Notes on the Brown–Douglas–Fillmore Theorem

The Brown–Douglas–Fillmore (BDF) theorem provides a remarkable solution to a long-standing open question in operator theory. The solution is obtained by first finding an equivalent formulation of the problem involving extensions of C^*-algebras. This does not make the original problem any simpler. None the less, Brown, Douglas, and Fillmore used new tools from algebraic topology and homological algebra to solve the problem of extensions of commutative C^*-algebras. This new approach also inspired the study of extensions of non-commutative C^*-algebras. The spurt of activity around the BDF theorem in the seventies was the precursor to the important fields of non-commutative geometry and K-theory.

The book presents the original proof of BDF in its entirety. Some parts of this proof, which were only briefly indicated in the BDF paper "Unitary equivalence modulo the compact operators and extensions of C*-algebras" published in the Springer Lecture Notes 345, appear here in complete detail. These notes have attempted to preserve the crucial connection of the BDF theorem with concrete problems in operator theory. This connection, for instance, is manifest in the study of the Arevson–Douglas conjecture.

Sameer Chavan is Professor in the Department of Mathematics and Statistics at the Indian Institute of Technology, Kanpur. His publications in single- and multi-variable operator theory use a variety of tools from graph theory and classical analysis. He is an Associate Editor of the *Indian Journal of Pure and Applied Mathematics* since 2020.

Gadadhar Misra is Professor in the Department of Mathematics at the Indian Institute of Science, Bangalore. His research in operator theory is mainly focused on operators in the Cowen–Douglas class and the study of homogeneous operators. He has been the Editor of the *Proceedings of Mathematical Sciences* during the period 2005–2012 and of the *Indian Journal of Pure and Applied Mathematics* during the period 2015–2020.

CAMBRIDGE–IISc SERIES

Cambridge–IISc Series aims to publish the best research and scholarly work in different areas of science and technology with emphasis on cutting-edge research.

The books aim at a wide audience including students, researchers, academicians and professionals and are being published under three categories: research monographs, centenary lectures and lecture notes.

The editorial board has been constituted with experts from a range of disciplines in diverse fields of engineering, science and technology from the Indian Institute of Science, Bangalore.

IISc Press Editorial Board:

Titles in print in this series:

- *Continuum Mechanics: Foundations and Applications of Mechanics* by C. S. Jog
- *Fluid Mechanics: Foundations and Applications of Mechanics* by C. S. Jog
- *Noncommutative Mathematics for Quantum Systems* by Uwe Franz and Adam Skalski
- *Mechanics, Waves and Thermodynamics* by Sudhir Ranjan Jain
- *Finite Elements: Theory and Algorithms* by Sashikumaar Ganesan and Lutz Tobiska
- *Ordinary Differential Equations: Principles and Applications* by A. K. Nandakumaran, P. S. Datti and Raju K. George
- *Lectures on von Neumann Algebras, 2nd Edition* by Serban Valentin Strătilă and László Zsidó
- *Biomaterials Science and Tissue Engineering: Principles and Methods* by Bikramjit Basu
- *Knowledge Driven Development: Bridging Waterfall and Agile Methodologies* by Manoj Kumar Lal
- *Partial Differential Equations: Classical Theory with a Modern Touch* by A. K. Nandakumaran and P. S. Datti
- *Modular Theory in Operator Algebras, 2nd Edition* by Şerban Valentin Strătilă

Cambridge–IISc Series

Notes on the
Brown–Douglas–Fillmore Theorem

Sameer Chavan
Gadadhar Misra

 CAMBRIDGE
UNIVERSITY PRESS

CAMBRIDGE
UNIVERSITY PRESS

University Printing House, Cambridge CB2 8BS, United Kingdom

One Liberty Plaza, 20th Floor, New York, NY 10006, USA

477 Williamstown Road, Port Melbourne, VIC 3207, Australia

314 to 321, 3rd Floor, Plot No.3, Splendor Forum, Jasola District Centre, New Delhi 110025, India

103 Penang Road, #05–06/07, Visioncrest Commercial, Singapore 238467

Cambridge University Press is part of the University of Cambridge.

It furthers the University's mission by disseminating knowledge in the pursuit of education, learning and research at the highest international levels of excellence.

www.cambridge.org
Information on this title: www.cambridge.org/9781316519301

First published 2021

Printed in India by Nutech Print Services, New Delhi 110020

A catalogue record for this publication is available from the British Library

Library of Congress Cataloging-in-Publication Data

Names: Chavan, Sameer, author. | Misra, Gadadhar, author.
Title: Notes on the Brown-Douglas-Fillmore theorem / Sameer Chavan, Gadadhar Misra.
Description: Cambridge ; New York, NY : Cambridge University Press, 2021. | Series: Cambridge-IISc series | Includes bibliographical references and index.
Identifiers: LCCN 2021013072 (print) | LCCN 2021013073 (ebook) | ISBN 9781316519301 (hardback) | ISBN 9781009023306 (ebook)
Subjects: LCSH: Operator theory. | C*-algebras. | Topological algebras. | BISAC: MATHEMATICS / General
Classification: LCC QA329 .C43 2021 (print) | LCC QA329 (ebook) | DDC 515/.724–dc23
LC record available at https://lccn.loc.gov/2021013072
LC ebook record available at https://lccn.loc.gov/2021013073

ISBN 978-1-316-51930-1 Hardback

To the memory of R. G. Douglas with great admiration,
much affection and deep respect.

The second named author is grateful to R. G. Douglas for the mathematical
training he has received from him, first as a doctoral student and then while
working together on several joint research projects.

Contents

Preface

A well-known theorem of Weyl says that every Hermitian operator on a separable complex Hilbert space is the sum of a diagonal and a compact operator. Halmos raised the question: Is every normal operator the sum of a diagonal and compact operator? Berg and Sikonia, independent of each other, proved that one may replace the Hermitain operators in Weyl's theorem by normal operators without changing the conclusion. However, this is not true if the operator is *essentially normal*, that is, its self-commutator is compact. The unilateral shift provides an example of such an operator. The obstruction to the decomposition of an essentially normal operator as a diagonal plus compact operator is a certain index data. This was established in a remarkable theorem by Brown, Douglas, and Fillmore that says that the index is the only obstruction to expressing, up to unitary equivalence, an essentially normal operator as the sum of a diagonal plus compact operator. They showed that the classification of essentially normal operators modulo compact is the same as an equivalence problem of *-monomorphisms of C^*-algebras. What has come to be known as the "BDF" theorem provides a solution to this problem. Their proofs are imaginative, novel and use a vast array of techniques and intuition from algebraic topology and homological algebra. This theory, cutting across several different areas, is quite sophisticated. Not surprisingly, therefore, many alternative proofs for different building blocks in the original proof were found soon afterward. So much so that the ingenious original proof in [26], which is a bit terse at some places, was all but forgotten.

These notes follow very closely the initial exposition of Brown, Douglas, and Fillmore [26]. In particular, we have resisted the temptation of giving simpler proofs to many of the arguments, which are now available. Our main goal was to occasionally fill in some details and explain few of the obscure points in the original proof of Brown, Douglas, and Fillmore, which appeared in [26], taking advantage of the exposition in the monograph [47]. In the process, we hope that the reader would develop an interest in modern operator theory and be exposed to some non-trivial techniques.

The first chapter provides standard preparatory material in basic operator theory; we largely follow the exposition of [44]. To get to the main points of the proof of the BDF theorem as fast as possible, a number of topics that aid in the understanding of the proof but are either not central to the proof or are expected to be known by the reader have been included in the appendices rather than in the main body of the text.

We make no claim to originality in preparing this material. All of it already exists in one form or the other. We reiterate that our goal is merely to make the original proof of Brown,

Douglas, and Fillmore available to beginning doctoral students and young researchers in its entirety.

These notes are an expanded version of the "lecture notes" prepared by the second named author on the Brown, Douglas, and Fillmore (BDF) theorem for the Instructional Conference on Operator Algebras, ISI-Bangalore, 1990. This deep and non-trivial theorem, from the early seventies, used tools from algebraic topology and homological algebra to solve an outstanding problem in operator theory. In the following years, many new and novel approaches to this topic have been discovered. These have considerably expanded the scope of the original theorem and have succeeded in making inroads into many other closely related topics. It is in the backdrop of this theorem that operator theory and operator algebras achieved their separate identities. Although new proofs have been found for different parts of the BDF theorem, the elegant and somewhat intriguing original proof remains an enigma.

There are several excellent expositions on the BDF theroem. However, these use many tools and advanced techniques from the theory of operator algebras. For instance, the exposition by Higson and Roe [77] provides a complete proof of the BDF theorem taking advantage of some of the later developments in operator theory and operator algebras. On the other hand, a number of intricate operator theoretic aspects of the original proof are apparent in the book of Davidson [40].

The goal we have set ourselves is to bring the original and beautiful proof of the BDF theorem to the notice of beginning doctoral students and researchers. Therefore, we have chosen to provide only brief explanations, where necessary, while following the original proof very closely. These notes attempt to provide an opportunity to those who might wish to learn what the BDF theorem says and what it takes to prove it. The proof of the BDF theorem in these notes occupies chapters two through four. The last chapter includes applications to operator theory. It contains, among other things, a discussion of essentially normal essentially homogeneous operators and reductive Hilbert modules which is at the heart of the Arveson–Douglas conjecture. We conclude with an Epilogue consisting of three short sections. In the first of these three sections, we discuss simpler proofs of different parts of the BDF theorem. The two main components here are the proof that Ext(X) is a group and Ext is a homotopy invariant functor. In the second section, we discuss some of the later developments related to the BDF theory. Here we discuss extensions of non-commutative C^*-algebras and describe the general notion of an index and relate it to K-theory. Finally, in the third section, we discuss some open problems concerning the complete set of invariants for essentially normal tuples and essential normality of essentially homogeneous tuples. We also discuss the Arveson–Douglas conjecture for semi-sub modules of a class of Hilbert modules.

The intended readership being the beginning doctoral students, we have added some preparatory material in the first chapter. To ensure that the original proof of the BDF theorem always remains the main focus of these notes, we have decided not to clutter the text with additional material except for occasionally filling in some details. Some of the topics which might give a greater understanding of the proof have been discussed in the Appendix. Again, the attempt here is not to be encyclopaedic but only to briefly recall some notions from topology and functional analysis that one may not easily find in one place. Consequently, the topics discussed in the Appendix appear in no particular order; they are somewhat unrelated to one another but serve the purpose of providing familiarity to a large variety of tools used in the proof and the flexibility of reading them independently or not at all.

The second named author would like to thank V. S. Sunder along with A. K. Vijayarajan and Uma Krishnan for several hours of discussion while preparing the initial notes. The idea of expanding the preliminary notes was conceived during the visit of the first author to the Indian Institute of Science in 2016. He lectured extensively during this visit on the BDF theorem to a lively audience consisting of Soumitra Ghara, Dinesh Kumar Keshari, Surjit Kumar, and Somnath Hazra. We thank all of them for their time and patience. We are also thankful to Rajeev Gupta, Shailesh Trivedi, Deepak Kumar Pradhan, Md. Ramiz Reza and Shubhankar Podder for several helpful comments on the initial draft of this book.

We thank the Indian Institute of Technology, Kanpur, and the Indian Institute of Science, Bangalore, for providing us an excellent opportunity to collaborate. The final version of this manuscript was prepared during a short visit of both the authors to the Department of Mathematics, University of Jammu. We thank Romesh Sharma for the invitation and the Department of Mathematics for the hospitality.

The work of the first author is supported through the P K Kelkar Research Fellowship received from IIT Kanpur for the duration 2017–2020. The second named author thanks the Department of Science and Technology, Government of India, for the very generous financial support he has received over the past decade through the J C Bose National Fellowship.

From: The Evolution of Modern Analysis, R. G. Douglas

Coburn had shown that there was more than one extension of the compact operators by the algebra of complex-valued continuous functions on the circle: a trivial one and the one I had exploited to calculate the index of continuous Toeplitz operators. Both Atiyah and Singer asked about the possibility of others and whether they could somehow be classified. Ultimately, investigating this question as well as several others, especially the classification of essentially normal operators, led to my joint work with Larry Brown and Peter Fillmore. ... The heart of the BDF results involves a functor from compact metric spaces to abelian groups defined by equivalence classes of C^*-extensions. Fundamentally, our work involved identifying this functor as defining the odd group in K-homology [48, pp. 283].

Overview

A normal operator on a finite dimensional inner product space can be diagonalised and the eigenvalues together with their multiplicities are a complete set of unitary invariants for the operator, while on an infinite dimensional Hilbert space the spectral theorem provides a model and a complete set of unitary invariants for such operators (refer to Appendix C). Thus, we view the theory of normal operators to be well understood. It is natural to study operators that may be thought of to be nearly normal in some sense. One hope is that it would be possible to provide canonical models and a complete set of invariants for such operators. Since an operator is normal if $[T, T^*] := TT^* - T^*T$ is 0, one may say an operator is nearly normal if $[T, T^*]$ is small in some appropriate sense, for example, finite rank, trace class, or compact. In these notes, we will take the last of these three measures of smallness for $[T, T^*]$ and make the following definition.

An operator T in $\mathcal{L}(\mathcal{H})$ is *essentially normal* if the self-commutator $[T, T^*]$ of T is compact. We say that T is *essentially unitary* if T is essentially normal and $T^*T - I$ is compact. Let $C(\mathcal{H})$ be the set of compact operators on a complex separable Hilbert space $\mathcal{H}$ and $\pi : \mathcal{L}(\mathcal{H}) \to \mathcal{L}(\mathcal{H})/C(\mathcal{H})$ be the natural quotient map. Set $Q(\mathcal{H}) := \mathcal{L}(\mathcal{H})/C(\mathcal{H})$. An operator T in $\mathcal{L}(\mathcal{H})$ is essentially normal if and only if $\pi(T)$ is normal in the C^*-algebra $Q(\mathcal{H})$. Further, an operator U in $\mathcal{L}(\mathcal{H})$ is essentially unitary if and only if $\pi(U)$ is unitary.

One of the main goals of these notes is to describe a complete set of invariants for the essentially normal operators with respect to a suitable notion of equivalence. As we are considering compact operators to be small, the correct notion of equivalence would seem to be the following.

Two operators T_1 and T_2 in $\mathcal{L}(\mathcal{H})$ are said to be *essentially equivalent* if there exist an essentially unitary operator U and a compact operator K such that $UT_1U^* = T_2 + K$. In this case, we write, $T_1 \sim T_2$. However, it turns out that one may replace the essentially unitary operator in this definition with a unitary operator without any loss of generality. The BDF theorem describes, among other things, the equivalence classes {essentially normal operators}/$\sim$.

The *essential spectrum* $\sigma_e(T)$, of an operator T in $\mathcal{L}(\mathcal{H})$ is the spectrum $\sigma(\pi(T))$ of $\pi(T)$ in the Calkin algebra $\mathcal{L}(\mathcal{H})/C(\mathcal{H})$. Let

$$\mathcal{N} + C = \{N + K : N \text{ is normal and } K \text{ is compact}\}.$$

For an operator T in $\mathcal{N}+C$ of the form $N+K$, note that $\sigma_e(T) = \sigma_e(N)$. The Weyl–von Neumann–Berg–Sikonia theorem states that the essential spectrum is a complete invariant for essential equivalence, that is, unitary equivalence modulo compact of operators in the class $\mathcal{N}+C$. Moreover, if X is any compact subset of the complex plane $\mathbb{C}$, then there is a normal operator N such that $\sigma_e(N) = X$.

Not all essentially normal operators are in $\mathcal{N}+C$. To give an example of an essentially normal operator not in $\mathcal{N}+C$, consider the Toeplitz operator T_z on the Hardy space $H^2(\mathbb{T})$. Note that $I - T_z T_z^* = P$ and $I - T_z^* T_z = 0$, where P is a rank one projection; therefore, T_z is an essentially unitary operator. An operator T is called *Fredholm* if it has a closed range and the dimension of its kernel and cokernel are finite. For a Fredholm operator T, the index $\operatorname{ind}(T)$ is an integer given by the formula

$$\operatorname{ind}(T) = \dim \ker(T) - \dim \ker(T^*).$$

If T is Fredholm and K is compact, then $\operatorname{ind}(T+K) = \operatorname{ind}(T)$. Moreover, the index of a Fredholm operator remains invariant under essential equivalence. Finally, if N is a normal operator that is also Fredholm, then its index is zero. It is easy to see that the Toeplitz operator T_z is Fredholm with $\operatorname{ind}(T_z) = -1$. If T_z is also in $\mathcal{N}+C$, then the index of T_z must be zero. This contradiction shows that the essentially normal operator T_z is not in $\mathcal{N}+C$.

First, the essential spectrum is a complete invariant modulo essential equivalence in the claas $\mathcal{N}+C$. Second, note that the normal operator M_z of multiplication by the coordinate function on $L^2(\mathbb{T})$ and the Toeplitz operator T_z both have the same essential spectrum, namely the unit circle $\mathbb{T}$. If these two operators were essentially equivalent via the unitary U, then we must have

$$-1 = \operatorname{ind}(T_z) = \operatorname{ind}(T_z + K) = \operatorname{ind}(U^* M_z U) = \operatorname{ind}(M_z) = 0.$$

This contradiction shows that the essential spectrum is not the only invariant for essential equivalence. The remarkable theorem of Brown, Douglas, and Fillmore says that the essential spectrum together with the index data is a complete set of invariants for essential equivalence.

Brown, Douglas, and Fillmore considered the problem of classifying the extension of compact operators by the C^*-algebra $C(X)$ of continuous functions on a compact set X induced by essentially normal operators with essential spectrum equal to X. Some of the details of this correspondence are provided in the following text. First, observe that if $\mathcal{S}_T$ is the C^*-algebra generated by the essentially normal operator T, the compact operators $C(\mathcal{H})$ and the identity operator I on the Hilbert space $\mathcal{H}$, then $\mathcal{S}_T/C(\mathcal{H})$ is isomorphic to the C^*-algebra generated by 1 and $\pi(T)$ in the Calkin algebra $Q(\mathcal{H})$. Since T is essentially normal, it follows that $\mathcal{S}_T/C(\mathcal{H})$ is commutative and we have

$$
\begin{array}{ccc}
\mathcal{S}_T & \longrightarrow & C(\sigma_e(T)) \\
\downarrow{\scriptstyle\pi} & & \uparrow{\scriptstyle\Gamma_{\mathcal{S}_T/C(\mathcal{H})}} \\
\mathcal{S}_T/C(\mathcal{H}) & \hookrightarrow & \mathcal{S}_T/C(\mathcal{H}) \subseteq Q(\mathcal{H})
\end{array}
$$

where $\Gamma_{S_T/C(\mathcal{H})}$ is the Gelfand map and we have an extension, that is

$$0 \to C(\mathcal{H}) \to S_T \overset{\varphi_T}{\to} C(\sigma_e(T)) \to 0$$

is exact. Conversely, if S is any C^*-algebra of operators on the Hilbert space $\mathcal{H}$ containing compact operators, that is $C(\mathcal{H}) \subseteq S \subseteq \mathcal{L}(\mathcal{H})$ and X is any compact subset of the complex plane $\mathbb{C}$ such that

$$0 \to C(\mathcal{H}) \overset{i}{\to} S \overset{\varphi}{\to} C(X) \to 0$$

is exact, then for any T in S, $\varphi(TT^* - T^*T) = 0$ and it follows that T is essentially normal. Fix any T in S such that $\varphi(T) = \mathrm{id}|_X$. Let S_T be the C^*-algebra generated by the operator T, the compact operators and the identity on $\mathcal{H}$. Now, $\varphi(S_T)$ is a C^*-subalgebra of $C(X)$ containing the identity function and therefore must be all of $C(X)$. If S is any operator in S, then there is always an operator S' in S_T such that $\varphi(S) = \varphi(S')$ so that $\varphi(S - S') = 0$, or equivalently, $S - S'$ is compact and hence, S is in S_T. Since $S_T \subseteq S$, it follows that $S_T = S$. Thus, there is a natural correspondence between essentially normal operators T with essential spectrum, a compact set $X \subseteq \mathbb{C}$ and extensions of $C(\mathcal{H})$ by $C(X)$.

Let us now relate unitary equivalence modulo the compacts of essentially normal operators to such extensions. If (S_1, φ_1) and (S_2, φ_2) are two extensions corresponding to equivalent essentially normal operators T_1 and T_2, that is, $U^*T_2U = T_1 + K$ for some unitary operator U and compact operator K, then $U^*S_2U = S_1$ by continuity of the map $T \mapsto U^*TU$ and $\varphi_2(T) = \varphi_1(U^*TU)$ for all T in S_2. It is therefore natural to say that two extensions (S_1, φ_1) and (S_2, φ_2) are *equivalent* if there exists a unitary operator U such that $U^*S_2U = S_2$ and $\varphi_2(T) = \varphi_1(U^*TU)$. Thus, if two essentially normal operators T_1 and T_2 are equivalent modulo the compacts, then the corresponding extensions are equivalent. Conversely, if the extensions are equivalent, then

$$\varphi_1(U^*T_2U) = \varphi_2(T_2) = \mathrm{id}|_X = \varphi_1(T_1)$$

and we see that $U^*T_2U - T_1$ is compact.

The classification problem for essentially normal operators and for extensions of $C(\mathcal{H})$ by $C(X)$ are identical for any compact subset X of $\mathbb{C}$. The extension point of view, of course, has many advantages. For any compact metrizable space X, let $\mathrm{Ext}(X)$ denote the *equivalence classes of the extensions* of $C(\mathcal{H})$ by $C(X)$. If X is a compact subset of the complex plane $\mathbb{C}$, then $\mathrm{Ext}(X)$ is just the *equivalence classes of essentially normal operators* N with $\sigma_e(N) = X$. Note that if Δ is a subset of the real line $\mathbb{R}$ and if S is any operator in $\mathcal{L}(\mathcal{H})$ such that $\pi(S)$ is normal with spectrum Δ, then $\pi(S)$ is self-adjoint,

$$\pi(S - S^*) = 0 \text{ implies that } S = \mathfrak{R}(S) + \text{compact},$$

where $\mathfrak{R}(S)$ denotes the real part of S. By the Weyl–von Neumann theorem, any two of these operators are equivalent modulo the compacts or in other words, $\mathrm{Ext}(\Delta) = 0$, for $\Delta \subseteq \mathbb{R}$.

The very deep and powerful theorem of Brown, Douglas, and Fillmore says that $\text{Ext}(X)$ is a group for any compact metric space X and in particular, if X is a planar set, then $\text{Ext}(X)$ is isomorphic to $\text{Hom}(\pi^1(X), \mathbb{Z})$. Here, $\pi^1(X)$ is the first cohomotopy group of X. These homomorphisms can be realized as follows: Let T be an essentially normal operator with essential spectrum X and $\{X_i\}_{i \in I}$ be the connected components of $\mathbb{C} \setminus X$. Suppose that the index of $T - \lambda$, $\lambda \in X_i$ be equal to n_i, $i \in I$. Define the map $\gamma_T : I \to \mathbb{Z}$ by $\gamma_T(i) = n_i$. This is the index data of the operator T and the BDF theorem says that the set $\{X, \gamma_T\}$ is a complete invariant for the operator T. In particular, taking $X = \mathbb{T}$, one obtains the following classification of essentially normal operators with essential spectrum $\mathbb{T}$. If T is an essentially normal operator and $\sigma_e(T) = \mathbb{T}$, then T is essentially equivalent to the bilateral shift, the n-fold direct sum of the unilateral forward shift or the m-fold direct sum of the unilateral backward shift according as the index of T is 0, $-n, n > 0$, or $m, m > 0$.

1

Spectral Theory for Hilbert Space Operators

The purpose of this chapter is two-fold. First, it serves as a rapid introduction to some of the basics of modern operator theory. Second, it has all the prerequisites needed to prove the Brown–Douglas–Fillmore theorem.

Unless otherwise stated, all Hilbert spaces considered in this text are assumed to be complex and separable. Whenever separability is not needed or it does not simplify the situation, the same is mentioned. Throughout this text, $\mathcal{H}$ denotes a complex Hilbert space and $\mathcal{L}(\mathcal{H})$ stands for the algebra of bounded linear operators from $\mathcal{H}$ into $\mathcal{H}$. Note that $\mathcal{L}(\mathcal{H})$ is a unital C^*-algebra, where the identity operator I is the unit, composition of operators is the multiplication and the uniquely defined adjoint T^* of a bounded linear operator T on Hilbert space $\mathcal{H}$ is the involution (the reader is referred to Appendix B for the definition of a C^*-algebra). To avoid ambiguity, whenever necessary, we let $I_{\mathcal{H}}$ denote the identity operator on $\mathcal{H}$. Given $T \in \mathcal{L}(\mathcal{H})$, the symbols $\ker T$ and $\operatorname{ran} T$ stand for the kernel and range of the operator T, respectively. As usual, we let $\|\cdot\|$ and $\langle \cdot, \cdot \rangle$ denote the norm and the inner product in the Hilbert space $\mathcal{H}$.

1.1 Partial Isometries and Polar Decomposition

If λ is a nonzero complex number, then $\lambda = |\lambda|e^{i\theta}$ for some real number θ; this is the polar decomposition of λ. Theorem 1.1.3 provides, for operators in $\mathcal{L}(\mathcal{H})$, a similar decomposition. The challenge is to find the two factors analogous to $|\lambda|$ and $e^{i\theta}$ in $\mathcal{L}(\mathcal{H})$. A natural choice for $|\lambda|$ quickly presents itself, namely, the operator $(T^*T)^{1/2}$. The choice for $e^{i\theta}$ would seem to be either a unitary or an isometry; however, none of these choices is quite correct for an operator on a Hilbert space of dimension greater than one as we will see here.

Definition 1.1.1

An operator $V \in \mathcal{L}(\mathcal{H})$ is a *partial isometry* if $\|Vf\| = \|f\|$ for all $f \in \mathcal{H}$ that are orthogonal to $\ker V$. If, in addition, the kernel of V is $\{0\}$, then V is said to be an *isometry*. The *initial space* of a partial isometry V is defined as the orthogonal complement $(\ker V)^\perp$ of $\ker V$, whereas the *final space* of V is the range $\operatorname{ran} V$ of V.

Remark 1.1.2 Let $V \in \mathcal{L}(\mathcal{H})$ be a partial isometry. Let $f \in \mathcal{H}$ and write $f = f_1 + f_2$, where $f_1 \in \ker V$ and $f_2 \in (\ker V)^\perp$. Then

$$\langle (I - V^*V)f, f \rangle = \|f\|^2 - \|Vf\|^2 = \|f_1\|^2.$$

Thus, $I - V^*V$ is a positive operator, and by Remark C.1.6, it has a unique positive square root, say, $(I - V^*V)^{1/2}$. Moreover, for any $f \in (\ker V)^\perp$,

$$\|(I - V^*V)^{1/2}f\|^2 = \langle (I - V^*V)f, f \rangle = 0.$$

Consequently, $(I - V^*V)^{1/2}f = 0$ or $V^*Vf = f$. In particular, V^*V is the orthogonal projection of $\mathcal{H}$ onto the initial space of V. Now, $VV^*(Vf) = Vf$ and if $g \perp \operatorname{ran} V$, then $VV^*g = 0$. So, VV^* is the orthogonal projection of $\mathcal{H}$ onto the final space of V. Thus, V^* is a partial isometry with initial space $\operatorname{ran} V$. It also follows that the range of a partial isometry is closed.

It can be seen that the correct analogy for $e^{i\theta}$ is a partial isometry. For $T \in \mathcal{L}(\mathcal{H})$, we let $|T|$ denote the positive square root $(T^*T)^{1/2}$ of the operator T^*T (see Remark C.1.6).

Theorem 1.1.3 (Polar Decomposition)

If $T \in \mathcal{L}(\mathcal{H})$, then there exists a positive operator P and a partial isometry V (with initial space $\overline{\operatorname{ran} P}$ and final space $\overline{\operatorname{ran} T}$) such that

$$T = VP \text{ and } \ker V = \ker P.$$

Moreover, such a pair (V, P) is uniquely determined. Furthermore, P can be chosen to be $|T|$.

Proof Note that for any $f \in \mathcal{H}$,

$$\||T|f\|^2 = \langle |T|f, |T|f \rangle = \langle |T|^2 f, f \rangle = \langle T^*Tf, f \rangle = \|Tf\|^2. \tag{1.1.1}$$

Thus, if we define $\tilde{V} : \operatorname{ran} |T| \to \mathcal{H}$ by

$$\tilde{V}(|T|f) = Tf, \quad f \in \mathcal{H},$$

then $\tilde{V}$ is well defined. Moreover, it is isometric and extends uniquely to an isometric mapping from $\overline{\operatorname{ran} |T|}$ to $\mathcal{H}$. If we further define $V : \mathcal{H} \to \mathcal{H}$ by

$$Vf = \begin{cases} \tilde{V}f & \text{if } f \in \overline{\operatorname{ran} |T|}, \\ 0 & \text{if } f \in (\operatorname{ran} |T|)^\perp, \end{cases}$$

then V is a partial isometry satisfying $T = V|T|$ and

$$\ker V = (\operatorname{ran} |T|)^\perp = \ker |T|.$$

Thus, choosing $P = |T|$, we have $T = VP$.

For uniqueness, suppose that $T = WQ$, where W is a partial isometry, Q is positive and $\ker W = \ker Q$. By Remark 1.1.2, W^*W is the orthogonal projection onto $(\ker W)^\perp = (\ker Q)^\perp = \overline{\operatorname{ran} Q}$. It follows that

$$|T|^2 = T^*T = QW^*WQ = Q^2.$$

Thus, by the uniqueness of the positive square root, we have $Q = |T|$. Consequently, $W|T| = V|T|$, and hence, $W = V$ on the range of $|T|$. However

$$(\operatorname{ran} |T|)^\perp = \ker |T| = \ker W = \ker V,$$

and hence, $W = V$ on $(\operatorname{ran} |T|)^\perp$. Therefore, $V = W$ and the proof is complete.   □

A polar decomposition in which the order of the factors are reversed is useful in certain instances.

Corollary 1.1.1

If $T \in \mathcal{L}(\mathcal{H})$, then there exist a positive operator Q and partial isometry W such that

$$T = QW \text{ and } \operatorname{ran} W = (\ker Q)^\perp.$$

Moreover, such a pair (Q, W) is uniquely determined.

Proof An application of Theorem 1.1.3 to the operator T^* shows that there exists a partial isometry V such that

$$T^* = V|T^*| \text{ and } \ker V = \ker |T^*|.$$

Setting $Q = |T^*|$, we have $T = |T^*|W$, where $W = V^*$ and $\ker W^* = \ker |T^*|$. By Remark 1.1.2, W is a partial isometry with closed range. Further, $\ker W^* = \ker |T^*|$ if and only if $\operatorname{ran} W = (\ker |T^*|)^\perp$. The uniqueness part now follows from the preceding theorem.   □

Remark **1.1.4** If W is chosen such that $\operatorname{ran} W = (\ker |T^*|)^\perp$, then $\operatorname{ran} W = \overline{\operatorname{ran} |T^*|}$, and hence,

$$\operatorname{ran} |T^*| = |T^*|(\mathcal{H}) = |T^*|(\overline{\operatorname{ran} |T^*|}) = |T^*|W(\mathcal{H}) = \operatorname{ran} T.$$

1.2 Compact and Fredholm Operators

In this section, we discuss some basic properties of compact and Fredholm operators on a Hilbert space. In particular, we show that the subset $C(\mathcal{H})$ of compact operators forms a closed two-sided $*$-ideal in the C^*-algebra $\mathcal{L}(\mathcal{H})$ of bounded linear operators on $\mathcal{H}$. Furthermore, we show that the quotient $\mathcal{L}(\mathcal{H})/C(\mathcal{H})$ is a C^*-algebra. The invertible elements of this quotient algebra naturally give rise to the notion of Fredholm operators.

Compact Operators

In this section, it is shown that an operator is compact if and only if it is the norm limit of a sequence of finite rank operators. Thus, compact operators are natural generalizations of finite dimensional operators in a topological sense. We first show that any closed subspace of the Hilbert space $\mathcal{H}$ in the range of a compact operator $T : \mathcal{H} \to \mathcal{H}$ must be finite dimensional. Moreover, any operator whether compact or not, possessing this property can be approximated in norm by a sequence of finite rank operators. Thus, we obtain another characterization of compact operators, namely an operator in $\mathcal{L}(\mathcal{H})$ is compact if and only if the only closed subspaces of the Hilbert space $\mathcal{H}$ in its range are finite dimensional.

Let $\mathbb{B} = \{x \in \mathcal{H} : \|x\| < 1\}$ denote the open unit ball in $\mathcal{H}$ and let $\overline{\mathbb{B}}$ denote the closure of $\mathbb{B}$ in $\mathcal{H}$.

Definition 1.2.1

A bounded linear transformation $T : \mathcal{H} \to \mathcal{K}$ is of *finite rank* if the dimension of its range is finite and *compact* if the image of the closed unit ball $\overline{\mathbb{B}}$ under T is relatively compact in $\mathcal{K}$. Let $\mathcal{T}(\mathcal{H})$ and $C(\mathcal{H})$ denote the set of all finite rank and compact operators on $\mathcal{H}$ respectively.

By the Heine–Borel theorem, finite rank operators are compact. The converse is certainly not true (see Exercise 1.7.11). Most of the elementary properties of finite rank operators are collected together in the following proposition.

Proposition 1.2.1

The collection $\mathcal{T}(\mathcal{H})$ is a minimal two-sided nonzero $*$-ideal in $\mathcal{L}(\mathcal{H})$.

Proof   Note that

$$\mathrm{ran}(S + T) \subseteq \mathrm{ran}\,S + \mathrm{ran}\,T \text{ and } \mathrm{ran}(ST) \subseteq \mathrm{ran}\,S.$$

These inclusions show that $\mathcal{T}(\mathcal{H})$ is a right ideal in $\mathcal{L}(\mathcal{H})$. Further, the equality

$$\mathrm{ran}\,T^* = T^*(\ker T^*)^\perp = T^*(\overline{\mathrm{ran}\,T})$$

shows that $T \in \mathcal{T}(\mathcal{H})$ if and only if $T^* \in \mathcal{T}(\mathcal{H})$. Finally, if $S \in \mathcal{L}(\mathcal{H})$ and $T \in \mathcal{T}(\mathcal{H})$, then $T^*S^* \in \mathcal{T}(\mathcal{H})$ and hence, $ST = (T^*S^*)^* \in \mathcal{T}(\mathcal{H})$. Therefore, $\mathcal{T}(\mathcal{H})$ is a two-sided $*$-ideal in $\mathcal{L}(\mathcal{H})$.

To show that $\mathcal{T}(\mathcal{H})$ is minimal, assume that $\mathcal{J}$ is a nonzero ideal in $\mathcal{L}(\mathcal{H})$. Thus, there exists a nonzero operator $T \in \mathcal{J}$, and hence, there is a nonzero vector f and a unit vector g in $\mathcal{H}$ such that $Tf = g$. For $k, h \in \mathcal{H}$, let $k \otimes h$ denote the rank one bounded linear operator defined by

$$k \otimes h(\ell) = \langle \ell, h \rangle k, \quad \ell \in \mathcal{H}. \tag{1.2.2}$$

Note that for any $\ell \in \mathcal{H}$,

$$(k \otimes g)T(f \otimes h)(\ell) = k \otimes g(\langle \ell, h \rangle Tf) = \langle \ell, h \rangle \langle Tf, g \rangle k = \langle \ell, h \rangle k = k \otimes h(\ell),$$

and therefore, $k \otimes h \in \mathcal{J}$ for any pair of vectors h and k in $\mathcal{H}$. However,

$$\{T \in \mathcal{L}(\mathcal{H}) : T \text{ is of rank one}\} = \{k \otimes h : h \text{ and } k \text{ belong to } \mathcal{H}\}.$$

Thus, $\mathcal{J}$ contains all rank one operators and hence, all finite rank operators. This completes the proof. □

Next, we obtain a very useful alternative characterization of compact operators.

Lemma 1.2.1

If T belongs to $\mathcal{L}(\mathcal{H})$, then T is compact if and only if for every sequence $\{f_n\}_{n \geqslant 0}$, which converges to f weakly, it is true that $\{Tf_n\}_{n \geqslant 0}$ converges to Tf in norm.

Proof Let $T \in \mathcal{L}(\mathcal{H})$ and suppose that T is compact. Let $\{f_n\}_{n \geqslant 0}$ be a sequence in $\mathcal{H}$ converging weakly to $f \in \mathcal{H}$. One may assume that $\{f_n\}_{n \geqslant 0}$ is contained in the unit ball after multiplying by a nonzero scalar if required (see Remark B.1.12). Assume contrary to the hypothesis that $\{Tf_n\}_{n \geqslant 0}$ does not converge in norm to Tf. After passing to a subsequence, if necessary, we may assume that for some $\epsilon > 0$,

$$\|Tf_n - Tf\| \geqslant \epsilon \text{ for all } n \geqslant 0. \qquad (1.2.3)$$

On the other hand, since $\{Tf_n\}_{n \geqslant 0}$ is a relatively compact subset of $\mathcal{H}$, it has a convergent subsequence $\{Tf_{n_k}\}_{k \geqslant 0}$ converging to $g \in \mathcal{H}$..By the uniqueness of the weak limit, we must have $g = Tf$. This is not possible in view of (1.2.3), and hence, this contradiction proves the necessary part.

To prove the converse, it suffices to check that every sequence in $T(\overline{\mathbb{B}})$ has a convergent subsequence. To verify this, let $\{f_n\}_{n \geqslant 0}$ be a sequence in $\mathbb{B}$. By Theorem B.1.13 $\{f_n\}_{n \geqslant 0}$ has a weakly convergent subsequence, and hence by assumption, $\{Tf_n\}_{n \geqslant 0}$ has a convergent subsequence. This completes the proof. □

In the following Lemma, we need not apply our standing assumption of separability to the Hilbert space $\mathcal{H}$. The proof below works for any complex Hilbert space.

Lemma 1.2.2

The closed unit ball in a Hilbert space $\mathcal{H}$ is compact if and only if $\mathcal{H}$ is finite dimensional.

Proof If $\mathcal{H}$ is finite dimensional, then it is isometrically isomorphic to $\mathbb{C}^n$ for some positive integer n, and hence, by the Heine–Borel theorem, its closed unit ball is compact. Conversely, if $\mathcal{H}$ is infinite dimensional, then there exists an infinite orthonormal sequence $\{e_n\}_{n \geqslant 0}$ in $\overline{\mathbb{B}}$. The fact that

$$\|e_n - e_m\| = \sqrt{2} \text{ for } n \neq m$$

shows that the sequence $\{e_n\}_{n \geqslant 0}$ has no convergent subsequence. Thus, the closed unit ball $\overline{\mathbb{B}}$ is not compact. □

The following proposition says that the range of a compact operator cannot contain a closed subspace of infinite dimension.

Proposition 1.2.2

If T is a compact operator on an infinite dimensional Hilbert space $\mathcal{H}$ and $\mathcal{M}$ is a closed subspace of $\mathcal{H}$ contained in the range of T, then $\mathcal{M}$ is finite dimensional.

Proof   Let T be a compact operator on an infinite dimensional Hilbert space $\mathcal{H}$ and let P_M denote the orthogonal projection of $\mathcal{H}$ onto $\mathcal{M}$. A simple application of Lemma 1.2.1 shows that $P_M T$ is compact. Define a linear transformation $A : \mathcal{H} \to \mathcal{M}$ by setting

$$Af = P_M Tf, \quad f \in \mathcal{H}.$$

Note that A is a bounded linear transformation that maps $\mathcal{H}$ onto $\mathcal{M}$. By the open mapping theorem (see Theorem B.1.9), A is an open map. Therefore,

$$A(\overline{\mathbb{B}}) \supseteq \mathbb{B}_\delta \text{ for some } \delta > 0,$$

where $\mathbb{B}_\delta$ denotes the ball $\{x \in \mathcal{M} : \|x\| < \delta\}$ in $\mathcal{M}$. As the relatively compact set $P_M T(\overline{\mathbb{B}})$ contains the ball $\mathbb{B}_\delta$, by Lemma 1.2.2, $\mathcal{M}$ is finite dimensional.   □

An important property of the set of compact operators is now evident.

Proposition 1.2.3

The set of compact operators $C(\mathcal{H})$ is norm closed in $\mathcal{L}(\mathcal{H})$. In particular, the norm closure of $\mathcal{T}(\mathcal{H})$ is contained in $C(\mathcal{H})$.

Proof   Assume that $\{K_n\}_{n \geqslant 0}$ is a sequence of compact operators that converges in norm to K. Let $\{f_n\}_{n \geqslant 0}$ be a sequence converging weakly to f and set $M = \max\{1, \|f_n\| : n \geqslant 0\}$. Choose a non-negative integer ℓ such that $\|K - K_\ell\| < \epsilon/3M$. As K_ℓ is compact, by Lemma 1.2.1, there exists an integer $n_0 \geqslant 0$ such that

$$\|K_\ell f_n - K_\ell f\| < \epsilon/3, \quad n \geqslant n_0.$$

Thus, for every integer $n \geqslant n_0$, we have

$$\|Kf_n - Kf\| \leqslant \|(K - K_\ell)f_n\| + \|K_\ell f_n - K_\ell f\| + \|(K_\ell - K)f\| < \epsilon,$$

and hence, by Lemma 1.2.1, K is a compact operator. The remaining part now follows from the fact that $\mathcal{T}(\mathcal{H})$ is contained in $C(\mathcal{H})$.   □

A diagonal operator on a separable infinite dimensional Hilbert space is compact if and only if the sequence of its diagonal entries converges to 0. This is a consequence of the following general fact:

Corollary 1.2.1

For a non-negative integer n, let T_n be a compact operator in $\mathcal{L}(\mathcal{H}_n)$. If $\|T_n\| \to 0$ as $n \to \infty$, then $T = \oplus_{n=0}^\infty T_n$ is a compact operator in $\mathcal{L}(\mathcal{H})$, where $\mathcal{H} = \oplus_{n=0}^\infty \mathcal{H}_n$.

Proof Recall that $\|\oplus_{n=0}^{\infty} T_n\| = \sup_{n \geq 0} \|T_n\|$. Thus, if $\lim_{n \to \infty} \|T_n\| = 0$, then $\oplus_{n=0}^{k} T_n \to T$ as $k \to \infty$. As the finite direct sum of compact operators is compact, by Proposition 1.2.3, T is compact. $\qquad\square$

The following theorem shows that compact operators can be approximated by finite rank operators in the operator norm topology. We point out that the proof of this theorem does not require the Hilbert space to be separable. In spite of our disclaimer to the contrary, we have included this proof to demonstrate the power of the spectral theorem, which is an essential tool for many arguments in these notes.

Theorem 1.2.2

If the range of a bounded linear operator T on the Hilbert space $\mathcal{H}$ does not contain any closed infinite dimensional subspace of $\mathcal{H}$, then T is in the norm closure of $\mathcal{T}(\mathcal{H})$. In particular, the norm closure of $\mathcal{T}(\mathcal{H})$ is $C(\mathcal{H})$.

Proof Let T in $\mathcal{L}(\mathcal{H})$ be such that the range of T does not contain any closed infinite dimensional subspace of $\mathcal{H}$ and let $T = PW$ be the polar decomposition for T, where $P = |T^*|$ and W is a partial isometry such that $\operatorname{ran} W = (\ker P)^{\perp}$ (see Corollary 1.1.1). Let $\chi_\epsilon = \chi_{(\epsilon, \|P\|]}$ be the characteristic function of the interval $(\epsilon, \|P\|]$. By the bounded Borel functional calculus of P (see Theorem C.1.4),

$$\mathbb{E}_\epsilon = \chi_\epsilon(P) \text{ is an orthogonal projection in } \mathcal{L}(\mathcal{H}).$$

If we define ψ_ϵ on $(0, \|P\|]$ by

$$\psi_\epsilon = \begin{cases} 1/x & \text{for } \epsilon < x \leq \|P\|, \\ 0 & \text{otherwise,} \end{cases}$$

then $Q_\epsilon = \psi_\epsilon(P)$ satisfies

$$Q_\epsilon P = \mathbb{E}_\epsilon = P Q_\epsilon.$$

Thus, we have

$$\operatorname{ran} \mathbb{E}_\epsilon = \operatorname{ran}(P Q_\epsilon) \subseteq \operatorname{ran} P = \operatorname{ran} T$$

(see Remark 1.1.4), and therefore the range of the projection $\mathbb{E}_\epsilon$ is finite dimensional by assumption. Hence, $P_\epsilon = P\mathbb{E}_\epsilon$ is in $\mathcal{T}(\mathcal{H})$ and $P_\epsilon W$ is also in $\mathcal{T}(\mathcal{H})$. Finally, since $\|W\| \leq 1$,

$$\begin{aligned} \|T - P_\epsilon W\| &= \|PW - P_\epsilon W\| \leq \|P - P_\epsilon\| = \|P - P\mathbb{E}_\epsilon\| \\ &= \sup_{0 \leq x \leq \|P\|} \|x - x\chi_\epsilon(x)\| < \epsilon. \end{aligned}$$

Therefore, T is in the norm closure of $\mathcal{T}(\mathcal{H})$. The proof of the remaining part about compact operators now follows from Proposition 1.2.2. $\qquad\square$

For emphasis, we separately record the following corollary, which is already contained in Lemma 1.2.1 and the second half of Theorem 1.2.2.

Corollary 1.2.2

If $\mathcal{H}$ is an infinite dimensional Hilbert space and T is an operator on $\mathcal{H}$, then T is compact if and only if range of T does not contain any closed infinite dimensional subspace.

Remark 1.2.3   It follows from the preceding corollary that an orthogonal projection is compact if and only if it is of finite rank.

Corollary 1.2.3

If $\mathcal{H}$ is an infinite dimensional Hilbert space, then $C(\mathcal{H})$ is the only proper closed nonzero two-sided $*$-ideal in $\mathcal{L}(\mathcal{H})$.

Proof   Let $\mathcal{H}$ be an infinite dimensional separable Hilbert space. Since $C(\mathcal{H})$ is the norm closure of $\mathcal{T}(\mathcal{H})$ and $\mathcal{T}(\mathcal{H})$ is a minimal two-sided nonzero $*$-ideal, it follows that $C(\mathcal{H})$ is itself a minimal closed two-sided nonzero $*$-ideal. Next, we prove that it is the only such ideal. Suppose that $T \in \mathcal{L}(\mathcal{H})$ is not compact. Then, by Corollary 1.2.2, the range of T contains a closed infinite dimensional subspace $\mathcal{M}$. Note that the operator $P_{\mathcal{M}}T$ maps $\mathcal{H}$ onto the subspace $\mathcal{M}$. Hence, by the open mapping theorem (see Theorem B.1.9), we find an operator $S : \mathcal{M} \to \mathcal{H}$ such that $P_{\mathcal{M}}TS = I_{\mathcal{M}}$. Now if $U : \mathcal{H} \to \mathcal{M}$ is a unitary map, that is, an isometric surjection (which exists since the Hilbert space dimensions of $\mathcal{H}$ and $\mathcal{M}$ are the same in view of the separability of $\mathcal{H}$), then

$$(U^*P_{\mathcal{M}})T(SU) = U^*I_{\mathcal{M}}U = I,$$

where $U^*P_{\mathcal{M}}$ and SU are operators in $\mathcal{L}(\mathcal{H})$. Hence, any two-sided ideal containing T must also contain I. This completes the proof.   □

Fredholm Operators

The Fredholm operators are those operators that have a finite dimensional kernel and co-kernel. Here co-kernel of an operator $T \in \mathcal{L}(\mathcal{H})$ is the quotient $\mathcal{H}/\operatorname{ran}T$. It then follows that the finite dimensionality of the co-kernel implies that the range of T is closed (see Exercise 1.7.31), and in the context of Hilbert spaces, it also implies that the kernel of T^* is finite dimensional. Taken together with the finite dimensionality of the kernel, this is the theorem of Atkinson.

Let B be a Banach algebra and let $\mathscr{I}$ be a proper, closed two-sided ideal of B. Then $B/\mathscr{I}$ is a quotient algebra with unity $1 + \mathscr{I}$, which is complete as a normed linear space with the quotient norm denoted by $\|\cdot\|_Q$ (see (B.1.2)). Further, $\|1 + \mathscr{I}\|_Q = 1$ and

$$\|(x + \mathscr{I})(y + \mathscr{I})\|_Q \leqslant \|x + \mathscr{I}\|_Q \|y + \mathscr{I}\|_Q. \tag{1.2.4}$$

In particular, $B/\mathscr{I}$ is a Banach algebra (see Lemma B.1.6).

Definition 1.2.4

Let $\mathcal{H}$ be an infinite dimensional Hilbert space. The Banach (quotient) algebra $Q(\mathcal{H}) = \mathcal{L}(\mathcal{H})/C(\mathcal{H})$ is referred to as the *Calkin algebra*.

The following basic result says that if $\mathscr{I}$ is a closed two sided involutive ideal, that is, closed under the involution, of $\mathcal{L}(\mathcal{H})$, then the quotient algebra is actually a C^*-algebra.

Theorem 1.2.5

If $\mathscr{I}$ is a closed two-sided involutive ideal in $\mathcal{L}(\mathcal{H})$, then the quotient $\mathcal{L}(\mathcal{H})/\mathscr{I}$ is a C^*-algebra.

Proof Let $\mathscr{I}$ be a closed two-sided involutive ideal in $\mathcal{L}(\mathcal{H})$. The proof that the quotient is an algebra possessing an involution inherited from $\mathcal{L}(\mathcal{H})$ is a routine verification and is omitted. The non-trivial part is to show that the quotient norm is a C^* norm. The following steps facilitate the proof of this claim.

Step I

Consider the sequence $\{u_n\}_{n\geqslant 1}$ of continuous functions on $[0, \infty)$ given by

$$u_n(t) = \begin{cases} nt & \text{if } 0 \leqslant t \leqslant 1/n, \\ 1 & \text{otherwise.} \end{cases}$$

For $T \in \mathscr{I}$ and a positive integer n, set $T_n = u_n(T^*T)$. Note that for every positive integer n,

(a) T_n is a self-adjoint operator of norm at most 1,

(b) T_n belongs to $\mathscr{I}$,

(c) $T_n \leqslant T_{n+1}$.

Indeed, since $\|u_n\|_\infty = 1$, u_n can be approximated by polynomials, $u_n(t) \in \mathbb{R}$ and $u_{n+1}(t) - u_n(t) \geqslant \min\{t, 0\}$ for every $t \in [0, \infty)$, parts (a)–(c) follow from the functional calculus of self-adjoint operators. We claim that

$$\lim_{n \to \infty} \|TT_n - T\| = 0. \tag{1.2.5}$$

It is easy to see using the C^*-identity that

$$\|TT_n - T\|^2 = \|T_n T^* TT_n - T^* TT_n - T_n T^* T + T^* T\|, \quad n \geqslant 1.$$

Another application of the functional calculus for T^*T shows that (1.2.5) holds provided $\lim_{n \to \infty} \|f_n\|_\infty = 0$, where

$$f_n(t) = t(u_n(t) - 1)^2, \quad t \in [0, \infty), \, n \geqslant 1.$$

Step II

For $S \in \mathcal{L}(\mathcal{H})$, let

$$\alpha = \inf\{\|S - ST\| : T \in \mathbb{B}_{\mathscr{I}}^+\},$$

where $\mathbb{B}^+_{\mathscr{I}}$ denotes the positive elements in the unit ball in $\mathscr{I}$. Since $S\mathscr{I} \subseteq \mathscr{I}$, we clearly have $\|S + \mathscr{I}\|_Q \leqslant \alpha$. We claim that $\|S + \mathscr{I}\|_Q = \alpha$. To see the claim, fix $T \in \mathscr{I}$ and let $\{T_n\}_{n \geqslant 1}$ be the so-called approximate identity for $\mathscr{I}$ as ensured in Step I. Since $0 \leqslant I - T_n \leqslant I$,

$$
\begin{aligned}
\|S + T\| &\geqslant \liminf_{n \geqslant 1} \|(S + T)(I - T_n)\| \\
&= \liminf_{n \geqslant 1} \|S(I - T_n) + T(I - T_n)\| \\
&= \liminf_{n \geqslant 1} \|S(I - T_n)\|,
\end{aligned}
$$

where we used the fact that $\lim_{n \to \infty} \|T(I - T_n)\| = 0$. It follows that the quotient norm $\|S + \mathscr{I}\|_Q$ is at least α, and the claim stands verified.

We now complete the proof of the theorem. By Step I, for any $T \in \mathscr{I}$, there exists a sequence $\{T_n\}_{n \geqslant 1}$ of self-adjoint operators such that (1.2.5) holds. Since $T_n \in \mathscr{I}$ and involution is continuous, $T^* \in \mathscr{I}$ whenever $T \in \mathscr{I}$. It is now easy to see using (1.2.4) that for any $S \in \mathcal{L}(\mathcal{H})$,

$$
\begin{aligned}
\|S^* + \mathscr{I}\|_Q &= \|S + \mathscr{I}\|_Q, \\
\|S^*S + \mathscr{I}\|_Q &= \|(S^* + \mathscr{I})(S + \mathscr{I})\|_Q \leqslant \|S + \mathscr{I}\|_Q^2.
\end{aligned}
$$

On the other hand, by Step II,

$$
\begin{aligned}
\|S + \mathscr{I}\|_Q^2 &= \inf\{\|S(I - T)\|^2 : T \in \mathbb{B}^+_{\mathscr{I}}\} \\
&= \inf\{\|(I - T)S^*S(I - T)\| : T \in \mathbb{B}^+_{\mathscr{I}}\} \\
&\leqslant \inf\{\|S^*S(I - T)\| : T \in \mathbb{B}^+_{\mathscr{I}}\} \\
&= \|S^*S + \mathscr{I}\|_Q.
\end{aligned}
$$

This completes the proof of the theorem. $\qquad\qquad\qquad\qquad\qquad\qquad\qquad\qquad\qquad\square$

***Remark* 1.2.6**   By Theorem 1.2.5, the Calkin algebra $Q(\mathcal{H})$ is a C^*-algebra. Hence, by the Gelfand–Naimark–Segal construction (refer to [36, Chapter VIII]), there exists a Hilbert space $\mathcal{K}$ and an injective unital $*$-representation $\sigma : Q(\mathcal{H}) \to B(\mathcal{K})$.

The quotient map from $\mathcal{L}(\mathcal{H})$ onto $Q(\mathcal{H})$, to be referred to as the *Calkin map*, is denoted by π. The Calkin map π is multiplicative.

We prove the basic spectral properties of compact operators after obtaining some elementary facts pertaining to Fredholm operators. The following definition of the Fredholm operator is equivalent to the classical one via Atkinson's theorem (see Theorem 1.2.8).

Definition 1.2.7

Let $\pi : \mathcal{L}(\mathcal{H}) \to Q(\mathcal{H})$ be the Calkin map. An operator T in $\mathcal{L}(\mathcal{H})$ is *Fredholm* if $\pi(T)$ is invertible in the Calkin algebra $Q(\mathcal{H})$. The collection of Fredholm operators in $\mathcal{L}(\mathcal{H})$ is denoted by $\mathcal{F}(\mathcal{H})$.

Since the Calkin map sends invertible elements to invertible ones, every invertible operator is Fredholm. By definition, compact perturbation of invertible operators is Fredholm. As expected, not every Fredholm operator is of this form (see Exercise 1.7.23).

Some elementary properties of Fredholm operators are immediate from the definition; they are as follows.

Proposition 1.2.4

For any Hilbert space $\mathcal{H}$, $\mathcal{F}(\mathcal{H})$ is an open subset of $\mathcal{L}(\mathcal{H})$, which is self-adjoint, closed under multiplication and invariant under compact perturbations. In particular, $\mathcal{F}(\mathcal{H})$ is a group and multiplication modulo compact operators serves as the group operation. Moreover,

$$\mathcal{G}(\mathcal{H}) + C(\mathcal{H}) \subsetneq \mathcal{F}(\mathcal{H}),$$

where $\mathcal{G}(\mathcal{H})$ denotes the group of invertible operators in $\mathcal{L}(\mathcal{H})$.

Proof Since $\pi : \mathcal{L}(\mathcal{H}) \to \mathcal{Q}(\mathcal{H})$ is continuous and $\mathcal{F}(\mathcal{H})$ is the inverse image of the group of invertible elements in $\mathcal{Q}(\mathcal{H})$, it follows that $\mathcal{F}(\mathcal{H})$ is open (see Corollary B.1.6). Again the fact that π is multiplicative implies that $\mathcal{F}(\mathcal{H})$ is closed under multiplication. It is obvious that $\mathcal{F}(\mathcal{H})$ is invariant under compact perturbations. If T is in $\mathcal{F}(\mathcal{H})$, then there exists an operator S and compact operators K_1 and K_2 such that

$$ST = I + K_1 \text{ and } TS = I + K_2.$$

Taking adjoints, we see that $\pi(T^*)$ is invertible in the Calkin algebra $\mathcal{Q}(\mathcal{H})$ and $\mathcal{F}(\mathcal{H})$ is self-adjoint. The remaining part is now immediate from the discussion following Definition 1.2.7. $\square$

Although the vector sum of two closed subspaces M and M_0 of a Hilbert space $\mathcal{H}$ is not closed in general (see Exercise 1.7.28), it is true that the sum is closed if one of the subspaces, say, M_0 is finite dimensional. To prove this, let E be the projection onto the subspace $M^\perp$ and note that $E(M_0)$ being finite dimensional is closed and

$$M + M_0 = E^{-1}(E(M_0)).$$

Theorem 1.2.8 (Atkinson's theorem)

An operator T in $\mathcal{L}(\mathcal{H})$ is Fredholm if and only if the range of T is closed, $\dim \ker T$ and $\dim \ker T^*$ are finite.

Proof Assume that T is a Fredholm operator. By the definition, there exists an operator S in $\mathcal{L}(\mathcal{H})$ and a compact operator K such that $ST = I + K$. If f is a vector in the kernel of $I + K$, then $Kf = -f$, and hence, f is in the range of K. Thus,

$$\ker T \subseteq \ker ST = \ker(I + K) \subseteq \operatorname{ran} K$$

and therefore, by Proposition 1.2.2, $\dim \ker T$ is finite. Similarly, $\dim \ker T^*$ is finite (provided $TS = I + K$ for some $S \in \mathcal{L}(\mathcal{H})$ and $K \in C(\mathcal{H})$). Further, by Theorem 1.2.2, there exists a finite rank operator F such that $\|K - F\| < \frac{1}{2}$. Hence, for $f \in \ker F$, we have

$$\|S\| \|Tf\| \;\geqslant\; \|STf\| = \|f + Kf\| = \|f + Kf - Ff\|$$

$$\geqslant\; \|f\| - \|Kf - Ff\| \geqslant \|f\|/2.$$

Therefore, T is bounded below on $\ker F$, which implies that $T(\ker F)$ is a closed subspace of $\mathcal{H}$. To show that range of T is closed, observe that $(\ker F)^{\perp}$ is finite dimensional and

$$\operatorname{ran} T = T(\ker F) + T(\ker F)^{\perp}.$$

Hence, by the discussion prior to the theorem, $\operatorname{ran} T$ is closed.

Conversely, assume that the range of T is closed and both kernel and cokernel of T are finite. Define $T_0 : (\ker T)^{\perp} \to \operatorname{ran} T$ by

$$T_0 f = Tf, \quad f \in (\ker T)^{\perp},$$

and note that T_0 is one to one and onto. Hence, it is invertible by the open mapping theorem (see Theorem B.1.9). If we define the operator S on $\mathcal{H}$ by

$$S f = \begin{cases} T_0^{-1} f & f \in \operatorname{ran} T, \\ 0 & f \in (\operatorname{ran} T)^{\perp}, \end{cases}$$

then S is bounded, $ST = I - P_1$ and, $TS = I - P_2$, where P_1 is the projection onto $\ker T$ and P_2 is the projection onto $(\operatorname{ran} T)^{\perp} = \ker T^*$. Therefore, $\pi(S)$ is the inverse for $\pi(T)$ in the Calkin algebra $Q(\mathcal{H})$ and the proof is complete. $\qquad\square$

Remark 1.2.9 An examination of the proof of Atkinson's theorem shows that if $T \in \mathcal{L}(\mathcal{H})$ is *left-Fredholm* (that is, there exists an operator S in $\mathcal{L}(\mathcal{H})$ and a compact operator K such that $ST = I + K$), then the range of T is closed and the kernel is finite dimensional.

An operator $N \in \mathcal{L}(\mathcal{H})$ is said to *normal* if the self-commutator $[T, T^*] := TT^* - T^*T$ of T is the zero operator.

Corollary 1.2.4

Let $N \in \mathcal{L}(\mathcal{H})$ be a normal operator. If N is Fredholm, then N invertible if and only if N is injective.

Proof Assume that N is a normal and injective operator. Note that N^* is also an injective operator. If, in addition, N is Fredholm, then by Atkinson's theorem, the range of N is closed and dense, and hence, N is invertible. $\qquad\square$

1.3 Fredholm Index and Abstract Index

In this section, we discuss two notions of index for Fredholm operators. The first of these is called the Fredholm index, whereas the second is the abstract index. We first show that the Fredholm index is a continuous homomorphism from the set of Fredholm operators onto $\mathbb{Z}$. It can be shown that the abstract index is a different manifestation of the Fredholm index (see Theorem 1.3.14). We also discuss the spectral theory of compact and normal operators.

Fredholm Index

If $\mathcal{H}$ is a Hilbert space and $T : \mathcal{H} \to \mathcal{H}$ is a Fredholm operator, then the two numbers

$$\alpha_T = \dim \ker T \text{ and } \beta_T = \dim \ker T^*$$

would seem to contain important information concerning the operator T. However, their difference, to be referred to as the *Fredholm index or index* of T,

$$\text{ind}(T) = \alpha_T - \beta_T$$

is of even greater importance. If $T : V \to W$ is a finite dimensional operator (that is, T is linear and V, W are finite dimensional spaces), then by the rank-nullity theorem [17],

$$\dim V - \alpha_T = \text{rank } T = \text{rank } T^* = \dim W - \beta_T.$$

Thus, we obtain the following basic fact:

Lemma 1.3.1

For any finite dimensional operator $T : V \to W$,

$$ind(T) = \dim V - \dim W.$$

Let L and L' be any two invertible operators on a given Hilbert space $\mathcal{H}$. By Proposition 1.2.4, an operator $T : \mathcal{H} \to \mathcal{H}$ is Fredholm if and only if LTL' is Fredholm. Moreover, since $\dim \ker(LTL') = \dim \ker T$, we obtain

$$\alpha_{LTL'} = \alpha_T, \quad \beta_{LTL'} = \beta_T \quad \text{and} \quad \text{ind}(LTL') = \text{ind}(T). \tag{1.3.6}$$

Finally, if $T = T_1 \oplus T_2$ (direct sum), then $\ker T = \ker T_1 \oplus \ker T_2$, and consequently, $\alpha_T = \alpha_{T_1} + \alpha_{T_2}$. Similarly, $\beta_T = \beta_{T_1} + \beta_{T_2}$ and hence,

$$\text{ind}(T_1 \oplus T_2) = \text{ind}(T_1) + \text{ind}(T_2). \tag{1.3.7}$$

Lemma 1.3.2

Let $\mathcal{H}_1 \oplus \mathcal{H}_2 = \mathcal{H} = \mathcal{H}'_1 \oplus \mathcal{H}'_2$ be any two direct sum decompositions of the Hilbert space $\mathcal{H}$. Consider the matrix representation of T with respect to this decomposition:

$$T = \begin{bmatrix} T_{11} & T_{12} \\ T_{21} & T_{22} \end{bmatrix} : \mathcal{H}_1 \oplus \mathcal{H}_2 \to \mathcal{H}'_1 \oplus \mathcal{H}'_2.$$

Assume that $T_{22} : \mathcal{H}_2 \to \mathcal{H}'_2$ is invertible and set $\tilde{T} = T_{11} - T_{12}T_{22}^{-1}T_{21} : \mathcal{H}_1 \to \mathcal{H}'_1$. Then,

$$\alpha_{\tilde{T}} = \alpha_T, \quad \beta_{\tilde{T}} = \beta_T \quad and \quad ind(T) = \alpha_{\tilde{T}} - \beta_{\tilde{T}}.$$

Proof The ingenious proof shown here is from [84]:

$$\begin{bmatrix} T_{11} & T_{12} \\ T_{21} & T_{22} \end{bmatrix} \longrightarrow \begin{bmatrix} T_{11} & T_{12} \\ T_{21} & T_{22} \end{bmatrix}\begin{bmatrix} I & 0 \\ 0 & T_{22}^{-1} \end{bmatrix}$$

$$= \begin{bmatrix} T_{11} & T_{12}T_{22}^{-1} \\ T_{21} & I \end{bmatrix}$$

$$\longrightarrow \begin{bmatrix} I & -T_{12}T_{22}^{-1} \\ 0 & I \end{bmatrix}\begin{bmatrix} T_{11} & T_{12}T_{22}^{-1} \\ T_{21} & I \end{bmatrix}$$

$$= \begin{bmatrix} T_{11} - T_{12}T_{22}^{-1}T_{21} & 0 \\ T_{21} & I \end{bmatrix}$$

$$\longrightarrow \begin{bmatrix} T_{11} - T_{12}T_{22}^{-1}T_{21} & 0 \\ T_{21} & I \end{bmatrix}\begin{bmatrix} I & 0 \\ -T_{21} & T_{22} \end{bmatrix}$$

$$= \begin{bmatrix} \tilde{T} & 0 \\ 0 & T_{22} \end{bmatrix}.$$

Thus, we obtain invertible operators L and L' such that $LTL' = \tilde{T} \oplus T_{22}$. Since T_{22} is invertible, $\ker LTL' = \ker \tilde{T} \oplus \{0\}$ and it follows that $\alpha_T = \alpha_{LTL'} = \alpha_{\tilde{T}}$. Similarly, $\beta_T = \beta_{LTL'} = \beta_{\tilde{T}}$. This completes the proof of the Lemma. □

Most of the basic properties of index are contained in the following theorem.

Theorem 1.3.1

The index of a Fredholm operator is

(1) *locally constant, that is, for any Fredholm operator T, there exists $\epsilon > 0$ such that $ind(S) = ind(T)$ for every Fredholm operator S such that $\|S - T\| < \epsilon$,*

(2) *invariant under compact perturbations, that is, for any Fredholm operator T and a compact operator K,*

$$ind(T + K) = ind(T),$$

(3) a homomorphism, that is, if S, T are any two Fredholm operators, then ST is Fredholm and

$$\text{ind}(ST) = \text{ind}(S) + \text{ind}(T).$$

Proof Let $T : \mathcal{H} \to \mathcal{H}$ be any Fredholm operator. We decompose the Hilbert space $\mathcal{H}$ as

$$\ker T \oplus (\ker T)^{\perp} = \mathcal{H} = \ker T^* \oplus (\ker T^*)^{\perp}.$$

Write T as a 2×2 block matrix with respect to this decomposition:

$$T = \begin{bmatrix} T_{11} & T_{12} \\ T_{21} & T_{22} \end{bmatrix} : \ker T \oplus (\ker T)^{\perp} \longrightarrow \ker T^* \oplus (\ker T^*)^{\perp}.$$

Since T is Fredholm, ran T is closed and it follows that the operator $T_{22} : (\ker T)^{\perp} \to (\ker T^*)^{\perp} =$ ran T is invertible. If S is any other Fredholm operator such that $\|S - T\| < \epsilon$ and S is written as a 2×2 block matrix with respect to the same decomposition of $\mathcal{H}$ as earlier, then $\|S_{22} - T_{22}\| < \epsilon$, and hence, S_{22} is invertible for sufficiently small ϵ (see Corollary B.1.6). By Lemma 1.3.2,

$$\text{ind}(S) = \alpha_{\tilde{S}} - \beta_{\tilde{S}},$$

where $\tilde{S} = S_{11} - S_{12}S_{22}^{-1}S_{21}$. However the operator $\tilde{S} : \ker T \to \ker T^*$ is finite dimensional. Therefore, by Lemma 1.3.1, we have

$$\alpha_{\tilde{S}} - \beta_{\tilde{S}} = \alpha_T - \beta_T = \text{ind}(T).$$

Thus, $\text{ind}(T)$ is locally constant and (1) is established.

To prove (2), note that for any K in $C(\mathcal{H})$, the map $\psi_K : [0, 1] \to \mathbb{Z}$ given by

$$\psi_K(t) = \text{ind}(T + tK), \quad t \in [0, 1]$$

is locally constant, and therefore constant on any connected set. In particular,

$$\text{ind}(T) = \psi(0) = \psi(1) = \text{ind}(T + K).$$

To prove (3), let S be a Fredholm operator and note that

$$ST \oplus I = LQ_{\epsilon}L', \tag{1.3.8}$$

where L, Q_{ϵ} and L' are given by

$$L = \begin{bmatrix} I & -\epsilon^{-1}S \\ 0 & I \end{bmatrix}, \quad Q_{\epsilon} = \begin{bmatrix} S & 0 \\ \epsilon I & T \end{bmatrix}, \quad \text{and} \quad L' = \begin{bmatrix} T & \epsilon^{-1}I \\ -\epsilon I & 0 \end{bmatrix}.$$

The two operators S and T are Fredholm and hence, $S \oplus T$ is Fredholm. Thus, by Proposition 1.2.4, for sufficiently small ϵ, Q_{ϵ} is Fredholm. Since L and L' are invertible, by Proposition

1.2.4 and (1.3.8), the operator $ST \oplus I$ is Fredholm, and hence, ST is Fredholm. By part (1), index is locally constant. Therefore, for sufficiently small ϵ,

$$\mathrm{ind}(ST) = \mathrm{ind}(ST \oplus I) = \mathrm{ind}(Q_\epsilon) = \mathrm{ind}(S \oplus T) = \mathrm{ind}(S) + \mathrm{ind}(T),$$

where we used (1.3.6), (1.3.7) and (1.3.8). This yields (3). □

A convenient way of calculating the Fredholm index (see [16, (1.2)]) is to first fix the dimension of the cokernel; then, there is only the dimension of the kernel that needs to be determined.

Corollary 1.3.1

Let $T \in \mathcal{L}(\mathcal{H})$ be a Fredholm operator and let $\{e_n\}_{n \geqslant 0}$ be an orthonormal basis of $\mathcal{H}$. For a non-negative integer n, let $\mathcal{H}_n$ be the closed subspace of $\mathcal{H}$ spanned by $\{e_j : j \geqslant n\}$. Set $T_n = P_n \circ T$, where P_n is the orthogonal projection onto $\mathcal{H}_n$. Suppose that $\{e_0, \ldots, e_{n_0-1}\}$ spans $\ker T^*$ for some positive integer n_0. Then,

$$\mathrm{ind}(T) = \dim \ker T_{n_0} - n_0.$$

Proof For any integer $n \geqslant 0$, P_n is a rank n perturbation of the identity operator on $\mathcal{H}$. Thus, P_n is Fredholm with index equal 0, and hence, by Theorem 1.3.1, T_n is Fredholm such that

$$\mathrm{ind}(T_n) = \mathrm{ind}(T), \quad n \geqslant 0. \tag{1.3.9}$$

Note that $T_{n_0}(\mathcal{H}) = P_{n_0}(T(\mathcal{H})) = \mathcal{H}_{n_0}$, and hence,

$$\mathrm{ind}(T_{n_0}) = \dim \ker T_{n_0} - \dim \ker T_{n_0}^* = \dim \ker T_{n_0} - n_0.$$

The desired conclusion is now immediate from (1.3.9). □

We explicitly compute the Fredholm index of the forward unilateral shift operator.

***Example* 1.3.2** Let $\mathbb{N}$ denote the set of non-negative integers and let $\ell^2(\mathbb{N})$ denote the complex Hilbert space of square-summable sequences indexed by $\mathbb{N}$. The *shift operator* $U_+ : \ell^2(\mathbb{N}) \to \ell^2(\mathbb{N})$ is defined as

$$U_+(a_0, \ldots, a_n, \ldots) = (0, a_0, \ldots, a_n, \ldots).$$

Note that U_+ is an isometry, that is, $\|U_+ a\| = \|a\|$ for every $a \in \ell^2(\mathbb{N})$. Thus, $\ker U_+ = \{0\}$ and $\mathrm{ran}\, U_+$ is closed in $\ell^2(\mathbb{N})$ (see Remark 1.1.2). Further, a simple computation shows that

$$U_+^* e_n = \begin{cases} 0 & \text{if } n = 0 \\ e_{n-1} & \text{if } n \geqslant 1, \end{cases}$$

where $\{e_n\}_{n \geqslant 0}$ is the standard orthonormal basis for $\ell^2(\mathbb{N})$. Here e_n denotes the sequence in $\ell^2(\mathbb{N})$ with 1 at the $(n+1)$th place and 0 elsewhere. It is now clear that $\dim \ker U_+^* = 1$. Thus, U_+ is Fredholm such that

$$\mathrm{ind}(U_+) = -1 \quad \text{and} \quad \mathrm{ind}(U_+^*) = 1.$$

This may also be derived from Corollary 1.3.1 (the case in which $n_0 = 1$). Since index is a homomorphism (Theorem 1.3.1(3)), it follows that

$$\text{ind}(U_+^n) = -n \quad \text{and} \quad \text{ind}(U_+^{*n}) = n.$$

This shows in particular that ind is a surjective homomorphism from the group of Fredholm operators in $\mathcal{L}(\ell^2(\mathbb{N}))$ onto $\mathbb{Z}$.   ∎

As any complex infinite dimensional Hilbert space contains $\ell^2(\mathbb{N})$ (up to an isometric isomorphism), we obtain the following general fact.

Corollary 1.3.2

For any complex infinite dimensional Hilbert space $\mathcal{H}$, the Fredholm index is a surjective homomorphism from the group $\mathcal{F}(\mathcal{H})$ of Fredholm operators in $\mathcal{L}(\mathcal{H})$ onto $\mathbb{Z}$.

Proof By Theorem 1.3.1, it suffices to check that ind is surjective. Let $\{f_n\}_{n \geqslant 0}$ be an infinite orthonormal subset of $\mathcal{H}$ and let $\mathcal{K}$ be the closed linear span of $\{f_n\}_{n \geqslant 0}$. Let $U : \mathcal{K} \to \ell^2(\mathbb{N})$ be the isometric isomorphism governed by $U f_n = e_n$ for integers $n \geqslant 0$. Let $U_n \in \mathcal{L}(\mathcal{K})$ be defined as

$$U_n = \begin{cases} U^* U_+^n U & \text{if } n \geqslant 0 \\ U^* U_+^{*n} U & \text{if } n \leqslant 0, \end{cases}$$

and let I denote the identity operator on $\mathcal{K}^\perp$. Decompose $\mathcal{H}$ as $\mathcal{K} \oplus \mathcal{K}^\perp$. In view of Example 1.3.2, for every integer n, the operator $U_n \oplus I$ in $\mathcal{L}(\mathcal{H})$ has index equal to $-n$.   □

By Theorem 1.3.1, ind is constant on any connected component of $\mathcal{F}(\mathcal{H})$ of Fredholm operators. For an integer n, let $\mathcal{F}_n(\mathcal{H})$ denote the set of Fredholm operators in $\mathcal{L}(\mathcal{H})$ of index n. As ind is also constant on $\mathcal{F}_n(\mathcal{H})$, the following natural questions arise. What are the connected components of $\mathcal{F}(\mathcal{H})$? Are these precisely $\mathcal{F}_n(\mathcal{H})$, $n \in \mathbb{Z}$? Before we answer these questions, we present a preliminary result.

Lemma 1.3.3

If $A \in \mathcal{F}_0(\mathcal{H})$, then there exists a finite rank partial isometry $F \in \mathcal{L}(\mathcal{H})$ with initial space $\ker A$ and final space $\ker A^*$. Moreover, the partial isometry F can be chosen so that $A + F$ is invertible.

Proof Let $A \in \mathcal{F}_0(\mathcal{H})$. As A is Fredholm of index 0, there exists an unitary $U : \ker A \to \ker A^*$. The desired partial isometry $F \in B(\mathcal{H})$ is given by $F = JU P_{\ker A}$, where J is the inclusion of $\ker A^*$ into $\mathcal{H}$ and $P_{\mathcal{M}}$ denotes the orthogonal projection of $\mathcal{H}$ onto the closed subspace $\mathcal{M}$ of $\mathcal{H}$. Clearly, F is of finite rank since $\ker A$ is finite dimensional. Further, as $A P_{\ker A} = 0$ and $F P_{(\ker A)^\perp} = 0$, we have

$$A + F = (A + F)(P_{\ker A} + P_{(\ker A)^\perp}) = A P_{(\ker A)^\perp} + JU P_{\ker A}. \tag{1.3.10}$$

Let $x \in \mathcal{H}$ be such that $(A + F)x = 0$. Then, by (1.3.10), $AP_{(\ker A)^\perp}x$ and $JUP_{\ker A}x$ being orthogonal are equal to 0. As J, U are injective, $x \in (\ker A)^\perp$, and hence, $Ax = 0$ implying that $x = 0$. Thus, $A + F$ is injective. Further, by (1.3.10),

$$\operatorname{ran}(A + F) = \operatorname{ran}(AP_{(\ker A)^\perp}) \oplus \operatorname{ran}(JUP_{\ker A}) = \operatorname{ran}A \oplus \ker A^* = \mathcal{H}.$$

Thus, $A + F$ is one–one and surjective. The invertibility of $A + F$ now follows from the bounded inverse theorem (see Theorem B.1.5). □

The following result describes the connected components of $\mathcal{F}(\mathcal{H})$.

Corollary 1.3.3

The connected components of the collection $\mathcal{F}(\mathcal{H})$ of Fredholm operators in $\mathcal{L}(\mathcal{H})$ are precisely $\mathcal{F}_n(\mathcal{H})$, $n \in \mathbb{Z}$.

Proof   We will first show that the set $\mathcal{F}_0(\mathcal{H})$ is path-connected. The path-connectedness of $\mathcal{F}_n(\mathcal{H})$ for $n > 0$ will be then derived from that of $\mathcal{F}_0(\mathcal{H})$. The proof is divided into the following steps.

Step I *As before, let $\mathcal{G}(\mathcal{H})$ denote the group of invertible operators in $\mathcal{L}(\mathcal{H})$ and $\mathcal{T}(\mathcal{H})$, the ideal of finite rank operators in $\mathcal{L}(\mathcal{H})$. Note that*

$$\mathcal{F}_0(\mathcal{H}) = \mathcal{G}(\mathcal{H}) + \mathcal{T}(\mathcal{H}). \tag{1.3.11}$$

An application of Theorem 1.3.1 shows that $\mathcal{G}(\mathcal{H}) + \mathcal{T}(\mathcal{H}) \subseteq \mathcal{F}_0(\mathcal{H})$. To see the reverse inclusion, let $A \in \mathcal{F}_0(\mathcal{H})$. By Lemma 1.3.3, there exists a finite rank, partial isometry $F \in \mathcal{L}(\mathcal{H})$ (with initial space $\ker(A)$ and final space $\ker(A^*)$) such that $A + F$ is invertible. Thus, A is a finite rank perturbation of an invertible operator, and the claim stands verified.

Step II $\mathcal{F}_0(\mathcal{H})$ *is path-connected.*

It suffices to check that any $A \in \mathcal{F}_0(\mathcal{H})$ is path-connected to I. By (1.3.11), $A + F$ is invertible for some finite rank operator F. Moreover, as $I + F \in \mathcal{F}_0(\mathcal{H})$ (see Theorem 1.3.1(2)), it suffices to show that $A + F$ can be connected to I. The desired conclusion now follows from the fact that the group $\mathcal{G}(\mathcal{H})$ of invertible operators forms a path-connected subset of $\mathcal{F}_0(\mathcal{H})$ (see Corollary C.1.7).

We now complete the proof of the corollary. Let $A, B \in \mathcal{F}_n(\mathcal{H})$. Then there exists $C \in \mathcal{F}_{-n}(\mathcal{H})$ such that $CB = I + K$, which belongs to $\mathcal{F}_0(\mathcal{H})$ by Step II. By Theorem 1.3.1, $\operatorname{ind}(A) = \operatorname{ind}(B) = -\operatorname{ind}(C)$ and AC belongs to $\mathcal{F}_0(\mathcal{H})$. Let γ be a path connecting I and AC as ensured by Step III. It is easy to see that

$$\delta(t) = \gamma(t)B - tAK, \quad t \in [0, 1]$$

is the desired path in $\mathcal{F}(\mathcal{H})$, which connects A and B. □

Spectral Properties of Compact Operators

Let $\mathcal{H}$ be a nonzero complex Hilbert space. The *spectrum* $\sigma(T)$ of $T \in \mathcal{L}(\mathcal{H})$ is the subset

$$\{\lambda \in \mathbb{C} : T - \lambda I \text{ is not invertible in } \mathcal{L}(\mathcal{H})\}$$

of the complex plane $\mathbb{C}$. For a finite dimensional operator, the spectrum is the set of eigenvalues. The reader is referred to Appendix B for elementary properties of the spectrum. Recall the fact that eigenvectors corresponding to distinct eigenvalues of a bounded linear operator in $\mathcal{L}(\mathcal{H})$ are linearly independent. In case $\mathcal{H}$ is separable and the eigenvectors of T corresponding to distinct eigenvalues are orthogonal (for normal operators this is the case), the set of eigenvalues is countable.

The following contains basic spectral properties of a compact operator.

Theorem 1.3.3

Suppose that $K \in \mathcal{L}(\mathcal{H})$ is a compact operator. Then, $\sigma(K)$ is countable with 0 being the only possible limit point. If λ is a nonzero point in $\sigma(K)$, then λ is an eigenvalue of finite multiplicity and $\bar{\lambda}$ is an eigenvalue of K^* with the same multiplicity.

Proof Let λ be a nonzero complex number in $\sigma(K)$. One may use the argument of Step II of the proof of Corollary 1.3.3 to conclude that $K - \lambda$ is Fredholm and $\text{ind}(K - \lambda) = 0$. As $\lambda \in \sigma(K)$, it now follows from Theorem 1.2.8 that $\ker(K - \lambda)$ is nonzero and finite dimensional. Thus, λ is an eigenvalue of K of finite multiplicity. Moreover, as $\text{ind}(K - \lambda) = 0$, we see that $\bar{\lambda}$ is an eigenvalue of K^* of the same multiplicity. Thus, it suffices to check that 0 is the only possible limit point of $\sigma(K)$.

We may assume that $\mathcal{H}$ is infinite dimensional. Let $\{\lambda_n\}_{n \geqslant 1}$ be a sequence of distinct eigenvalues of K with corresponding eigen vector $\{f_n\}_{n \geqslant 1}$. For $n \in \mathbb{N}$, let $\mathcal{M}_0 = \{0\}$ and $\mathcal{M}_n = \text{span}\{f_1, \ldots, f_n\}$, $n \geqslant 1$. Note that $\mathcal{M}_1 \subsetneq \mathcal{M}_2 \subsetneq \ldots$ (see the discussion prior to Theorem 1.3.3). For every positive integer n, let g_n be a unit vector in $\mathcal{M}_n$ orthogonal to $\mathcal{M}_{n-1}$ (as ensured by the Gram–Schmidt orthogonalization process). For any h in $\mathcal{H}$, $h = \sum_{n \geqslant 1} \langle h, g_n \rangle g_n + g$, where g is orthogonal to g_n for all integers $n \geqslant 1$. As $\|h\|^2 = \sum_{n \geqslant 1} |\langle h, g_n \rangle|^2 + \|g\|^2$, it follows that $\lim_{n \to \infty} \langle h, g_n \rangle = 0$. Therefore, the sequence $g_n \to 0$ weakly and hence, by Lemma 1.2.1,

$$\|Kg_n\| \to 0 \text{ as } n \to \infty. \tag{1.3.12}$$

As $g_n \in \mathcal{M}_n$, there exists scalars $\alpha_1, \ldots, \alpha_n$ such that $g_n = \sum_{k=1}^{n} \alpha_k f_k$. Moreover, for some $h_n \in \mathcal{M}_{n-1}$,

$$
\begin{aligned}
Kg_n &= \sum_{k=1}^{n} \alpha_k \lambda_k f_k \\
&= \lambda_n \sum_{k=1}^{n} \alpha_k f_k + \sum_{k=1}^{n-1} \alpha_k (\lambda_k - \lambda_n) f_k \\
&= \lambda_n g_n + h_n.
\end{aligned}
$$

Therefore, by (1.3.12),

$$|\lambda_n|^2 \leqslant |\lambda_n|^2 \|g_n\|^2 + \|h_n\|^2 = \|Kg_n\|^2 \to 0 \text{ as } n \to \infty.$$

This completes the proof of the theorem.   □

Remark 1.3.4   If $\mathcal{H}$ is infinite dimensional, then the spectrum of any compact operator on $\mathcal{H}$ is countable, 0 is a limit point of this set and it is the only limit point.

Let us see some applications of Theorem 1.3.3.

Corollary 1.3.4

Let K be a compact operator on an infinite dimensional Hilbert space $\mathcal{H}$. If K is normal, then there exists an orthonormal basis of $(\ker K)^{\perp}$ consisting of eigenvectors corresponding to nonzero eigenvalues.

Proof   Assume that K is normal. It is easy to see that eigenvectors of K corresponding to distinct eigenvalues are orthogonal. By Theorem 1.3.3 and Remark 1.3.4, $\sigma(K) \backslash \sigma_p(K) = \{0\}$ and the eigenspaces of K are invariant for K and K^* (recall that $\mathcal{M}$ is an *invariant subspace* for T if $T\mathcal{M} \subseteq \mathcal{M}$). Let $\mathcal{M}$ denote the orthogonal direct sum of all eigenspaces of K corresponding to nonzero eigenvalues and note that $\ker K \subseteq \mathcal{M}^{\perp}$. We contend that $\mathcal{M}^{\perp} = \ker K$. We note that $K|_{\mathcal{M}^{\perp}}$ is a normal and compact operator. If $K|_{\mathcal{M}^{\perp}}$ is nonzero, then by Theorem 1.3.3, it has a nonzero eigenvalue. However, eigenvalue of $K|_{\mathcal{M}^{\perp}}$ is also an eigenvalue of K, which is not possible. Thus, $K|_{\mathcal{M}^{\perp}} = 0$, and hence, $\mathcal{M}^{\perp} = \ker K$. This completes the proof.   □

As the second application of Theorem 1.3.3, we derive the following decomposition theorem for compact operators.

Corollary 1.3.5 (Singular Value Decomposition)

For a Hilbert space $\mathcal{H}$, let $K \in \mathcal{L}(\mathcal{H})$ be a compact operator. If $\mathcal{H}$ is infinite dimensional, then there exists a decreasing sequence $\{\mu_j(K)\}_{j \geqslant 0}$ of non-negative real numbers converging to 0 and orthonormal families $\{\varphi_j\}_{j \geqslant 0}$ and $\{\psi_j\}_{j \geqslant 0}$ in $\mathcal{H}$ such that

$$K = \sum_{j=0}^{\infty} \mu_j(K)\varphi_j \otimes \psi_j, \quad |K| = \sum_{j=0}^{\infty} \mu_j(K)\varphi_j \otimes \varphi_j \qquad (1.3.13)$$

(see (1.2.2)).

Proof   Assume that $\mathcal{H}$ is infinite dimensional. By Proposition 1.2.1, K^*K is a compact positive operator. Hence, by Corollary 1.3.4, there exists an orthonormal basis for $(\ker K)^{\perp}$ such that K^*K is a diagonal operator with diagonal entries $\{\mu_j^2(K)\}_{j \geqslant 0}$, where $\{\mu_j(K)\}_{j \geqslant 0}$ is a decreasing sequence of non-negatives real numbers. Further, by Theorem 1.3.3, $\{\mu_j^2(K)\}_{j \geqslant 0}$ converges to 0. This gives the desired representation for $|K|$. To see the required representation of K, consider the polar decomposition $K = V|K|$, where V is a partial isometry, isometric on the range of $|K|$

(see the proof of Theorem 1.1.3). In particular, $\psi_n = V\varphi_n$, $n \geqslant 0$ forms an orthonormal family. Further, by the continuity of V,

$$K = V|K| = \sum_{j=0}^{\infty} \mu_j(K)V(\varphi_j \otimes \varphi_j) = \sum_{j=0}^{\infty} \mu_j(K)\psi_j \otimes \varphi_j,$$

where we used the fact that $V(\varphi_j \otimes \varphi_j) = \psi_j \otimes \varphi_j$, $j \geqslant 0$. □

The eigenvalues $\{\mu_j(K)\}_{j\geqslant 0}$ of $|K|$ as appearing in the singular value decomposition of compact operators are referred to as *singular values* of K. The action of K completely determines these values in the following sense:

$$\mu_0(K) = \|K\|, \quad \mu_j(K) = \inf_{\psi_1,\dots,\psi_{j-1}} \Big(\sup_{\substack{\varphi \perp \psi_1,\dots,\psi_{j-1} \\ \|\varphi\|=1}} \|K\varphi\| \Big), \quad j \geqslant 1. \tag{1.3.14}$$

These formulae in turn rely on the following fact.

Corollary 1.3.6 (Max–Min Criterion)

Let $\mathcal{H}$ be an infinite dimensional complex Hilbert space and let $K \in \mathcal{L}(\mathcal{H})$ be a positive compact operator. If $\lambda_0(K) \geqslant \lambda_1(K) \geqslant \dots$ are the eigenvalues of K (ordered with counting multiplicities). then, for any positive integer n,

$$\lambda_n(K) = \inf_{\psi_1,\dots,\psi_{n-1}} \Big(\sup_{\substack{\varphi \perp \psi_1,\dots,\psi_{n-1} \\ \|\varphi\|=1}} \langle K\varphi, \varphi \rangle \Big). \tag{1.3.15}$$

Proof By Corollary 1.3.4, there exists an orthonormal basis $\{\varphi_j\}_{j\geqslant 0}$ for $(\ker K)^\perp$ such that $K\varphi_j = \lambda_j(K)\varphi_j$, $j \in \mathbb{N}$. Let $\psi_1,\dots,\psi_{n-1} \in \mathcal{H}$. Clearly, the intersection $\text{span}\{\varphi_1,\dots,\varphi_n\} \cap \{\psi_1,\dots,\psi_{n-1}\}^\perp$ is non-trivial. Hence, for any unit vector φ in this intersection, we have

$$\langle K\varphi, \varphi \rangle = \sum_{j=0}^{\infty} \lambda_j(K)|\langle \varphi, \varphi_j \rangle|^2 = \sum_{j=0}^{n} \lambda_j(K)|\langle \varphi, \varphi_j \rangle|^2$$

$$\geqslant \lambda_n(K) \sum_{j=0}^{n} |\langle \varphi, \varphi_j \rangle|^2 = \lambda_n(K).$$

This shows that for any $\psi_1,\dots,\psi_{n-1}$,

$$\sup_{\substack{\varphi \perp \psi_1,\dots,\psi_{n-1} \\ \|\varphi\|=1}} \langle K\varphi, \varphi \rangle \geqslant \lambda_n(K).$$

Further, the supremum above is attained at $\psi_j = \varphi_j$, $j = 1,\dots,n-1$ and $\varphi = \varphi_n$, and hence, infimum is attained in (1.3.15). □

Let us now complete the verification of (1.3.14). By (1.3.15),

$$\mu_j(K)^2 = \lambda_j(K^*K) = \inf_{\psi_1,\dots,\psi_{n-1}} \Big(\sup_{\substack{\varphi \perp \psi_1,\dots,\psi_{n-1} \\ \|\varphi\|=1}} \langle K^*K\varphi, \varphi \rangle \Big),$$

which after taking positive square root on both sides yields (1.3.14).

Essential Spectrum of a Normal Operator

In this subsection, we discuss the relationship between spectrum and essential spectrum of a normal operator. First let us define the essential spectrum.

Definition 1.3.5

The *essential spectrum* $\sigma_e(T)$ of $T \in \mathcal{L}(\mathcal{H})$ is given by

$$\sigma_e(T) = \sigma(\pi(T)),$$

that is, the spectrum of the image of T under the Calkin map π in the Calkin algebra $Q(\mathcal{H})$.

Remark 1.3.6   As the spectrum is non-empty (Appendix B), so is the essential spectrum. Moreover, by Atkinson's theorem, the complement $\mathbb{C}\backslash\sigma_e(T)$ of the essential spectrum $\sigma_e(T)$ of T is given by

$$\{\lambda \in \mathbb{C} : \mathrm{ran}(T - \lambda) \text{ is closed, } \dim \ker(T - \lambda) < \infty \text{ and } \dim \ker(T^* - \overline{\lambda}) < \infty\}.$$

The essential spectrum of a compact operator on an infinite-dimensional Hilbert space is equal to $\{0\}$.

For a normal operator, all points in the spectrum except isolated eigenvalues of finite multiplicity form its essential spectrum.

Theorem 1.3.7

If $N \in \mathcal{L}(\mathcal{H})$ is a normal operator, then

$$\sigma(N) = \sigma_e(N) \cup \pi_{00}(N), \tag{1.3.16}$$

where $\pi_{00}(N)$ denotes the set of isolated eigenvalues of N of finite multiplicity (repeated as often as their multiplicity).

Proof   Let $N \in \mathcal{L}(\mathcal{H})$ be a normal operator with the spectral measure $E(\cdot)$ and let $\mathcal{K} = E(\sigma(N)\backslash\{\lambda\})\mathcal{H}$ for $\lambda \in \pi_{00}(N)$. Note that $\mathcal{K}$ is a reducing proper closed subspace of $\mathcal{H}$. As $\mathcal{K}^\perp = \ker(N - \lambda)$, the range of $N - \lambda$ equals $(N - \lambda)\mathcal{K}$. However, $(N - \lambda)|_{\mathcal{K}}$ is invertible, and hence, the range of $N - \lambda$ is closed. This shows that

$$\pi_{00}(N) \subseteq \sigma(N)\backslash\sigma_e(N).$$

Thus, it is sufficient to check that any $\lambda \in \sigma(N)\backslash\sigma_e(N)$ is an isolated eigenvalue of finite multiplicity. To see that, let $\lambda \in \sigma(N)\backslash\sigma_e(N)$. If $\ker(N - \lambda) = \{0\}$, then by Corollary 1.2.4, we arrive at the contradiction that $N - \lambda$ is invertible. Hence, λ must be an eigenvalue of N of finite multiplicity. Thus, $\ker(N - \lambda)$ is a nonzero reducing subspace for N, and hence, N decomposes into $N_1 \oplus N_2$, where $N_1 = \lambda I$ and N_2 is a normal operator such that $\ker(N_2 - \lambda I) = \{0\}$. Moreover, as $\sigma_e(N_2) \subseteq \sigma_e(N)$ (see Exercise 1.7.38), $\lambda \notin \sigma_e(N_2)$. By another application of Corollary 1.2.4, $\lambda \notin \sigma(N_2)$. Since $\sigma(N) = \{\lambda\} \cup \sigma(N_2)$, λ must be an isolated point of $\sigma(N)$.   □

Let us construct a normal operator with essential spectrum equal to a given compact subset of the plane.

***Example* 1.3.8**   Let X be any compact subset of $\mathbb{C}$ and let $\{\mu_n\}_{n\geqslant1}$ denote the set of isolated points of X. Note that $X\backslash\{\mu_n\}_{n\geqslant1}$ is compact. Thus, there exists a diagonal operator D_1 such that $\sigma(D_1) = X\backslash\{\mu_n\}_{n\geqslant1}$. Let $D_2 = \oplus_{n\geqslant1}\mu_nI$, where μ_nI acts on $\ell^2(\mathbb{N})$, and consider $D = D_1 \oplus D_2$. Then, D is a block diagonal operator. Moreover, by the previous theorem, $\sigma_e(D) = \sigma(D) = X$. ∎

There is a simple way to obtain a diagonal perturbation of a normal operator with the property that the essential spectrum is equal to the spectrum after the perturbation. The precise statement follows.

Corollary 1.3.7

Let $D \in \mathcal{L}(\mathcal{H})$ be a normal operator such that

$$\sigma(D) = \sigma_e(D) \cup \{\mu_n\}_{n\geqslant1}$$

(cf. (1.3.16)), where $\{\mu_n\}_{n\geqslant1}$ is a countable set of isolated eigenvalues of D of finite multiplicity (repeated as often as their multiplicity). If $h_1,h_2,\ldots$ are orthonormal eigenvectors corresponding to $\mu_1,\mu_2,\ldots$ respectively, then $D_1 := D - \sum_{j=1}^{\infty}\mu_jh_j\otimes h_j$ is a normal operator that satisfies $\sigma(D_1) = \sigma_e(D_1)$.

Proof   Note first that $\sum_{j=1}^{\infty}\mu_jh_j\otimes h_j$ defines a bounded linear operator as $\{\mu_j\}_{j\geqslant1}$ is a bounded sequence. Choose $\{e_m\}_{m\geqslant1} \subseteq \mathcal{H}$, so that $\{e_m\}_{m\geqslant1} \cup \{h_n\}_{n\geqslant1}$ forms an orthonormal basis for $\mathcal{H}$. Since $\bigvee\{h_n\}_{n\geqslant1}$ is a reducing subspace for D, it is easy to see that D_1 is a normal operator. To prove the remaining part, suppose that there exists $\mu \in \sigma(D_1)\backslash\sigma_e(D_1)$. By Theorem 1.3.7, μ is an isolated eigenvalue of D_1 of finite multiplicity. Thus, there exists non-zero $h \in \mathcal{H}$ such that $D_1h = \mu h$. As $\ker D_1$ is infinite dimensional, $\mu \neq 0$. Write $h = \sum_{m=1}^{\infty}x_me_m + \sum_{n=1}^{\infty}y_nh_n$. Since $(D_1 - \mu)h = 0$, $D_1h_n = 0$, and $De_m = D_1e_m$ for all integers $m,n \geqslant 1$, we obtain

$$(D-\mu)\sum_{m=1}^{\infty}x_me_m = (D_1-\mu)\sum_{m=1}^{\infty}x_me_m$$
$$= -(D_1-\mu)\sum_{n=1}^{\infty}y_nh_n$$
$$= \mu\sum_{n=1}^{\infty}y_nh_n.$$

As $\bigvee\{e_m\}$ is reducing for D and $\mu \neq 0$, each $y_n = 0$. Thus, μ is an eigenvalue of D with eigenvector $\sum_{m=1}^{\infty}x_me_m$. Clearly, $\mu \neq \mu_n$ for every integer $n \geqslant 1$. Moreover, since μ is an isolated eigenvalue of finite multiplicity for D_1, μ is an isolated eigenvalue of D and its multiplicity is also finite. This is not possible as $\{\mu_n\}$ lists all isolated eigenvalues of D of finite multiplicity. This contradiction shows that $\sigma(D_1) = \sigma_e(D_1)$. □

An alternate and conceptually simpler proof of Corollary 1.3.7 can be given using orthogonal decomposition of D (see Exercise 1.7.39).

Abstract Index

Let B be a unital Banach algebra and let G denote the group of invertible elements in B. Let G_0 be the connected component of G containing the identity of G. Note that any connected component of G is open because all connected neighborhoods of its element are contained in it. It follows immediately that any connected component of G is closed in G, as well. In particular, G_0 is open as well as closed subset of G.

Example **1.3.9**   Let X be a compact Hausdorff space. Consider the Banach algebra $C(X)$ of continuous functions from X into $\mathbb{C}$. Note that the group G of invertible elements in $C(X)$ is given by

$$G = \{f \in C(X) : f \text{ is nowhere vanishing on } X\}.$$

Moreover, 1 is the identity for $C(X)$. Let $f \in C(X)$ belong to the connected component G_0 of G containing 1. Since G_0 is locally path-connected (G_0 is open and $C(X)$ is a vector space, therefore locally path-connected), there exists a collection $\{f_\lambda\}_{\lambda \in [0,1]}$ of continuous functions in $C(X)$ such that $f_0 = 1$ and $f_1 = f$. Define $\Gamma : X \times [0,1] \to G$ by $\Gamma(x, \lambda) = f_\lambda(x)$, and note that $\Gamma(x, 0) = 1$ and $\Gamma(x, 1) = f(x)$ for every $x \in X$. In other words, G_0 consists of precisely those elements in G that are *homotopic* to 1.   ∎

Recall that $\mathcal{F}_n(\mathcal{H})$ is the set of Fredholm operators in $\mathcal{L}(\mathcal{H})$ of index n.

Lemma 1.3.4

Let $\mathscr{G}(\mathcal{H})$ be the group of invertible elements in $Q(\mathcal{H})$. If $\mathscr{G}_0(\mathcal{H})$ denotes the connected component of $\mathscr{G}(\mathcal{H})$ containing the identity element in $\mathcal{G}(\mathcal{H})$, then

$$\mathscr{G}(\mathcal{H}) = \{\pi(T) : T \in \mathcal{F}(\mathcal{H})\} \text{ and } \mathscr{G}_0(\mathcal{H}) = \{\pi(T) : T \in \mathcal{F}_0(\mathcal{H})\},$$

where $\pi : \mathcal{L}(\mathcal{H}) \to Q(\mathcal{H})$ is the Calkin map.

Proof   Since $\mathcal{F}_n(\mathcal{H})$ is a connected component of $\mathcal{F}(\mathcal{H})$ (see Corollary 1.3.3) and since π is an open map, $\mathscr{G}(\mathcal{H})$ is the union of clopen sets $\{\pi(T) : T \in \mathcal{F}_0(\mathcal{H})\}$ and $\{\pi(T) : T \in \mathcal{F}(\mathcal{H}) \backslash \mathcal{F}_0(\mathcal{H})\}$. Moreover, this union is disjoint. Indeed, if $S_n \in \mathcal{F}_n(\mathcal{H})$ and $S_m \in \mathcal{F}_m(\mathcal{H})$, then $\pi(S_n) = \pi(S_m)$ implies $S_n - S_m$ is a compact operator. Since the Fredholm index is invariant under compact perturbations (Theorem 1.3.1), we have $m = n$ in this case.   □

Let B be a unital Banach algebra and let G denote the group of invertible elements in B. Let G_0 be the connected component of G containing the identity of G. For $f, g \in G_0$, we record the following observations:

- Since $L_g : G_0 \to gG_0$ given by $L_g(f) = gf$ is surjective and continuous, every coset of G_0 is connected.

- Since fg and f belongs to the connected subset fG_0 of B, $fG_0 \cup G_0$ is connected and non-empty. In particular, $fG_0 \cup G_0 \subseteq G_0$, and hence, G_0 is a subgroup.

- For any $h \in G$, hG_0h^{-1} is a connected subset containing identity, and hence, $hG_0h^{-1} = G_0$. That is, G_0 is normal.

- The cosets of G_0 are connected components of G. Since G_0 is a clopen subset of G, the quotient G/G_0 is a discrete group.

The previous discussion forms the basis of the following definition.

Definition 1.3.10

Let B be a unital Banach algebra. Let G denote the group of invertible elements in B and let G_0 denote the connected component of G containing the identity of G. The *abstract index group* for B is the quotient group G/G_0.

In what follows, we will be particularly interested in the abstract index group for the Banach algebra $C(X)$.

***Example* 1.3.11 (Example 1.3.9 continued)** Recall that the group G of invertible elements in $C(X)$ is the set of nowhere vanishing continuous functions in $C(X)$. It turns out that the connected component G_0 of $C(X)$ containing the identity element 1 in $C(X)$ is the group $\exp(C(X))$ of functions of the form e^f for some $f \in C(X)$. To see this, note that if f lies in an open unit ball in $C(X)$ around 1, then $f = e^g$, where $g \in B$ is given by

$$g = -\sum_{n=1}^{\infty} \frac{1}{n}(1-f)^n.$$

It is now easy to see that $\exp(C(X))$ forms a connected subgroup of G_0. Since G_0 is connected, we must have $G_0 = \exp(C(X))$.

In case X is of special form, the connected component G_0 can be described in geometric terms. Recall that any nonzero complex number z takes the form $z = |z|e^{i \arg(z)}$, where $\arg(z) \in [0, 2\pi)$ denotes the argument of z counted in the counter-clockwise direction. Suppose that X is a compact subset of $\mathbb{C}$ enclosed by a smooth Jordan curve C with parametrization $\gamma(t)$, $t \in [0,1]$, and let $f \in G$. Then, $f \in G_0$ if and only if the variation of $\arg(f)$ around C is zero. Indeed, if $f \in G_0$, then $f = e^g$ for some $g \in C(X)$, and hence,

$$\arg(f(\gamma(1))) - \arg(f(\gamma(0))) = \Im(g(\gamma(1))) - \Im(g(\gamma(0))) = 0,$$

where $\Im(z)$ denotes the imaginary part of the complex number z. Conversely, if the variation of $\arg(f)$ around C is zero, then g given by

$$g(z) = \log|f(z)| + i \arg(f(z))$$

is a well-defined continuous function that satisfies $e^g = f$. ∎

The abstract index group for the Banach algebra $C(X)$, denoted by $\Lambda(X)$, is also known as the *Bruschlinsky group*. We see in the next result that the abstract index group for $C(X)$ can be identified with the *first cohomotopy group* $\pi^1(X)$ of X. Recall that $\pi^1(X)$ is the group of (pointed) homotopy classes $[f]$ of continuous mappings f from X into the unit circle $\mathbb{T}$.

Proposition 1.3.1

For any compact Hausdorff space X, the abstract index group $\Lambda(X)$ for $C(X)$ and the first cohomotopy group $\pi^1(X)$ of X are isomorphic.

Proof   Let G_0 denote the connected component of $C(X)$ containing the identity element 1 in $C(X)$. Define $\Phi : \pi^1(X) \to \Lambda(X)$ by

$$\Phi([f]) = fG_0, \quad [f] \in \Lambda(X).$$

If f, g are homotopic, then $g^{-1}f$ is path-connected to 1, and hence, $g^{-1}f \in G_0$. Thus, Φ is well-defined. Clearly, Φ is a group homomorphism:

$$\Phi([f][g]) = \Phi([f])\Phi([g]), \quad [f], [g] \in \pi^1(X).$$

To see that Φ is onto, let $f \in C(X)$ be nowhere vanishing, and define $F : X \times [0,1] \to \mathbb{C}\backslash\{0\}$ by

$$F(x,t) = \frac{f(x)}{|f(x)|^t}, \quad x \in X, \, t \in [0,1].$$

Note that F is continuous, $F(x,0) = f(x)$ and $|F(x,1)| = 1$ for every $x \in X$. It is now easy to see that $g = F(\cdot,1) : X \to \mathbb{T}$ satisfies $\Phi([g]) = fG_0$. To see that Φ is injective, let $f, g : X \to \mathbb{T}$ be continuous functions such that $fG_0 = gG_0$. Then, $g^{-1}f \in G_0$, and hence, there exists a continuous function $H : X \times [0,1] \to \mathbb{C}\backslash\{0\}$ such that $H(\cdot,0) = g^{-1}f$ and $H(\cdot,1) = 1$. If we define $F : X \times [0,1] \to \mathbb{T}$ by

$$F(x,t) = \frac{g(x)H(x,t)}{|H(x,t)|}, \quad x \in X, \, t \in [0,1],$$

then F is continuous and establishes that f and g are homotopic in the class of continuous functions from X into $\mathbb{T}$. This shows that $[f] = [g]$, and hence, Φ is injective. $\square$

Remark **1.3.12**   Since the first cohomotopy group of $\mathbb{T}$ is same as the first homotopy group of $\mathbb{T}$, by Proposition 1.3.1, the abstract index group for the Banach algebra $C(\mathbb{T})$ is isomorphic to the group $\mathbb{Z}$ of integers.

For future reference, we record an immediate consequence of the previous proposition.

Corollary 1.3.8

Let X be a compact subset of $\mathbb{C}$. If the complement $\mathbb{C}\backslash X$ of X in $\mathbb{C}$ is connected, then the first cohomotopy group $\pi^1(X)$ of X is trivial.

Proof   We need the fact that any nowhere vanishing continuous function f on X is of the form e^g for some $g \in C(X)$ (see Lemma 4.5.1 for a generalization). In particular, the abstract index group for $C(X)$ is trivial (see Example 1.3.11). Now apply Proposition 1.3.1 to conclude that $\pi^1(X)$ is trivial. $\square$

We now introduce the second notion of index relevant to the Fredholm theory.

Definition 1.3.13

Let $\mathscr{G}(\mathcal{H})$ denote the group of invertible elements in the Calkin algebra $Q(\mathcal{H})$ and let $\mathscr{G}_0(\mathcal{H})$ be the connected component of $G(\mathcal{H})$ containing the identity. The *abstract index* $\iota : \mathcal{F}(\mathcal{H}) \to \mathscr{G}(\mathcal{H})/\mathscr{G}_0(\mathcal{H})$ is the homomorphism given by $\iota = q_0 \circ \pi|_{\mathcal{F}(\mathcal{H})}$, where $\mathcal{F}(\mathcal{H})$ denotes the group of Fredholm operators, $q_0 : \mathscr{G}(\mathcal{H}) \to \mathscr{G}(\mathcal{H})/\mathscr{G}_0(\mathcal{H})$, the quotient homomorphism and $\pi : \mathcal{L}(\mathcal{H}) \to Q(\mathcal{H})$, the Calkin map.

Lemma 1.3.5

The abstract index $\iota : \mathcal{F}(\mathcal{H}) \to \mathcal{G}(\mathcal{H})/\mathcal{G}_0(\mathcal{H})$ *is a continuous surjection with* $\ker(\iota) = \mathcal{F}_0(\mathcal{H})$. *Further, it satisfies the following:*

(1) $\iota(ST) = \iota(S)\iota(T)$ *for every* $S, T \in \mathcal{F}(\mathcal{H})$.

(2) $\iota(T + K) = \iota(T)$ *for every* $T \in \mathcal{F}(\mathcal{H})$ *and every compact operator* $K \in \mathcal{L}(\mathcal{H})$.

Proof   All the statements follow from the definition of the abstract index *i*. $\qquad\square$

The following result says that the abstract index is a manifestation of the Fredholm index.

Theorem 1.3.14

There exists an isomorphism $\alpha : \mathbb{Z} \to \mathcal{G}(\mathcal{H})/\mathcal{G}_0(\mathcal{H})$ *such that the following diagram commutes:*

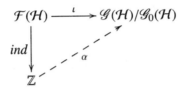

that is, $\alpha \circ \mathrm{ind} = \iota$, *where* ind *denotes the Fredholm index.*

Proof   We define $\alpha : \mathbb{Z} \to \mathcal{G}(\mathcal{H})/\mathcal{G}_0(\mathcal{H})$ by

$$\alpha(n) = \iota(T), \quad n \in \mathbb{Z}, \ T \in \mathcal{F}_n(\mathcal{H}),$$

where $\mathcal{F}_n(\mathcal{H})$ is the connected component of $\mathcal{F}(\mathcal{H})$ consisting of Fredholm operators of index *n*. Note that α is well-defined since ι being continuous, is constant on the connected component $\mathcal{F}_n(\mathcal{H})$ of $\mathcal{F}(\mathcal{H})$ (see Lemma 1.3.5). Since ι is surjective, so is α.

Fix $L \in \mathcal{F}_1(\mathcal{H})$ (see Corollary 1.3.2), and note that by Theorem 1.3.1, $L^n \in \mathcal{F}_n(\mathcal{H})$. It follows that $\alpha(n) = \iota(L^n)$, $n \in \mathbb{Z}$. By Lemma 1.3.5(1), $\iota(ST) = \iota(S)\iota(T)$, and hence,

$$\alpha(m + n) = \iota(L^{m+n}) = \iota(L^m)\iota(L^n) = \alpha(m)\alpha(n).$$

Further, by Lemma 1.3.5(2), $\alpha(n) = 0$ if and only if $\iota(T) = 0$ if and only if $T \in \mathcal{F}_0(\mathcal{H})$ if and only if $n = 0$. Finally, note that for any $T \in \mathcal{F}(\mathcal{H})$, $\alpha \circ \mathrm{ind}(T) = \iota(T)$. This completes the proof. $\qquad\square$

It is worth noting that Theorem 1.3.14 together with Lemma 1.3.5 recovers the facts that the Fredholm index is continuous, multiplicative, and invariant under compact perturbations.

1.4   Schatten Classes

For any compact operator $A \in \mathcal{L}(\mathcal{H})$ on an infinite dimensional Hilbert space $\mathcal{H}$, the set $\{\mu_j(A)\}_{j \geqslant 0}$ denotes the set of singular values of *A* (see (1.3.13)).

Definition 1.4.1

For $1 \leqslant p < \infty$, the *Schatten p-class* $\mathscr{I}_p$ is defined as the collection of those compact operators $A \in \mathcal{L}(\mathcal{H})$ for which the sequence $\{\mu_j(A)\}_{j \geqslant 0}$ of singular values of A belongs to $\ell^p(\mathbb{N})$ (the complex Banach space of p-summable sequences indexed by $\mathbb{N}$). The operators in the classes $\mathscr{I}_1$ and $\mathscr{I}_2$ are respectively known as the *trace-class operators* and the *Hilbert–Schmidt operators*.

Remark 1.4.2   Let $A \in \mathcal{L}(\mathcal{H})$ and $B \in \mathscr{I}_p$. Then A^* and AB belongs to $\mathscr{I}_p$. Indeed, it is easy to conclude from (1.3.13) and (1.3.14) that

$$\mu_j(A) = \mu_j(A^*), \quad \mu_j(AB) \leqslant \|A\|\mu_j(B), \quad j \in \mathbb{N}. \tag{1.4.17}$$

This in turn implies that for any $j \in \mathbb{N}$,

$$\mu_j(BA) = \mu_j(A^*B^*) \leqslant \|A^*\|\mu_j(B^*) = \|A\|\mu_j(B).$$

It follows that $BA \in \mathscr{I}_p$.

     The remark above shows that $\mathscr{I}_p$ is a two-sided $*$-ideal. Further, $\mathscr{I}_p$ carries a natural norm induced from $\ell^p(\mathbb{N})$.

Theorem 1.4.3

The Schatten class $\mathscr{I}_p$ is a two-sided $*$-ideal with norm

$$\|A\|_p = \Big(\sum_{j=0}^{\infty} \mu_j(A)^p \Big)^{1/p}, \quad A \in \mathscr{I}_p.$$

Proof   It suffices to check that

$$\|A + B\|_p \leqslant \|A\|_p + \|B\|_p, \quad A, B \in \mathscr{I}_p.$$

In view of the Minkowski inequality, it is sufficient to verify that

$$\|A\|_p^p = \sup_{\varphi,\psi \in \mathcal{B}} \sum_{j=0}^{\infty} |\langle A\psi_n, \varphi_n \rangle|^p, \tag{1.4.18}$$

where $\mathcal{B}$ denotes the family of all orthonormal sets $\{\varphi_n\}_{n \geqslant 0}$ of $\mathcal{H}$. By (1.3.13), there exists $\varphi, \psi \in \mathcal{B}$ such that

$$A = \sum_{j=0}^{\infty} \mu_j(A)\varphi_j \otimes \psi_j. \tag{1.4.19}$$

It follows that $|\langle A\psi_n, \varphi_n \rangle| = \mu_j(A)$, $n \in \mathbb{N}$, and hence,

$$\|A\|_p^p \leqslant \sup_{\varphi,\psi \in \mathcal{B}} \sum_{j=0}^{\infty} |\langle A\psi_n, \varphi_n \rangle|^p.$$

To see the reverse inequality, let $\kappa, \eta \in \mathcal{B}$. By (1.4.19),

$$\langle A\eta_n, \kappa_n \rangle = \sum_{m=0}^{\infty} \mu_m(A) \langle \eta_n, \psi_m \rangle \langle \varphi_m, \kappa_n \rangle, \quad n \in \mathbb{N}.$$

Consider the linear operator B in $\ell^p(\mathbb{N})$ induced by the matrix $(b_{mn})_{m,n=0}^{\infty}$, where $b_{mn} = \langle \eta_n, \psi_m \rangle \langle \varphi_m, \kappa_n \rangle$, $m, n \in \mathbb{N}$. We claim that

(1) B is *doubly substochastic* in the sense that

$$\sum_{m=0}^{\infty} |b_{mn}| \leqslant 1, \; n \in \mathbb{N}, \quad \sum_{n=0}^{\infty} |b_{mn}| \leqslant 1, \; m \in \mathbb{N}. \tag{1.4.20}$$

(2) $B : \ell^p(\mathbb{N}) \to \ell^p(\mathbb{N})$ is well-defined with the operator norm at most 1.

Assuming the claim, we can see that

$$\sum_{j=0}^{\infty} |\langle A\eta_n, \kappa_n \rangle|^p = \|B(\mu_m(A))_{m=0}^{\infty}\|_p^p \leqslant \|B\|^p \|(\mu_m(A))_{m=0}^{\infty}\|_p^p \leqslant \sum_{j=0}^{\infty} \mu_j(A)^p,$$

which completes the verification of (1.4.18).

First applying the Cauchy-Schwarz inequality and then the Bessel's inequality to $\sum_{m=0}^{\infty} |\langle \eta_n, \psi_m \rangle| |\langle \varphi_m, k_n \rangle|$ and $\sum_{n=0}^{\infty} |\langle \eta_n, \psi_m \rangle| |\langle \varphi_m, k_n \rangle|$, we have (1). To see that $\|B\| \leqslant 1$, let $\alpha \in \ell^p(\mathbb{N})$. In case $p = 1$, by (1.4.20),

$$\|B\alpha\|_1 = \sum_{m=0}^{\infty} \left| \sum_{n=0}^{\infty} b_{mn} \alpha_n \right| \leqslant \sum_{n=0}^{\infty} \left(\sum_{m=0}^{\infty} |b_{mn}| \right) |\alpha_n| \leqslant \|\alpha\|_1.$$

Assume now that $1 < p < \infty$ and let $q = p/(p-1)$. For any $\beta \in \ell^q(\mathbb{N})$, by Hölder's inequality and (1.4.20),

$$\sum_{m,n=0}^{\infty} |b_{mn} \alpha_n \beta_m| \leqslant \left(\sum_{m,n=0}^{\infty} |b_{mn}| |\alpha_n|^p \right)^{1/p} \left(\sum_{m,n=0}^{\infty} |b_{mn}| |\beta_m|^q \right)^{1/q} \leqslant \|\alpha\|_p \|\beta\|_q.$$

Hence, by the Riesz representation theorem for $\ell^p(\mathbb{N})$ (see Example B.1.11), $\|B\alpha\|_p \leqslant \|\alpha\|_p$. This completes the verification of (2). $\qquad\square$

The following corollary lists some applications of Theorem 1.4.3 to the trace-class operators. It turns out that any trace-class operator can be expressed as a product of two Hilbert–Schmidt operators.

Corollary 1.4.1

Let $A, T \in \mathcal{L}(\mathcal{H})$. If $A \in \mathcal{L}(\mathcal{H})$ be of trace-class, then

(1) there exists two Hilbert–Schmidt operators C and B such that $A = CB$,

(2) trace $A := \sum_{n=0}^{\infty} \langle A\varphi_n, \varphi_n \rangle$ converges absolutely to the same complex number for any orthonormal basis $\{\varphi_n\}_{n \geqslant 0}$ of $\mathcal{H}$,

(3) the operators AT and TA are of trace-class with $trace(AT) = trace(TA)$.

Proof   One may conclude from (1.4.17) that

$$\|S_1 S_2\|_2 \leqslant \|S_1\| \|S_2\|_2, \quad S_1 \in \mathcal{L}(\mathcal{H}), \ S_2 \in \mathscr{I}_2. \tag{1.4.21}$$

Let A be a trace-class operator. By polar decomposition, $A = C^* B$ with $C = |A|^{1/2} W^*$ and $B = |A|^{1/2}$ for some partial isometry W. Hence, by Theorem 1.4.3, A is a product of two Hilbert–Schmidt operators C^* and B. This yields (1), while the absolute convergence of the series in (2) follows from (1.4.18). The first part in (3) follows from Theorem 1.4.3. To see the remaining parts in (2) and (3), consider the singular value decomposition $A = \sum_{j=0}^{\infty} \mu_j(A)\varphi_j \otimes \psi_j$ of A for some orthonormal sets $\{\varphi_j\}_{j \geqslant 0}$ and $\{\psi_j\}_{j \geqslant 0}$ of $\mathcal{H}$ (see Theorem 1.3.5). For any orthonormal basis $\{\xi_j\}_{j \geqslant 0}$ of $\mathcal{H}$,

$$
\begin{aligned}
\sum_{m=0}^{\infty} \langle AT\xi_m, \xi_m \rangle &= \sum_{m=0}^{\infty} \sum_{n=0}^{\infty} \mu_n(A)\langle T\xi_m, \psi_n \rangle \langle \varphi_n, \xi_m \rangle \\
&= \sum_{n=0}^{\infty} \mu_n(A)\Big\langle \varphi_n, \sum_{m=0}^{\infty} \langle T^* \psi_n, \xi_m \rangle \xi_m \Big\rangle \\
&= \sum_{n=0}^{\infty} \mu_n(A)\langle \varphi_n, T^* \psi_n \rangle, \tag{1.4.22}
\end{aligned}
$$

where the interchange in the order of infinite sums is justified in view of

$$
\begin{aligned}
&\sum_{m,n=0}^{\infty} \mu_n(A)|\langle T\xi_m, \psi_n \rangle \langle \varphi_n, \xi_m \rangle| \\
&= \sum_{n=0}^{\infty} \mu_n(A) \sum_{m=0}^{\infty} |\langle T\xi_m, \psi_n \rangle \langle \varphi_n, \xi_m \rangle| \\
&\leqslant \sum_{n=0}^{\infty} \mu_n(A)\Big(\sum_{m=0}^{\infty} |\langle \xi_m, T^* \psi_n \rangle|^2 \Big)^{1/2} \Big(\sum_{m=0}^{\infty} |\langle \varphi_n, \xi_m \rangle|^2 \Big)^{1/2} \\
&= \sum_{n=0}^{\infty} \mu_n(A)\|T^* \psi_n\| \|\varphi_n\| \\
&\leqslant \|T\| \sum_{n=0}^{\infty} \mu_n(A).
\end{aligned}
$$

Note that (1.4.22) shows that the trace of AT is independent of the choice of the orthonormal basis $\{\xi_j\}_{j \geqslant 0}$. By letting $T = I$, this also shows that trace is independent of the choice of the orthonormal basis completing the verification of (2). Almost the same argument as earlier shows that

$$\sum_{m=0}^{\infty} \langle TA\xi_m, \xi_m \rangle = \sum_{n=0}^{\infty} \mu_n(A)\langle T\varphi_n, \psi_n \rangle,$$

which completes the verification of (3).   □

The following provides a useful criterion for the membership of trace-class and Hilbert–Schmidt class operators.

Corollary 1.4.2

Let $A \in \mathcal{L}(\mathcal{H})$ and let $1 \leqslant p \leqslant 2$ be given. Assume that there exists an orthonormal basis $\{e_n\}_{n \geqslant 0}$ of $\mathcal{H}$ such that $\sum_{n=0}^{\infty} \||A|^{p/2} e_n\|^2 < \infty$. Then A belongs to the Schatten class $\mathscr{I}_p$. In particular, A is compact.

Proof   Let $h \in \mathcal{H}$ and write $h = \sum_{n=0}^{\infty} \alpha_n e_n$ for some sequence $\{\alpha_n\}$ of scalars. By (1.1.1) and the Cauchy–Schwarz inequality,

$$
\begin{aligned}
\|Ah\| &\leqslant \sum_{n=0}^{\infty} |\alpha_n| \|Ae_n\| = \sum_{n=0}^{\infty} |\alpha_n| \||A|e_n\| \\
&\leqslant \||A|^{1-p/2}\| \sum_{n=0}^{\infty} |\alpha_n| \||A|^{p/2} e_n\| \\
&\leqslant \||A|^{1-p/2}\| \Big(\sum_{n=0}^{\infty} \||A|^{p/2} e_n\|^2 \Big)^{1/2} \|h\|.
\end{aligned}
$$

This shows that for some positive constant M,

$$
\|A\| \leqslant M \Big(\sum_{n=0}^{\infty} \||A|^{p/2} e_n\|^2 \Big)^{1/2}. \tag{1.4.23}
$$

For an integer $k \geqslant 0$, define A_k by $A_k e_n = A e_n$ for $n \leqslant k$, and zero otherwise. Then each A_k is a finite rank operator. Further, by arguing as earlier, we may conclude that

$$
\|A_k - A\| \leqslant M \sum_{n>k} \||A|^{p/2} e_n\|^2 \to 0 \text{ as } k \to \infty.
$$

Since each A_k is compact, by Proposition 1.2.3, so is A. On the other hand, it is easy to see that $\sum_{n=0}^{\infty} \||A|^{p/2} e_n\|^2$ is independent of the choice of orthonormal basis $\{e_n\}_{n \geqslant 0}$ of $\mathcal{H}$ (see Exercise 1.7.17). Hence, by letting $\{e_n\}_{n \geqslant 0}$ be the orthonormal basis $\{\varphi_n\}_{n \geqslant 0}$ of eigenvectors of $|A|$ (see (1.3.13)), one can see that $\big(\sum_{n=0}^{\infty} \||A|^{p/2} e_n\|^2 \big)^{1/2} = \|A\|_p$, and hence, the singular values $\{\mu_j(A)\}_{j \geqslant 0}$ of A are p-summable.   □

The cases $p = 1$ and $p = 2$ (corresponding to trace-class and Hilbert–Schmidt operators) of the aforementioned result will be used in the sequel.

1.5   Isometries and von Neumann–Wold Decomposition

In this section, we discuss the spectral theory for the class of isometries. In particular, we present a decomposition theorem for isometries obtained independently by von Neumann and Wold.

Let T be a bounded linear operator on a complex Hilbert space $\mathcal{H}$. Recall that T is an isometry if

$$\|Th\| = \|h\| \text{ for all } h \in \mathcal{H}.$$

Equivalently, T is an isometry if $T^*T = I$.

Remark 1.5.1   Let T be an isometry. Then T^n is isometric for all integers $n \geqslant 1$. Moreover, if $\mathcal{M}$ is an invariant subspace for T, that is, $T\mathcal{M} \subseteq \mathcal{M}$, then, the restriction $T|_{\mathcal{M}}$ of T to $\mathcal{M}$ is also an isometry.

Every unitary is trivially an isometry. Here is a basic example of an isometry, which is not unitary.

Example 1.5.2   The *Hardy space* $H^2(\mathbb{D})$ of the unit disc $\mathbb{D}$ is a normed linear space of complex-valued functions f holomorphic on $\mathbb{D}$ for which

$$\|f\|^2_{H^2(\mathbb{D})} = \sup_{0<r<1} \int_0^{2\pi} |f(re^{i\theta})|^2 \frac{d\theta}{2\pi} < \infty.$$

We claim that for any holomorphic function $f(z) = \sum_{n=0}^{\infty} a_n z^n$ in $H^2(\mathbb{D})$,

$$\|f\|^2_{H^2(\mathbb{D})} = \sum_{n=0}^{\infty} |a_n|^2. \tag{1.5.24}$$

An application of Parseval's identity shows that

$$\int_0^{2\pi} |f(re^{i\theta})|^2 \frac{d\theta}{2\pi} = \sum_{n=0}^{\infty} |a_n|^2 r^{2n}, \quad 0 \leqslant r < 1$$

(see Corollary B.1.4). Since the series on the right-hand side is non-decreasing in r, for any sequence $\{r_m\}_{m\geqslant0} \subseteq (0,1)$ increasing to 1,

$$\sup_{0<r<1} \int_0^{2\pi} |f(re^{i\theta})|^2 \frac{d\theta}{2\pi} = \lim_{m\to\infty} \sum_{n=0}^{\infty} |a_n|^2 r_m^{2n}.$$

An application of dominated convergence theorem immediately yields (1.5.24). It is now easy to see that $W : H^2(\mathbb{D}) \to \ell^2(\mathbb{N})$ given by $W(z^n) = e_n$ extends isometrically from $H^2(\mathbb{D})$ onto $\ell^2(\mathbb{N})$, where e_n denotes the sequence in $\ell^2(\mathbb{N})$ with only nonzero entry equal to 1 at $(n+1)$th stage. In particular, $H^2(\mathbb{D})$ is a separable Hilbert space. Since $zf \in H^2(\mathbb{D})$ for any $f \in H^2(\mathbb{D})$, we may define the *Szegö shift* $S : H^2(\mathbb{D}) \to H^2(\mathbb{D})$ by

$$(Sf)(z) = zf(z), \quad f \in H^2(\mathbb{D}).$$

Note that S is an isometry. Indeed, if $f(z) = \sum_{n\geqslant0} a_n z^n \in H^2(\mathbb{D})$, then by (1.5.24),

$$\|Sf\|^2_{H^2(\mathbb{D})} = \Big\|\sum_{n\geqslant0} a_n z^{n+1}\Big\|^2_{H^2(\mathbb{D})} = \sum_{n\geqslant0} |a_n|^2 = \|f\|^2_{H^2(\mathbb{D})}.$$

Finally, we note that S is unitarily equivalent to the shift operator U_+ on $\ell^2(\mathbb{N})$ (see Example 1.3.2). Indeed, a routine calculation shows that $WS = U_+W$. ∎

For ready reference, we state an elementary but important fact pertaining to isometries (the proof is implicitly given in Remark 1.1.2).

Proposition 1.5.1

If $T \in \mathcal{L}(\mathcal{H})$ is an isometry, then T has a trivial kernel and closed range. In particular, the final space of any partial isometry is closed.

Here are some useful consequences of Proposition 1.5.1.

Corollary 1.5.1

If $E \in \mathcal{L}(\mathcal{H})$ is an infinite rank projection, then there exists an isometry $U \in \mathcal{L}(\mathcal{H})$ such that $UU^* = E$. In this case, $U(\mathcal{H}) = E(\mathcal{H})$ and $EU : \mathcal{H} \to E(\mathcal{H})$ is a unitary.

Proof Let $E \in \mathcal{L}(\mathcal{H})$ be an infinite rank projection. Since the Hilbert space dimensions of $\mathcal{H}$ and $E(\mathcal{H})$ are the same (for $\mathcal{H}$ is separable), there exists a unitary $W : E(\mathcal{H}) \to \mathcal{H}$. Then $U = (WE)^* : \mathcal{H} \to \mathcal{H}$ is an isometry. Indeed, since W^* maps $\mathcal{H}$ onto $E(\mathcal{H})$,

$$WE(WE)^* = WEW^* = WW^* = I.$$

Moreover, $(WE)^*WE = E$. This completes the verification of the first part. To see the remaining part, note that by Proposition 1.5.1, the range of U^* is dense in $\mathcal{H}$. Hence, in view of $UU^* = E$, the range of U (being closed by Proposition 1.5.1) is equal to the range of E. Clearly, EU is a unitary. $\qquad\qquad\square$

The following is immediate from Remark 1.5.1 and Proposition 1.5.1.

Corollary 1.5.2

For $T \in \mathcal{L}(\mathcal{H})$, set

$$\mathcal{H}_u = \bigcap_{n \geqslant 0} T^n \mathcal{H}. \tag{1.5.25}$$

If T is an isometry, then $\mathcal{H}_u$ is a closed subspace of $\mathcal{H}$.

We say $T \in \mathcal{L}(\mathcal{H})$ is *analytic* if $\mathcal{H}_u = \{0\}$, where $\mathcal{H}_u$ is as given in (1.5.25).

***Example* 1.5.3 (Example 1.5.2 continued)** Let us see that the shift operator S on $\mathcal{H} := H^2(\mathbb{D})$ is analytic. Indeed, if $f \in \mathcal{H}_u$, then for every positive integer n, there exists a holomorphic function $f_n \in H^2(\mathbb{D})$ such that $f(z) = z^n f_n(z)$, $z \in \mathbb{D}$. However, then all derivatives of f vanish at 0, and hence, f is identically zero. $\qquad\blacksquare$

The von Neumann–Wold decomposition theorem says that the example of the isometry discussed in Example 1.5.3 is representative of the class of non-unitary isometries. We need a sequence of lemmas to prove this fact.

Lemma 1.5.1

For $T \in \mathcal{L}(\mathcal{H})$, let $\mathcal{H}_u = \cap_{n \geqslant 0} T^n \mathcal{H}$. If T is an isometry, then

(1) $T\mathcal{H}_u = \mathcal{H}_u$ and $T^*\mathcal{H}_u = \mathcal{H}_u$.

(2) $T|_{\mathcal{H}_u}$ is unitary.

(3) $T|_{\mathcal{H}_u^\perp}$ is analytic.

Proof   Assume that T is an isometry. Clearly, $T\mathcal{H}_u \subseteq \mathcal{H}_u$. To see the reverse inclusion, let $x \in \mathcal{H}_u$. Then there exists $y_n \in \mathcal{H}$ such that $x = T^n y_n$ for all integers $n \geqslant 0$. In particular, $Ty_1 = T^n y_n$. Since T is injective, $y_1 = T^{n-1} y_n$ for all integers $n \geqslant 1$. Thus, $y_1 \in \mathcal{H}_u$ and $x = Ty_1 \in T\mathcal{H}_u$. It follows that $T\mathcal{H}_u = \mathcal{H}_u$, and hence, $T^*\mathcal{H}_u = T^*(T\mathcal{H}_u) = \mathcal{H}_u$. This proves (1). Since $T|_{\mathcal{H}_u}$ is onto and isometric, it is unitary proving (2). To prove (3), set $S = T|_{\mathcal{H}_u^\perp}$, and note that $\cap_{n \geqslant 0} S^n \mathcal{H}_u^\perp \subseteq \mathcal{H}_u \cap \mathcal{H}_u^\perp = \{0\}$. $\square$

Lemma 1.5.2

If $\{\mathcal{M}_k\}_{k \geqslant 1}$ is a sequence of closed subspaces of $\mathcal{H}$, then

$$\left(\bigcap_{k \geqslant 1} \mathcal{M}_k \right)^\perp = \bigvee_{k \geqslant 1} \mathcal{M}_k^\perp.$$

Proof   Let $\mathcal{N}_1 = \bigvee_{k \geqslant 1} \mathcal{M}_k^\perp$, $\mathcal{N}_2 = \{\cap_{k \geqslant 1} \mathcal{M}_k\}^\perp$, and note that $\mathcal{N}_1 \subseteq \mathcal{N}_2$. Suppose that the strict inclusion $\mathcal{N}_1 \subsetneq \mathcal{N}_2$ holds, so that there exists $x \in \mathcal{N}_2$ such that $\langle y, x \rangle = 0$ for every $y \in \mathcal{N}_1$. It follows that $\langle y, x \rangle = 0$ for every $y \in \mathcal{M}_k^\perp$, $k \geqslant 1$. Thus, $x \in (\mathcal{M}_k^\perp)^\perp = \overline{\mathcal{M}_k} = \mathcal{M}_k$, $k \geqslant 1$. It follows that $x \in \mathcal{N}_2^\perp$, and hence, $x = 0$. This completes the proof. $\square$

Lemma 1.5.3

Let $T \in \mathcal{L}(\mathcal{H})$ and let n be a positive integer. If T is an isometry, then the following statements are true:

(1) $\ker T^{*n} = \bigvee_{k=0}^{n-1} T^k(\ker T^*)$.

(2) $(\cap_{n \geqslant 0} T^n \mathcal{H})^\perp = \bigvee_{n \geqslant 0} T^n(\ker T^*)$.

(3) The family $\{T^n(\ker T^*)\}_{n \geqslant 0}$ of closed subspaces of $\mathcal{H}$ is mutually orthogonal.

Proof   Assume that T is an isometry. Note the following identity:

$$I - T^n T^{*n} = \sum_{k=0}^{n-1} T^k (I - TT^*) T^{*k}.$$

Since $I - TT^*$ is an orthogonal projection onto $\ker T^*$, for any $x \in \ker T^{*n}$,

$$x = \sum_{k=0}^{n-1} T^k y_k, \text{ where } y_k = (I - TT^*) T^{*k} x \in \ker T^*.$$

This proves the first part. To prove the remaining part, note that by Lemma 1.5.2,

$$(\cap_{n \geqslant 0} T^n \mathcal{H})^\perp = \bigvee_{n \geqslant 1} (T^n \mathcal{H})^\perp = \bigvee_{n \geqslant 1} \ker(T^{*n}) = \bigvee_{n \geqslant 0} T^n(\ker T^*),$$

where the last equality follows from part (1). We leave the routine verification of (3) to the reader. □

Finally, putting these together, we obtain the following theorem revealing the structure of an arbitrary isometry.

Theorem 1.5.4 (von Neumann–Wold Decomposition)

The following statements are true for any isometry T on the Hilbert space $\mathcal{H}$.

(1) *There exists a unitary operator U on $\mathcal{H}_u = \cap_{n \geqslant 0} T^n \mathcal{H}$ and an analytic operator A on $\mathcal{H}_a = \bigvee_{n \geqslant 0} T^n(\ker T^*)$ such that*

$$T = U \oplus A \quad \text{on} \quad \mathcal{H} = \mathcal{H}_u \oplus \mathcal{H}_a.$$

(2) *If $k = \dim \ker T^*$ (possibly infinite), then A is unitarily equivalent to the k-fold inflation $S^{(k)} = S \oplus \cdots \oplus S$ of S, where S is the Szegö shift on the Hardy space $H^2(\mathbb{D})$ (see Example 1.5.2).*

In particular, any isometry is a direct sum of unitary and shift operator.

Proof The first part follows from Lemmas 1.5.1 and 1.5.3. To see the second part, let $\{f_1, \ldots, f_k\}$ be an orthonormal basis for $\ker T^*$, where $k \in \mathbb{N} \cup \{\infty\}$. Set

$$E_j = \{A^l f_j : l \geqslant 0\}, \quad j = 1, \ldots, k,$$

and note that by Lemma 1.5.3(3), $\{E_j\}_{j=1}^k$ forms a mutually orthogonal family of closed subspaces of $\mathcal{H}$. If

$$\mathcal{H}_j = \bigvee E_j, \quad j = 1, \ldots, k,$$

then $\mathcal{H}_a$ is the orthogonal direct sum $\mathcal{H}_1 \oplus \cdots \oplus \mathcal{H}_k$. Note that each $\mathcal{H}_j$ is invariant under A and $A|_{\mathcal{H}_j}$ is unitarily equivalent to the shift operator S on $H^2(\mathbb{D})$ implemented by the unitary $U_j : \mathcal{H}_j \to H^2(\mathbb{D})$ governed by

$$U_j(T^l f_j) = z^l, \quad l \in \mathbb{N}, \ j = 1, \ldots, k,$$

that is, $S U_j = U_j A|_{\mathcal{H}_j}, j = 1, \ldots, k$. □

The number k appearing in the statement of Theorem 1.5.4, is referred to as the *multiplicity* of T.

We now give some applications of the von Neumann–Wold decomposition.

Corollary 1.5.3

If $T \in \mathcal{L}(\mathcal{H})$ is an isometry, then $T^*T - TT^*$ is compact if and only if the multiplicity of T is finite.

Proof   Assume that T is an isometry and let k be the multiplicity of T. By Theorem 1.5.4, $T^*T - TT^* = A^*A - AA^*$. However, A is unitarily equivalent to the direct sum of k copies of the shift operator S on $H^2(\mathbb{D})$. Let E_0 denote the orthogonal projection of $H^2(\mathbb{D})$ onto the closed subspace $\{f \in H^2(\mathbb{D}) : f(0) = 0\}$, and note that $S^*S - SS^* = I - E_0$. It follows that $T^*T - TT^*$ is an orthogonal projection of rank k, and hence, $T^*T - TT^*$ is compact if and only if k is finite (see Remark 1.2.3). $\qquad\square$

Corollary 1.5.4

For every invariant subspace M of the shift operator S on the Hardy space $H^2(\mathbb{D})$, one has $M = \bigvee_{n \geqslant 0} S^n(M \ominus S M)$.

Proof   Since S is an analytic isometry (see Example 1.5.3), so is $S|_M$. The desired conclusion now follows from von Neumann–Wold decomposition. $\qquad\square$

Remark **1.5.5**   Notice that the conclusion of the corollary above for the invariant subspace $M := H^2(\mathbb{D})$ is equivalent to the density of polynomials in $H^2(\mathbb{D})$. Moreover, if M is nonzero, then so is $M \ominus S M$.

We discuss here another application of von Neumann–Wold decomposition.

Example **1.5.6**   Consider the shift operator S on $H^2(\mathbb{D})$ described in Example 1.5.2 and set

$$S_\lambda = S - \lambda I, \quad |\lambda| < 1.$$

Recall that the analytic polynomials are dense in $H^2(\mathbb{D})$ (see Remark 1.5.5). If P_λ denotes the orthogonal projection onto $\ker S_\lambda^*$, then

$$\ker S_\lambda^* = P_\lambda H^2(\mathbb{D}) = P_\lambda \bigvee_{n \geqslant 0} \{z^n\} = P_\lambda \bigvee_{n \geqslant 0} \{(z - \lambda)^n\} = P_\lambda \bigvee_{n \geqslant 0} \{S_\lambda^n 1\},$$

and hence, $\ker S_\lambda^*$ equals the space $P_\lambda\{f \in H^2(\mathbb{D}) : f \text{ is a constant function}\}$ of dimension at most 1. Actually, it can be seen that the dimension of $\ker S_\lambda^*$ is exactly 1 (see Exercise 1.7.40). Next, note that for any $f \in H^2(\mathbb{D})$,

$$\|S_\lambda f\| \geqslant \Big|\|S f\| - |\lambda|\|f\|\Big| = (1 - |\lambda|)\|f\|.$$

Thus, S_λ has a trivial kernel and closed range. Hence, by Atkinson's theorem, S_λ is Fredholm. In particular,

$$\sigma_e(S) \subseteq \sigma(S) \cap \{z \in \mathbb{C} : |z| \geqslant 1\}.$$

However, $\|S\| = 1$ and hence $\sigma(S)$ is contained in the closed unit disc (see (B.1.3)). Consequently, $\sigma_e(S)$ is a subset of the unit circle $\mathbb{T}$. We show that $\sigma_e(S) = \mathbb{T}$. To see that, define the unitary operator $U_\theta : H^2(\mathbb{D}) \to H^2(\mathbb{D})$ by

$$(U_\theta f)(z) = f(e^{i\theta}z), \quad |z| < 1, \, \theta \in \mathbb{R},$$

Note that $U_\theta S = e^{i\theta} S U_\theta$. It follows that all the spectral parts of S have circular symmetry. In particular, $z \in \sigma_e(S)$ if and only if $e^{i\theta}z \in \sigma_e(S)$. Since the essential spectrum is always non-empty (Remark 1.3.6), $\sigma_e(S)$ must be the entire unit circle. It is now easy to deduce from the von Neumann–Wold decomposition that for any analytic isometry T, $\sigma_e(T) = \mathbb{T}$ if and only if the multiplicity of T is finite. We leave it to the reader to check that $\sigma_e(T) = \overline{\mathbb{D}}$ in case the multiplicity of T is infinite. ∎

1.6 Toeplitz Operators with Continuous Symbols

Consider the Hilbert space $L^2(\mathbb{T})$ of square-integrable functions on the unit circle $\mathbb{T}$ with respect to the normalized arc-length measure $d\theta/2\pi$. Note that $\{z^n : n \in \mathbb{Z}\}$ is an orthonormal subset of $L^2(\mathbb{T})$. Since continuous functions are dense in $L^2(\mathbb{T})$, by the Stone–Weierstrass theorem (see Appendix B), the set $\{z^n : n \in \mathbb{Z}\}$ forms an orthonormal basis for $L^2(\mathbb{T})$ (see Exercise 1.7.15 for an alternate deduction based on Corollary 1.3.4). In particular, any $f \in L^2(\mathbb{T})$ admits the Fourier representation

$$f(z) = \sum_{n \in \mathbb{Z}} \hat{f}(n) z^n,$$

where $\hat{f}(n) = \frac{1}{2\pi} \int_0^{2\pi} f(e^{i\theta}) e^{-in\theta} d\theta$, $n \in \mathbb{Z}$.

Consider the closed, z-invariant subspace $H^2(\mathbb{T}) = \bigvee \{z^n : n \geqslant 0\}$ of $L^2(\mathbb{T})$. Note that

$$H^2(\mathbb{T}) = \{f \in L^2(\mathbb{T}) : \hat{f}(n) = 0 \text{ for all integers } n < 0\}.$$

Further, $H^2(\mathbb{D})$ (as discussed in Example 1.5.2) and $H^2(\mathbb{T})$ are isomorphic. Indeed, the mapping $U : H^2(\mathbb{D}) \to H^2(\mathbb{T})$ governed by $U(z^n) = e^{i\theta n}$ is a unitary mapping such that $US = M_z U$, where S is the Szegö shift and M_z denotes the operator of multiplication by z on $H^2(\mathbb{T})$.

Definition 1.6.1

For $\varphi \in C(\mathbb{T})$, let M_φ denote the bounded linear operator of multiplication by φ in $\mathcal{L}(L^2(\mathbb{T}))$. The *Toeplitz operator* T_φ on $H^2(\mathbb{T})$ with *symbol* $\varphi \in C(\mathbb{T})$ is defined by

$$T_\varphi = PM_\varphi|_{H^2(\mathbb{T})},$$

where $P : L^2(\mathbb{T}) \to H^2(\mathbb{T})$ is the orthogonal projection of $L^2(\mathbb{T})$ onto $H^2(\mathbb{T})$.

Remark **1.6.2** Note that the identity operator I and the multiplication operator M_z are Toeplitz operators with symbols 1 and z respectively. Note further that M_z^* is the Toeplitz operator with symbol $\bar{z}$.

Definition 1.6.3

Let $\mathcal{A}$ be a unital C^*-algebra with unit 1. For a subset $\mathbb{A}$ of $\mathcal{A}$, the C^*-*algebra* $C^*(\mathbb{A})$ *generated by* $\mathbb{A}$ is the smallest unital C^*-algebra containing $\mathbb{A}$. In case $\mathbb{A} = \{a\}$ for some $a \in \mathcal{A}$, then, by abuse of notation, we write $C^*(a)$. In case $\mathcal{A}$ is the C^*-algebra of bounded linear operators $\mathcal{L}(\mathcal{H})$, we let $\mathcal{S}_T$ denote the C^*-subalgebra of $\mathcal{L}(\mathcal{H})$ generated by the operator $T \in \mathcal{L}(\mathcal{H})$. Following the usual convention, we let $C^*(T_z)$ denote the C^*-algebra generated by the Toeplitz operator T_z. The C^*-algebra $C^*(T_z)$ is commonly known as the *Toeplitz* C^*-*algebra*.

***Remark* 1.6.4**   The existence of $C^*(\mathbb{A})$ can be ensured by taking intersection of all unital C^*-algebras containing $\mathbb{A}$ ($\mathcal{A}$ belongs to this collection; hence, it is non-empty). Concretely, it can be realized as follows. Consider the unital, free associative algebra $\mathcal{P}$ of complex polynomials in the variables x, y. For each $p \in \mathcal{P}$, one may associate uniquely an element $p(a, a^*) \in \mathcal{A}$. It can be easily seen that $C^*(\mathbb{A})$ is the closure of $\{p(a, a^*) : p \text{ in } \mathcal{P} \text{ and } a \in \mathbb{A}\}$ in $\mathcal{A}$.

Although the Toeplitz C^*-algebra is not abelian (since $T_z^* T_z - T_z T_z^* \neq 0$), it is abelian modulo the ideal of compact operators. This is contained in the following structure theorem for the Toeplitz C^*-algebra.

Theorem 1.6.5

If $C(H^2(\mathbb{T}))$ denotes the algebra of compact operators on $H^2(\mathbb{T})$, then the C^*-algebra $C^*(T_z)/C(H^2(\mathbb{T}))$ is isomorphic to the C^*-algebra $C(\mathbb{T})$ via the mapping $\Gamma(\pi(T_\varphi)) = \varphi$, where $\pi : \mathcal{L}(H^2(\mathbb{T})) \to Q(H^2(\mathbb{T}))$ is the Calkin map.

Proof   We have already observed that $T_z^* T_z - T_z T_z^*$ is a rank one operator (see the proof of Corollary 1.5.3). In particular, the C^*-algebra $C^*(\pi(T_z))$ generated by the normal element $\pi(T_z)$ is abelian. By the spectral theorem (see Corollary C.1.2 and Remark 1.2.6) $C^*(\pi(T_z))$ may be identified with

$$C(\sigma(\pi(T_z))) = C(\sigma_e(T_z)) = C(\mathbb{T}),$$

where we used the fact that the essential spectrum of T_z is the unit circle $\mathbb{T}$ (see Example 1.5.6). Note that the Gelfand map Γ implementing the isometric $*$-isomorphism from $C^*(\pi(T_z))$ onto $C(\mathbb{T})$ is governed by

$$\Gamma(p(\pi(T_z), \pi(T_z^*))) = p(z, \bar{z})$$

for any polynomial p in two variables. Since π is a $*$-homomorphism, we have $p(\pi(T_z), \pi(T_z^*)) = \pi(p(T_z, T_z^*))$, which implies that $\Gamma(\pi(p(T_z, T_z^*))) = p(z, \bar{z})$. Moreover, since $T_z^* T_z = I$ and $T_z T_z^* - I \in C(H^2(\mathbb{T}))$ (see the proof of Corollary 1.5.3), for any $\varphi(z) = p(z, \bar{z})$ for some two variable polynomial p, we have

$$\Gamma(\pi(T_\varphi)) = \varphi. \tag{1.6.26}$$

To see that (1.6.26) holds for any $\varphi \in C(\mathbb{T})$, let $\{p_n\}$ be a sequence of polynomials in z and $\bar{z}$ converging to φ in the sup norm as ensured by the Stone–Weierstrass theorem (see Appendix B). By the continuous functional calculus for the multiplication operators on $L^2(\mathbb{T})$ (see Theorem C.1.4),

$$\|T_{p_n} - T_\varphi\| \leqslant \|M_{p_n} - M_\varphi\| = \|p_n - \varphi\|_\infty \to 0.$$

It follows that $\Gamma \circ \pi(T_{p_n})$ converges to $\Gamma \circ \pi(T_\varphi)$, and hence, $\Gamma(\pi(T_\varphi)) = \varphi$.   □

Remark 1.6.6   Note that T_φ is compact if and only if $\varphi = 0$.

For a map φ, let $\text{Im}\,\varphi$ denote its image. We reserve the notation ran for the range of linear transformations between Hilbert spaces and use Im for the image of maps between C^*-algebras. We now discuss some important applications of the last theorem.

Corollary 1.6.1

Let $\pi : \mathcal{L}(H^2(\mathbb{T})) \to Q(H^2(\mathbb{T}))$ *be the Calkin map. Consider the map* $\Gamma : C^*(T_z)/C(H^2(\mathbb{T})) \to C(\mathbb{T})$ *as defined in (1.6.26) and let* $\varphi = \Gamma \circ \pi$. *Then,* $\ker\varphi = C(H^2(\mathbb{T}))$ *and* $\text{Im}\,\varphi = C(\mathbb{T})$. *In particular, we have the short exact sequence*

$$0 \xrightarrow{0} C(H^2(\mathbb{T})) \overset{i}{\hookrightarrow} C^*(T_z) \xrightarrow{\varphi} C(\mathbb{T}) \xrightarrow{0} 0, \tag{1.6.27}$$

that is, $\ker i = \{0\}$, $\text{Im}\,i = \ker\varphi$ *and* $\text{Im}\,\varphi = C(\mathbb{T})$, *where* $i : C(H^2(\mathbb{T})) \hookrightarrow C^*(T_z)$ *denotes the inclusion map.*

Proof   In view of Theorem 1.6.5, it suffices to check that $C^*(T_z)$ contains $C(H^2(\mathbb{T}))$. Since finite rank operators on $H^2(\mathbb{T})$ are dense in $C(H^2(\mathbb{T}))$, it is sufficient to see that $C^*(T_z)$ contains an arbitrary finite rank operator. Recall that an arbitrary rank one operator $x \otimes y$ in $\mathcal{L}(\mathcal{H})$ is given by

$$x \otimes y(h) = \langle h, y \rangle x, \quad h \in \mathcal{H},$$

where $x, y \in \mathcal{H}$. In this terminology, we have $T_z^* T_z - T_z T_z^* = 1 \otimes 1$, where 1 is the constant function in $H^2(\mathbb{T})$. Now if $f, g \in H^2(\mathbb{T})$ are polynomials in $e^{i\theta}$, then

$$T_f(T_z^* T_z - T_z T_z^*)T_g^* = T_f(1 \otimes 1)T_g^* = f \otimes g.$$

Since polynomials in $e^{i\theta}$ are dense in $H^2(\mathbb{T})$, by a routine approximation argument, $C^*(T_z)$ contains all rank one operators in $\mathcal{L}(H^2(\mathbb{T}))$. Moreover, since $C^*(T_z)$ is a vector space, $C^*(T_z)$ contains all finite rank operators on $H^2(\mathbb{T})$. □

Corollary 1.6.2
For any $\varphi, \psi \in C(\mathbb{T})$, $T_{\varphi\psi} - T_\varphi T_\psi$ *is compact.*

Proof   In view of Corollary 1.6.1, it suffices to check that $\pi(T_\varphi T_\psi - T_{\varphi\psi}) = 0$. Note that by (1.6.26),

$$\Gamma \circ \pi(T_\varphi T_\psi) = \Gamma \circ \pi(T_\varphi)\Gamma \circ \pi(T_\psi) = \varphi\psi = \Gamma \circ \pi(T_{\varphi\psi}),$$

and hence, $\Gamma \circ \pi(T_\varphi T_\psi - T_{\varphi\psi}) = 0$. Since Γ is injective (see Corollary 1.6.1), $\pi(T_\varphi T_\psi - T_{\varphi\psi}) = 0$. □

Let us characterize Fredholm Toeplitz operators with continuous symbols. First, we need a result on spectral permanence, which is the lemma stated below.

Lemma 1.6.1

Suppose that B is a unital C^*-subalgebra of a C^*-algebra $\mathcal{A}$. Then for all $x \in B$, $\sigma_B(x) = \sigma_{\mathcal{A}}(x)$.

Proof   Clearly, $\sigma_{\mathcal{A}}(x) \subseteq \sigma_B(x)$. To see the reverse inclusion, let $x \in B$ and $\lambda \notin \sigma_{\mathcal{A}}(x)$. Thus $y = x - \lambda$ is invertible with inverse in $\mathcal{A}$. Note that y^*y is invertible in $\mathcal{A}$. We must show that y is invertible in B. Since $y^{-1} = (y^*y)^{-1}y^*$, it suffices to check that $(y^*y)^{-1}$ belongs to B. Since $\sigma_{\mathcal{A}}(y^*y)$ is a compact subset of the open interval $(0, \infty)$, by the Stone-Weierstrass Theorem, $\frac{1}{t}$ can be approximated uniformly by polynomials in t on $\sigma_{\mathcal{A}}(y^*y)$. Hence, by the Spectral theorem, one may approximate $(y^*y)^{-1}$ by polynomials in y^*y, and hence y^*y is invertible in B. This completes the proof.   □

Corollary 1.6.3

If $\varphi \in C(\mathbb{T})$, then T_φ is Fredholm if and only if φ is nowhere vanishing. In particular, $\sigma_e(T_\varphi) = \varphi(\mathbb{T})$.

Proof   Note that T_φ is Fredholm if and only if $\pi(T_\varphi)$ is invertible in $Q(H^2(\mathbb{T}))$. This holds, by the spectral permanence, if and only if $\pi(T_\varphi)$ is invertible in $C^*(T_z)/C(H^2(\mathbb{T}))$. However, by Corollary 1.6.1, the later one holds if and only if φ is invertible in $C(\mathbb{T})$. The desired conclusion now follows from the fact that φ is invertible in $C(\mathbb{T})$ if and only if φ is nowhere vanishing.   □

The last result prompts the following question: If $\varphi \in C(\mathbb{T})$ is nowhere vanishing, then what is the Fredholm index of T_φ ? To guess the answer, note that the Fredholm index of T_{z^n} equals $-n$, which is negative of the winding number of z^n. This statement, usually known as the *index theorem*, holds for any Toeplitz operator T_φ on $H^2(\mathbb{T})$ with (continuous) symbol φ.

Before we state the index theorem, we briefly recall the definition and some elementary properties of the so-called *winding number $wind_a(\gamma)$ of a curve $\gamma : \mathbb{T} \to \mathbb{C}$ that encloses the point $a \in \mathbb{C}$.* Recall that a *closed curve* is just a continuous mapping from the unit circle into the complex plane. If γ is piecewise smooth, then $wind_a(\gamma)$ is given by

$$\text{wind}_a(\gamma) = \frac{1}{2\pi i} \int_\gamma \frac{dz}{z - a}.$$

Otherwise, since $\mathbb{C}$ is simply connected, any continuous closed curve γ is homotopic to a smooth curve $\tilde{\gamma}$ (cf. [132]), and hence, one may set $wind_a(\gamma) = wind_a(\tilde{\gamma})$. This is well-defined since the fundamental group of the unit circle $\mathbb{T}$ is the additive group $\mathbb{Z}$ of all integers [96]. Recall that $a \mapsto wind_a(\gamma)$ is an integer-valued continuous function from $\mathbb{C}\backslash\gamma$ onto the set of integers. Thus, winding number and Fredholm index share the same properties. This is anticipated in the following theorem.

Theorem 1.6.7

Suppose that $\varphi \in C(\mathbb{T})$ is nowhere vanishing. Then the Toeplitz operator T_φ is Fredholm and

$$ind(T_\varphi) = -wind_0(\varphi).$$

Proof   Let $\psi \in C(\mathbb{T})$ be nowhere vanishing and homotopic to φ. Let

$$\Lambda : [0, 1] \times \mathbb{T} \to \mathbb{C}\backslash\{0\}$$

be a homotopy between φ and ψ. Then the mapping $t \mapsto T_{\Lambda(t,\cdot)}$ from $[0,1]$ into the set of Fredholm operators in $\mathcal{L}(H^2(\mathbb{T}))$ is continuous. Since the Fredholm index ind is locally constant (Theorem 1.3.1), $\text{ind}(T_\varphi) = \text{ind}(T_\psi)$. Now φ is homotopic to $\psi = z^{\text{wind}_0(\varphi)}$, so that

$$\text{ind}(T_\varphi) = \text{ind}(T_\psi) = \text{wind}_0(\varphi)\,\text{ind}(T_z) = -\text{wind}_0(\varphi),$$

where we used the fact that $\text{ind}(T_z) = -1$ (see Example 1.5.6). This completes the proof. $\square$

1.7   Notes and Exercises

- The treatment of polar decomposition, compact operators, and abstract index follows closely the classic text [44]. A part of the Fredholm theory and the proof of Segal's theorem (see Theorem 1.2.5) are taken from [35, 84].

- The presentation on Schatten classes is borrowed from the exposition [131].

- The exposition of the section on isometries including Wold–von Neumann decomposition is inspired by Shimorin's approach to the wandering subspace problem for operators close to isometries, see [128].

- The discussion of the Toeplitz operators in these notes are from [44] and [118]. Toeplitz operators with the continuous symbol defined on a strongly pseudo-convex domain are a precursor to the BDF theory (see [47, pp. 3]). Keeping this in mind, we have limited our discussion to Toeplitz operators with continuous symbol defined on the Hardy space of the unit disc. For the spectral and index theory of quotient modules obtained from Beurling type submodules of the Hardy module for the Euclidean ball, see [69].

Exercises

Exercise 1.7.1 Show that every (separable) Hilbert space has (countable) orthonormal basis.

Hint: Let $\mathcal{P}$ be the collection of all orthonormal subsets of H ordered by inclusion. Conclude from Zorn's lemma that $\mathcal{P}$ contains a maximal element $B \in \mathcal{P}$.

Exercise 1.7.2 (Jordan and von Neumann) Let X denote a normed linear space with the norm $\|\cdot\|$. If X satisfies the *parallelogram law*

$$\|x+y\|^2 + \|x-y\|^2 = 2\|x\|^2 + 2\|y\|^2, \quad x,y \in X,$$

then show that the expression

$$\langle x, y \rangle = \frac{1}{4}\left(\|x+y\|^2 - \|x-y\|^2 + i\|x+iy\|^2 - i\|x-iy\|^2\right), \quad x,y \in X$$

(to be referred to as the *polarization identity*) satisfies $\sqrt{\langle \cdot, \cdot \rangle} = \|\cdot\|$ and defines an inner-product on X. Conclude that the norm on any normed linear space is induced by an inner-product if and only if it satisfies the parallelogram law.

Hint: Verify the following:

- $\langle x, y \rangle = \overline{\langle y, x \rangle}$, $x, y \in X$.

- $\langle x/2, y \rangle = 1/2 \langle x, y \rangle$, $x, y \in X$.

- $\langle x + y, z \rangle = \langle x, z \rangle + \langle y, z \rangle$, $x, y \in X$ (use $x + y + z = x + y/2 + y/2 + z$)

- $\langle \alpha x, y \rangle = \alpha \langle x, y \rangle$ for any $\alpha \in \mathbb{C}$ (use the density of $\{m/2^n : m \in \mathbb{Z}, \ n \geqslant 0\}$ in the real line $\mathbb{R}$ to conclude that $\langle \alpha x, y \rangle = \alpha \langle x, y \rangle$ for every $\alpha \in \mathbb{R}$).

Exercise 1.7.3 Let $T \in \mathcal{L}(\mathcal{H})$ be a left-invertible operator. Show that T^*T is invertible. Hence, conclude that T has unique polar decomposition given by $T = VP$, where $V = T(T^*T)^{-1/2}$ and $P = (T^*T)^{1/2}$.

Hint: A positive operator bounded from below is invertible.

Exercise 1.7.4 Show that linearly homeomorphic Hilbert spaces are isometrically isomorphic.

Hint: Exercise 1.7.3.

Exercise 1.7.5 For $U \in \mathcal{L}(\mathcal{H})$, show that the following are equivalent:

(i) U is a partial isometry.

(ii) $U = UU^*U$.

(iii) U^*U and UU^* are orthogonal projections.

Hint: Remark 1.1.2.

Exercise 1.7.6 Show that for $T \in \mathcal{L}(\mathcal{H})$, the partial isometries in the polar decompositions $T = U|T|$ and $T = |T^*|V$ of T can be chosen to be the same.

Hint: Use $|T^*| = U|T|U^*$ to conclude that $T^* = U^*|T^*|$.

Exercise 1.7.7 Show that any rank one bounded linear operator in $\mathcal{L}(\mathcal{H})$ is of the form $x \otimes y$ for some $x, y \in \mathcal{H}$ (see (1.2.2)). Does there exist a closed rank one linear operator on a separable Hilbert space $\mathcal{H}$, which is not bounded ?

Hint: If $T \in B(\mathcal{H})$ is a rank one operator with range spanned by a unit vector $y \in \mathcal{H}$, then for every $x \in \mathcal{H}$, $Tx = \alpha_x y$ for some scalar $\alpha_x \in \mathbb{C}$. Now verify that $T = y \otimes T^*y$. To prove the remaining part, try $x \otimes y$ for $x \in \mathcal{H}$ and $y \notin \mathcal{H}$ (which means that the Fourier coefficients of y are not square-summable).

Exercise 1.7.8 Let $x, y \in \mathcal{H}$ be unit vectors and let $x \otimes y$ be as defined in (1.2.2). Verify the following statements:

(1) $x \otimes y$ is self-adjoint if and only if $x = \pm y$.

(2) $x \otimes y$ is normal if and only if there exists a unimodular scalar $\alpha \in \mathbb{C}$ such that $x = \alpha y$.

Hint: This may be derived from

$$x \otimes \alpha y = \bar{\alpha} x \otimes y, \ \alpha \in \mathbb{C},$$
$$(x \otimes y)^* = y \otimes x, \quad (x \otimes y)(z \otimes w) = \langle z, y \rangle x \otimes w,$$

where $x, y, z, w \in \mathcal{H}$ and $\alpha \in \mathbb{C}$.

Exercise 1.7.9 Let $x, y \in \mathcal{H}$. Show that

$$\sigma(x \otimes y) = \{0, \langle x, y \rangle\} = \sigma_p(x \otimes y).$$

Hint: $x \otimes y$ is an *algebraic operator*:

$$p(x \otimes y) = 0 \text{ with } p(\lambda) = \lambda(\lambda - \langle x, y \rangle).$$

Now use the spectral mapping property for polynomials (see Lemma B.1.10).

Exercise 1.7.10 Let $k : [0,1] \times [0,1]$ be a continuous function. Define a linear operator $T : L^2([0,1]) \to L^2([0,1])$ by

$$(Tu)(x) = \int_{[0,1]} k(x,y)u(y)dy, \ u \in L^2([0,1]), \ x \in [0,1].$$

Show that T is a Hilbert–Schmidt operator.

Hint: For any orthonormal basis $\{f_n\}_{n \geqslant 0}$ of $L^2[0,1]$,

$$\sum_{n=0}^{\infty} \|Tf_n\|^2 = \int_{[0,1]} \int_{[0,1]} |k(x,y)|^2 dxdy.$$

Exercise 1.7.11 For $t \in C_0(\mathbb{R}^{n+m})$ (space of continuous functions with compact support), define a linear operator $T : L^2(\mathbb{R}^m) \to L^2(\mathbb{R}^n)$ by

$$(Tu)(x) = \int_{\mathbb{R}^m} t(x,y)u(y)dy, \ u \in L^2(\mathbb{R}^m), \ x \in \mathbb{R}^n.$$

Verify the following:

(1) T is a bounded, linear operator.

(2) Let $\{u_k\}$ be a bounded sequence with bound M. Then,

$$|(Tu_k)(x)| \leqslant \|t\|_{\infty} (2c)^{m/2} M,$$

where c is chosen so that $t(x,y) = 0$ if $\|x\|_2 > c$ or $\|y\|_2 > c$.

(3) Given $\epsilon > 0$, there exists $\delta \in (0,1)$ such that

$$|(Tu_k)(x) - (Tu_k)(x')| \leqslant \epsilon (2c)^{m/2} M$$

whenever $\|x - x'\|_2 < \delta$.

(4) T is a compact operator.

Exercise 1.7.12 Let $1 \leqslant p < \infty$. Set $(Tf)(x) = \int_0^x f(y)dy$ $(x \in [0,1])$ and

$$(T_n f)(x) = \sum_{k=0}^{n-1} \chi_{k/n,(k+1)/n}(x) \int_0^{k/n} f(y)dy \ (x \in [0,1]).$$

Verify the following:

(1) $T \in B(L^1(0,1), L^p(0,1))$.
(2) $T_n \in B(L^1(0,1), L^p(0,1))$ has n-dimensional range.
(3) $\|T - T_n\|_p \leqslant n^{-1/p}$.
(4) T is compact.

Exercise 1.7.13 Show that a positive operator $P \in \mathcal{L}(\mathcal{H})$ is compact if and only if the square root of P is compact.

Hint: Consult Corollary 1.3.4.

Exercise 1.7.14 Verify the following for $T : L^2(0,\infty) \to L^2(0,\infty)$ given by

$$(Tf)(x) = \frac{1}{x} \int_0^x f(t)dt \ (x \in (0,\infty)).$$

(1) T is bounded with $\|T\| \leqslant 2$.

(2) T is not compact.

Hint: To see (2), consider $f_n(t) = n$ if $0 < t \leqslant 1/n^2$, and 0 otherwise. Note that $\|Tf_n\| \leqslant 1$ for all n. However, as $n \to \infty$, $\langle Tf_n, g \rangle \to 0$ for all $g \in L^2(0,\infty)$.

Exercise 1.7.15 Let T be the linear operator of differentiation with domain $\mathcal{D}(T)$ in $L^2(0,1)$. Verify the following:

(1) If $\mathcal{D}(T) = \{f \in AC[0,1] : f' \in L^2(0,1), \ f(0) = f(1)\}$, then

$$\sigma(T) = \sigma_p(T) = \{2i\pi n : n \in \mathbb{Z}\}.$$

(2) For the operator T in part (1), $(T-1)^{-1}$ is compact.

(3) The eigenvectors $1, e^{2\pi ix}, e^{-2\pi ix}, e^{4\pi ix}, e^{-4\pi ix}, \ldots$ of iT form a complete orthonormal set in $L^2[0,1]$.

Hint: For $f \in \mathcal{D}(T)$, $Tf = \lambda f$ iff $f' = \lambda f$ a.e. iff $(e^{-\lambda t}f)' = 0$ a.e. By the fundamental theorem of calculus [119, Theorem 7.18], and the boundary conditions $f(0) = f(1)$, we must have $\lambda = 2i\pi n$ for some $n \in \mathbb{Z}$. If $\lambda \notin \sigma_p(T)$, then $R(\lambda)$ is given by

$$(R(\lambda)g)(x) = \int_{(0,1)} \frac{e^{\lambda(x-y)}}{1-e^{\lambda}} t(x,y)g(y)dy, \quad g \in L^2(0,1),$$

where $t(x,y) = e^{\lambda}$ if $x \leqslant y$; and $t(x,y) = 1$ if $x > y$. To see (2), note that

$$((T - I)^{-1}g)(x) = \int_{(0,1)} \frac{e^{(x-y)}}{1 - e} t(x,y)g(y)dy, \quad g \in L^2(0,1),$$

where $t(x,y) = e$ if $x \leqslant y$; and $t(x,y) = 1$ if $x > y$. For (3), apply Corollary 1.3.4.

Exercise 1.7.16 For an integer $n \geqslant 1$, let T denote the $n \times n$ matrix $(T_{ij})_{1 \leqslant i,j \leqslant n}$, where $T_{ij} \in \mathcal{L}(\mathcal{H})$, $i,j = 1,\ldots,n$. Show that T is a compact operator on n-fold inflation of $\mathcal{H}$ if and only if $T_{ij} \in \mathcal{L}(\mathcal{H})$ is compact for every $i,j = 1,\ldots,n$.

Hint: To prove the necessity, use Corollary 1.2.3.

Exercise 1.7.17 Let $A \in \mathcal{L}(\mathcal{H})$ be a compact operator and let p be a positive integer. Show that $\sum_{n=0}^{\infty} \||A|^p e_n\|^2$ is independent of the choice of orthonormal basis $\{e_n\}_{n \geqslant 0}$ of $\mathcal{H}$.

Exercise 1.7.18 Let $F \in \mathcal{L}(\mathcal{H})$ be a finite rank operator with $m = \dim \operatorname{ran} F$. Show that the Schatten p norm $\|F\|_p$ of F is at most $m^{1/p}\|T\|$.

Hint: We may assume that $F = \sum_{j=1}^{m} F(x_j) \otimes y_j$ for some $x_j, y_j \in \mathcal{H}$ such that $\{F(x_j)\}_{j=1}^{m}$ is orthonormal in $\mathcal{H}$. Note that

$$F^*F = \Big(\sum_{j=1}^{m} y_j \otimes F(x_j)\Big)\Big(\sum_{j=1}^{m} F(x_j) \otimes y_j\Big) = \sum_{j=1}^{m} y_j \otimes y_j = \sum_{k=1}^{\infty} \Big(\sum_{j=1}^{m} |\langle y_j, e_k \rangle|^2\Big) e_k \otimes e_k,$$

where $\{e_j\}_{j \geqslant 0}$ is an orthonormal basis for $\mathcal{H}$. Thus, the singular values of F are $\Big(\sum_{j=1}^{m} |\langle y_j, e_k \rangle|^2\Big)^{1/2}$, $j = 1,\ldots,m$.

Exercise 1.7.19 Let $\mathcal{H}$ be a complex separable Hilbert space and let $\mathcal{N}$ be the collection of normal operators in $B(\mathcal{H})$. Then the set of eigenvalues together with their multiplicities is a complete unitary invariant for $\mathcal{N}$ if and only if $\mathcal{H}$ is finite dimensional.

Hint: If $n < \infty$, then the conclusion follows from the spectral theorem for normal matrices. If n is infinite, then we may assume that $\mathcal{H} = L^2(\mathbb{T})$. Note that $N = M_z$ and $M = M_{z/2}$ both have empty eigenspectrum, but they are not unitarily equivalent: For any unitary U, $\|UM_z 1\| = 1$ and $\|M_{z/2}U 1\| = 1/2$.

Exercise 1.7.20 Let X be a compact subset of $\mathbb{C}$. For $f \in C(X)$, set

$$\rho(f) = \{\lambda \in \mathbb{C} : g(f - \lambda) = 1 \text{ for some } g \in C(X)\}.$$

Verify the following:

(i) $\mathbb{C} \backslash \rho(f) = \{f(\lambda) : \lambda \in X\}$.

(ii) The maximal ideal space (up to homeomorphism) of the C^*-algebra $C^*(f)$ generated by f is equal to the image of f.

Hint: If $\lambda \notin f(X)$, then $(f - \lambda)^{-1} \in C(X)$. Note that $C(X)$ can be embedded into $B(L^2(X))$ via the mapping $f \mapsto M_f$. Thus, $C^*(f)$ is isomorphic to $C^*(M_f)$.

Exercise 1.7.21 Let $N \in \mathcal{L}(\mathcal{H})$ be a normal operator. Show that N is invertible if and only if there exists a positive number c such that $N^*N > cI$. Conclude that $\sigma(N) = \sigma_{ap}(N)$ (see Remark B.1.16 for the definition of the approximate-point spectrum $\sigma_{ap}(\cdot)$).

Hint: The second part can be obtained by applying the first part to the normal operator $N - \lambda$, $\lambda \in \mathbb{C}$.

Exercise 1.7.22 Show that there exists two normal operators with the same spectrum, which are not unitarily equivalent.

Hint: The spectrum of the multiplication operator M_z on $L^2(\mathbb{T})$ equals the unit circle $\mathbb{T}$. If $\{\lambda_n\}$ is a countable dense subset of $\mathbb{T}$, then the spectrum of the diagonal operator with the nth diagonal entry λ_n equals $\mathbb{T}$. Clearly, these two operators cannot be unitarily equivalent as their eigen-spectra differ.

Exercise 1.7.23 Let U_+ be as defined in Example 1.3.2. Show that U_+ is a Fredholm operator, which is not a compact perturbation of any invertible operator.

Exercise 1.7.24 ([75], Weyl's Theorem) Let $A, B \in \mathcal{L}(\mathcal{H})$ be such that $A - B$ is a compact operator. Show that

$$\sigma(A) \backslash \sigma_p(A) \subseteq \sigma(B), \quad \sigma(B) \backslash \sigma_p(B) \subseteq \sigma(A).$$

Hint: By symmetry, we need to check only the first inclusion; after translation, it suffices to see the inclusion for $\lambda = 0$. Now apply Corollary 1.2.3 and Theorem 1.3.3.

Exercise 1.7.25 Let N be a normal operator. If, for some open neighborhood U of the origin, $\chi_U(N)$ is a finite rank orthogonal projection, then show that N is Fredholm.

Exercise 1.7.26 For any compact subset X of $\mathbb{C}$, show that there exists a diagonal operator D such that $\sigma(D) = X$.

Exercise 1.7.27 Let E be an orthogonal projection in $\mathcal{L}(\mathcal{H})$ and let $T \in \mathcal{L}(\mathcal{H})$ be such that $\pi(T)$ commutes with $\pi(E)$. Show that T is diagonal modulo compact in the following sense: $T = T_1 \oplus T_2 + K$ on $\ker E \oplus \ker(I - E)$ for some bounded linear operators T_1, T_2 and a compact operator K.

Exercise 1.7.28 Suppose that there exists an injective operator $T \in B(\mathcal{K})$ such that $\operatorname{ran} T \subsetneq \mathcal{K}$ is dense in $\mathcal{H}$. Prove that there exist a Hilbert space $\mathcal{H}$ and closed subspaces $\mathcal{M}$ and $\mathcal{N}$ of $\mathcal{H}$ such that $\mathcal{M} + \mathcal{N} \subsetneq \mathcal{H}$ is dense in $\mathcal{H}$ and $\mathcal{M} \cap \mathcal{N} = \{0\}$.

Hint: Let $\mathcal{H} = \mathcal{K} \oplus \mathcal{K}$ with inner-product

$$\langle (x \oplus y, x' \oplus y') \rangle = \langle x, x' \rangle + \langle y, y' \rangle,$$

$\mathcal{M} = \{x \oplus Tx : x \in \mathcal{K}\}$ and $\mathcal{N} = \mathcal{K} \oplus \{0\}$.

Exercise 1.7.29 Show that every compact operator $T \in \mathcal{L}(\mathcal{H})$ admits a separable reducing subspace with orthogonal complement contained in $\ker T$.

Hint: Let $\mathcal{M}$ denote the closed linear span of all eigenvectors of T^*T corresponding to nonzero eigenvalues and note that $\mathcal{M}^\perp = \ker T$. Consider now the closed linear span of $\{Sx : S \in C^*(T),\ x \in \mathcal{M}\}$.

Exercise 1.7.30 Let N be a normal operator such that $NM - MN$ is compact for some bounded linear operator M. Show that $f(N)M - Mf(N)$ is compact for any continuous $f : \sigma(N) \to \mathbb{C}$.

Exercise 1.7.31 ([99]) Let X be a Banach space and let T be in $\mathcal{L}(X)$ such that $X/\mathrm{ran}\,T$ is finite dimensional, i.e., there exists a finite dimensional subspace Z of X such that the (vector space) direct sum of $\mathrm{ran}\,T$ and Z is X. Show that $\mathrm{ran}\,T$ is closed in X.
Hint: After quotienting out by $\ker T$, we may assume that T is injective. Verify that $(x,z) \mapsto Tx + z$ is a Banach space isomorphism from $X \oplus Z$ onto X, and apply open mapping theorem.

Exercise 1.7.32 Let $A, B \in \mathcal{L}(\mathcal{H})$ be self-adjoint operators such that $AB = BA$. If AB is compact, then show that $\sigma_e(A + iB)$ is homeomorphic to a compact subset of the real line.
Hint: Use the spectral mapping property.

Exercise 1.7.33 Show that if two Banach algebras are isomorphic, then so are their abstract index groups.

Exercise 1.7.34 Show that abstract index group for $\mathcal{L}(\mathcal{H})$ is trivial.
Hint: The group of invertible operators in $\mathcal{L}(\mathcal{H})$ is path-connected.

Exercise 1.7.35 Show that the point evaluation on the $H^2(\mathbb{D})$ is bounded, that is, for any $w \in \mathbb{D}$, the mapping $E_w : H^2(\mathbb{D}) \to \mathbb{C}$ given by $E_w(f) = f(w)$ is bounded.
Hint: Apply the Cauchy integral formula to see that for any $f \in H^2(\mathbb{D})$,

$$|f(w)| \leqslant \frac{\|f\|_{H^2(\mathbb{D})}}{R - r}, \quad |w| \leqslant r < R < 1.$$

Exercise 1.7.36 Suppose that $T \in \mathcal{L}(\mathcal{H})$ is *finitely multicyclic*, that is, there exist finitely many vectors (to be referred to as *cyclic vectors*) $e_1, \ldots, e_k \in \mathcal{H}$ such that

$$\bigvee \{T^n e_i : i = 1, \ldots, k\} = \mathcal{H}.$$

Verify the following:

(1) If T is left-invertible, then it is Fredholm.

(2) The approximate-point spectrum

$$\sigma_{ap}(T) = \{\lambda \in \mathbb{C} : T - \lambda I \text{ is not bounded below}\}$$

of T contains its essential spectrum $\sigma_e(T)$.

Conclude that if T has no eigenvalues, then $\sigma_{ap}(T) = \sigma_e(T)$.

Exercise 1.7.37 For $j = 1,\ldots,n$, let $T_j \in \mathcal{L}(\mathcal{H}_j)$. Show that the spectrum of the orthogonal direct sum $\oplus_{j=1}^n T_j \in \mathcal{L}(\oplus_{j=1}^d \mathcal{H}_j)$ is union of spectra of $T_1,\ldots,T_d$. Show that this is no longer true for infinite direct sum, even after taking closure.

Exercise 1.7.38 For $j = 1,\ldots,n$, let $T_j \in \mathcal{L}(\mathcal{H}_j)$. Show that the essential spectrum of the orthogonal direct sum $\oplus_{j=1}^n T_j \in \mathcal{L}(\oplus_{j=1}^d \mathcal{H}_j)$ is a union of essential spectra of $T_1,\ldots,T_d$. Show that this is no longer true for the infinite direct sum, even after taking closure.

Hint: Consult Theorem 1.3.16.

Exercise 1.7.39 In the notation of Corollary 1.3.7, decompose D with respect to $\mathcal{H} = \mathcal{M} \oplus \mathcal{M}^\perp$, where $\mathcal{M}$ is the closed linear span of $\{h_n\}_{n \geqslant 1}$. Deduce Corollary 1.3.7 from this decomposition.

Hint: Apply Exercises 1.7.37 and 1.7.38 noting that $D_1 = 0 \oplus D|_{\mathcal{M}^\perp}$.

Exercise 1.7.40 Let S_λ be as introduced in Example 1.5.6. Verify that the dimension of $\ker S_\lambda^*$ equals 1. Conclude that $\sigma(S)$ equals the closed unit disc in the plane.

Exercise 1.7.41 Show that any isometry U has the following spectral dichotomy: Either $\sigma(U)$ is the closed unit disc or a closed subset of the unit circle. Conclude further that the spectrum of a non-normal isometry is necessarily the closed unit disc.

Hint: Use Theorem 1.5.4 (von Neumann-Wold Decomposition) and Exercise 1.7.40.

Exercise 1.7.42 Show that the multiplicity is a *complete unitary invariant* for the class of analytic isometries, that is, two analytic isometries are unitarily equivalent if and only if their multiplicities are the same.

Exercise 1.7.43 ([128]) Let T be left-invertible and let

$$L = (T^*T)^{-1} T^*$$

be a left inverse of T. Verify:

(1) $I - T^n L^n = \sum_{k=0}^{n-1} T^k (I - TL) L^k$.

(2) $I - TL$ is an orthogonal projection onto $\ker T^*$.

(3) $\ker L^n = \bigvee_{k=0}^{n-1} T^k (\ker T^*)$.

(4) $(\cap_{n \geqslant 0} L^{*n} \mathcal{H})^\perp = \bigvee_{n \geqslant 0} S^n (\ker T^*)$.

Hint: Imitate the argument of Lemma 1.5.3. The operator L^*, introduced by S. Shimorin, is known as the *Cauchy dual* of T.

Exercise 1.7.44 Let $\varphi \in C(\mathbb{T})$. Calculate the matrix representation of T_φ with respect to the orthonormal basis $\{e^{i\theta n}\}_{n \in \mathbb{N}}$ of $H^2(\mathbb{T})$.

Exercise 1.7.45 Let $T \in \mathcal{H}(\mathcal{H})$ be left-invertible. Show that $C^*(T)$ is same as $C_0^*(T)$, where $C_0^*(T)$ is equal to the

     closure of $\{p(T, T^*) : p$ is a polynomial in z, w with $p(0,0) = 0\}$ in $\mathcal{L}(\mathcal{H})$.

Hint: Note that T^*T is invertible. It now suffices to check that $(T^*T)^{-1}$ belongs to $C_0^*(T)$. To prove that, combine the Stone–Weierstrass theorem with the spectral theorem.

Exercise 1.7.46 Show that C^*-algebra generated by $\{T_\varphi : \varphi \in C(\mathbb{T})\}$ coincides with the Toeplitz C^*-algebra.

Hint: Use the Stone–Weierstrass Theorem.

2

Ext(X) as a Semigroup with Identity

In his very influential article [72], Halmos listed ten problems for Hilbert space operators. One of these (Problem 4) asked "Is every normal operator the sum of a diagonal operator and a compact one?" Soon after the question was raised, Berg [20] and Sikonia [129] independently of each other showed that the answer is "yes". Much of what follows reproduces their theorem generalizing the Weyl–von Neumann theorem. Moreover, we recast this theorem in the language of the 'Ext" group.

2.1 Essentially Normal Operators

An essentially normal operator $T \in \mathcal{L}(\mathcal{H})$ defines an extension of $C(X)$, where X is the essential spectrum of T, by the compact operators $C(\mathcal{H})$, or equivalently, a $*$-monomorphism of $C(X)$ into the Calkin algebra $Q(\mathcal{H})$ and conversely. One of the main goals of this chapter is to show the advantage of studying the class of essentially normal operators using the C^*-extensions they define. We begin by recalling the notion of an essentially normal operator and other notions closely related to it.

Definition 2.1.1

An operator T in $\mathcal{L}(\mathcal{H})$ is *essentially normal* if the self-commutator $[T, T^*]$ of T is compact. We say that T is *essentially unitary* if T is essentially normal and $T^*T - I$ is compact. Two operators T_1 and T_2 in $\mathcal{L}(\mathcal{H}_1, \mathcal{H}_2)$ are said to be *essentially equivalent* if there exists a unitary operator $U \in \mathcal{L}(\mathcal{H}_1, \mathcal{H}_2)$ and a compact operator $K \in \mathcal{L}(\mathcal{H}_1, \mathcal{H}_2)$ such that $UT_1U^* = T_2 + K$. In this case we write, $T_1 \sim T_2$.

Remark 2.1.2   Let T_1, T_2 in $\mathcal{L}(\mathcal{H})$ be essentially equivalent. Then,

(1) T_1 is essentially normal if and only if so is T_2.

(2) T_1 is Fredholm if and only if so is T_2. In this case, $\mathrm{ind}(T_1) = \mathrm{ind}(T_2)$.

In particular, T_1 and T_2 have the same essential spectra and index data.

  Note that any operator is essentially equivalent to its compact perturbation. Here is one concrete illustration of this general fact.

Example 2.1.3   Let $\beta = \{\beta_n\}_{n \geqslant 0}$ be a sequence of positive numbers such that

$$\sup_{n \geqslant 0} \frac{\beta_{n+1}}{\beta_n} < \infty.$$

The *weighted Hardy space* of the open unit disc $\mathbb{D}$, denoted by H_β^2, is given by

$$H_\beta^2 = \Big\{ f : \mathbb{D} \to \mathbb{C} \mid f(z) = \sum_{n=0}^\infty \hat{f}(n) z^n \text{ such that } \sum_{n=0}^\infty |\hat{f}(n)|^2 \beta_n^2 < \infty \Big\}.$$

H_β^2 is endowed with the inner-product

$$\langle f, g \rangle = \sum_{n=0}^\infty \beta_n^2 \hat{f}(n) \overline{\hat{g}(n)}, \quad f, g \in H_\beta^2.$$

The shift operator $M_z : H_\beta^2 \to H_\beta^2$ is given by

$$(M_z f)(z) = z f(z), \quad f \in H_\beta^2.$$

If $\beta_n = 1$ for all integers $n \geqslant 0$, then H_β^2 is nothing but the Hardy space $H^2(\mathbb{D})$ introduced in Example 1.5.2. In this case, M_z agrees with the shift operator S. In case $\beta_n = \frac{1}{\sqrt{n+1}}$, $n \geqslant 0$, the space H_β^2 is known as the *Bergman space* $L_a^2(\mathbb{D})$, and the associated multiplication operator is known as the *Bergman shift*, denoted by B. Note that

$$\|z^n\|_{H^2(\mathbb{D})} = 1, \quad \|z^n\|_{L_a^2(\mathbb{D})} = 1/\sqrt{n+1}, \quad n \geqslant 0.$$

For any unitary $U : L_a^2(\mathbb{D}) \to H^2(\mathbb{D})$, note that $\|S U 1\|_{H^2(\mathbb{D})} = 1$ and $\|UB1\| = \|z\|_{L_a^2(\mathbb{D})} = 1/\sqrt{2}$. It follows that S and B are not unitarily equivalent. Let us verify the following statements:

(1) S and B are essentially normal. We have seen earlier in Chapter 1 that S is essentially normal (see Corollary 1.5.3). To obtain the essential normality of B, note that $B^* B - BB^*$ is a diagonal operator with diagonal entries $\{1/(n+1)(n+2)\}_{n \in \mathbb{N}}$ with respect to the orthonormal basis $\{z^n/\|z^n\|\}_{n \in \mathbb{N}}$. Since the diagonal entries of $B^*B - BB^*$ converge to 0, B is essentially normal (see Corollary 1.2.1).

(2) S and B are essentially equivalent. In fact, if $U(z^n) = \frac{z^n}{\sqrt{n+1}}$, $n \geqslant 0$, then $U : L_a^2(\mathbb{D}) \to H^2(\mathbb{D})$ is a unitary operator that satisfies

$$UB - SU = -SUD,$$

where $D : L_a^2(\mathbb{D}) \to L_a^2(\mathbb{D})$ is the diagonal operator given by

$$D(z^n) = (1 - \sqrt{n+1/n+2})z^n, \quad n \geqslant 0.$$

Since D is compact, by Corollary 1.2.3, $UB - SU$ is also a compact operator.

Finally, note that $\text{ind}(S - \lambda) = \text{ind}(B - \lambda) = -1$ for $\lambda \in \mathbb{D}$, the bounded component of the complement of the essential spectrum $\sigma_e(S) = \sigma_e(B) = \mathbb{T}$ (see Example 1.5.6). ∎

Recall that $T \in \mathcal{L}(\mathcal{H})$ is *diagonalizable* if there exists an orthonormal basis for $\mathcal{H}$ consisting of eigenvectors of T. Note that a diagonalizable operator is necessarily normal. To classify a class of diagonalizable operators up to essential equivalence, we need the following lemma.

Lemma 2.1.1 (von Neumann)

Let (X, d) be a compact metric space with metric d. If two sequences $\{x_n\}_{n\geqslant 0}$ and $\{y_n\}_{n\geqslant 0}$ in X have the same set L of limit points and if $\{x_n\}_{n\geqslant 0} \subseteq L$ and $\{y_n\}_{n\geqslant 0} \subseteq L$, then for any $\epsilon > 0$, there exists a permutation $\eta : \mathbb{N} \to \mathbb{N}$ such that $\sum_{n=0}^{\infty} d(x_n, y_{\eta(n)}) < \epsilon$.

Proof   As pointed out by Halmos [74], the proof is due to Stampfli. We verify by a two-step induction on integers $n \geqslant 0$ that there exist permutations σ and τ such that

$$d(x_{\sigma(n)}, y_{\tau(n)}) < \epsilon/2^{n+1}, \quad n \geqslant 0. \tag{2.1.1}$$

Set $\sigma(0) = 1$. Since $x_{\sigma(0)} \in L$ (which is also a limit point of $\{y_n\}_{n\geqslant 0}$), there exists an integer $\tau(0) \geqslant 0$ such that $d(x_{\sigma(0)}, y_{\tau(0)}) < \epsilon/2$. Let $\tau(1) \geqslant 0$ be the smallest integer distinct from $\tau(0)$ (if $\tau(0) = 0$, let $\tau(1) = 1$, and if $\tau(0) > 0$, let $\tau(1) = 0$). Since $y_{\tau(1)} \in L$, arguing as earlier, there exists an integer $\sigma(1) \geqslant 0$ distinct from $\sigma(0)$ such that $d(x_{\sigma(1)}, y_{\tau(1)}) < \epsilon/4$. Thus, we obtain (2.1.1) for $n = 0, 1$. Assume now that (2.1.1) holds for $n, n+1$ for some integer $n \geqslant 0$. Without loss of generality, we may assume that n is even (in case n is odd, we need to choose $\tau(n+2), \sigma(n+2), \sigma(n+3), \tau(n+3)$ in order). Let $\sigma(n+2) \geqslant 0$ be the smallest integer distinct from $\{\sigma(0), \ldots, \sigma(n+1)\}$. Since $x_{\sigma(n+2)} \in L$, there exists an integer $\tau(n+2) \geqslant 0$ distinct from $\tau(0), \ldots, \tau(n+1)$ such that $d(x_{\sigma(n+2)}, y_{\tau(n+2)}) < \epsilon/2^{n+3}$. Let $\tau(n+3) \geqslant 0$ be the smallest integer distinct from $\tau(0), \ldots, \tau(n+2)$. Since $y_{\tau(n+3)} \in L$, there exists an integer $\sigma(n+3) \geqslant 0$ distinct from $\sigma(0), \ldots, \sigma(n+2)$ such that $d(x_{\sigma(n+3)}, y_{\tau(n+3)}) < \epsilon/2^{n+4}$. Clearly, this construction ensures that σ and τ are one to one. Furthermore, the choice of $\sigma(n)$, $n = 0, 2, 4, \ldots$, shows that σ is surjective. Let $\eta = \tau \circ \sigma^{-1}$ and note that

$$\sum_{n=0}^{\infty} d(x_n, y_{\eta(n)}) = \sum_{n=0}^{\infty} d(x_{\sigma(n)}, y_{\tau(n)}) < \epsilon$$

by (2.1.1) completing the proof. □

The following proposition provides a classification, up to essential equivalence, of the diagonalizable operators without isolated eigenvalues of finite multiplicity (see Theorem 1.3.7).

Proposition 2.1.1

Let $\mathcal{H}$ be a separable Hilbert space and let D_1, D_2 be two diagonalizable operators in $\mathcal{L}(\mathcal{H})$. Assume that

$$\sigma_e(D_1) = \sigma(D_1), \quad \sigma_e(D_2) = \sigma(D_2).$$

Then, D_1 and D_2 are essentially equivalent if and only if $\sigma_e(D_1) = \sigma_e(D_2)$.

Proof The necessity part follows from Remark 2.1.2(2). To see the sufficiency part, let $\{\mu_n\}_{n \in \mathbb{N}}$ (resp. $\{\nu_m\}_{m \in \mathbb{N}}$) be eigenvalues of D_1 (resp. D_2) with eigenvectors $\{e_n\}_{n \in \mathbb{N}}$ (resp. $\{f_m\}_{m \in \mathbb{N}}$). Note that

$$\sigma_e(D_1) = \overline{\{\mu_n\}}_{n \in \mathbb{N}} \text{ and } \sigma_e(D_2) = \overline{\{\nu_m\}}_{m \in \mathbb{N}}.$$

Note further that for each $j \in \mathbb{N}$, there is a subsequence of $\{\mu_m\}_{m \in \mathbb{N}}$ converging to ν_j. Indeed, if ν_j is an isolated point then by (1.3.16), it must be an eigenvalue of infinite multiplicity, and hence, $\nu_j = \mu_i$ for infinitely many values of i and a constant subsequence works; otherwise, ν_j is a limit point of $\{\mu_m\}_{m \in \mathbb{N}}$. Thus, $\{\mu_n\}_{n \in \mathbb{N}}$ and $\{\nu_m\}_{m \in \mathbb{N}}$ have the same set of limit points. By Proposition 2.1.1, there exists a permutation $\eta : \mathbb{N} \to \mathbb{N}$ such that $|\mu_n - \nu_{\eta(n)}| \to 0$ as $n \to \infty$. Thus, the unitary operator U on $\mathcal{H}$ defined by $U e_n = f_{\eta(n)}$, $n \in \mathbb{N}$ satisfies

$$\|(D_1 - U^* D_2 U)e_n\| = \|(\mu_n - \nu_{\eta(n)})e_n\| = |\mu_n - \mu_{\eta(n)}| \to 0 \text{ as } n \to \infty,$$

which together with Corollary 1.2.1 implies that D_1 and D_2 are essentially equivalent. □

In view of the preceding proposition, the natural question arises whether the earlier result extends to a pair of normal operators. The answer is affirmative and provided by the celebrated Weyl–von Neumann–Berg–Sikonia theorem to be discussed in the next section. This theorem is a special case of the Brown–Douglas–Fillmore theorem describing the class of mutually inequivalent essentially normal operators. In what follows, we will very closely follow the fundamental papers of Brown, Douglas, and Fillmore [26, 27] (we have also generously borrowed from the monograph [47]). Why should the problem of classifying essentially equivalent essentially normal operators be tractable at all? To answer this question, we have to look at some early history preceding the work of Brown, Douglas, and Fillmore [26, 27].

2.2 Weyl–von Neumann–Berg–Sikonia Theorem

In this section, we give a proof of the Weyl–von Neumann theorem which says that a bounded self-adjoint operator on a complex separable Hilbert space is essentially equivalent to a diagonal operator. Berg and Sikonia have shown that the theorem of Weyl–von Neumann extends to all normal operators. We reproduce a proof due to Halmos. The fact that every normal operator is a continuous function of a self-adjoint operator is a key ingredient in this proof.

In 1909, Weyl [141] defined the essential spectrum of a self-adjoint operator to be all points in its spectrum except the isolated eigenvalues of finite multiplicity (cf. Theorem 1.3.7). He proved that if two self-adjoint operators S and T are essentially equivalent then S and T have the same essential spectrum. Some twenty years later, von Neumann [101] proved a striking converse, that is, if the essential spectrum of two self-adjoint operators are equal, then they are essentially equivalent. In response to a question of Halmos [72], Berg and Sikonia, independent of each other, showed around 1971 that the Weyl von Neumann theorem actually holds for normal operators (refer to [20] and [129]).

It might be asked: What does this all have to do with essentially normal operators? The point is that if $C(\mathcal{H})$ is the set of compact operators and $\pi : \mathcal{L}(\mathcal{H}) \to Q(\mathcal{H})$ is the natural quotient map, then an operator T is essentially normal if and only if the class $\pi(T)$ is normal in the Calkin algebra $Q(\mathcal{H})$. The essential spectrum $\sigma_e(N)$ of a normal operator N is the same as the spectrum $\sigma(\pi(N))$ of the class $\pi(N)$ in $Q(\mathcal{H})$. Let

$$\mathcal{N} + C = \{N + K : N \in \mathcal{L}(\mathcal{H}) \text{ be normal and } K \in \mathcal{L}(\mathcal{H}) \text{ be compact}\}.$$

For an operator $T = N + K$ in $\mathcal{N} + C$, we see that

$$\sigma_e(T) = \sigma_e(N + K) = \sigma(\pi(N + K)) = \sigma(\pi(N)) = \sigma_e(N),$$

so that the Weyl–von Neumann–Berg–Sikonia theorem actually extends to operators in the class $\mathcal{N} + C$.

Theorem 2.2.1 (Weyl–von Neumann–Berg–Sikonia theorem)

Any two operators T_1 and T_2 in $\mathcal{N} + C$ are essentially equivalent if and only if $\sigma_e(T_1) = \sigma_e(T_2)$. Moreover, if X is any compact subset of the complex plane $\mathbb{C}$, then there is a normal operator N such that $\sigma_e(N) = X$.

The theorem above shows that the essential spectrum of an operator in $\mathcal{N} + C$ is *completely invariant* for unitary equivalence modulo compact and the classification problem for such operators is complete. It is natural to ask if all essentially normal operators belong to N + C ? We present here an example of an essentially normal operator not in $\mathcal{N} + C$.

Example 2.2.2   Consider the shift operator S on the Hardy space $H^2(\mathbb{D})$ (see Example 1.5.2). We have already seen that $S^*S = I$ and $I - SS^* = P$, where P is a rank one projection onto the kernel of S^* (see Example 1.5.3). It follows that S is an essentially unitary operator such that $\mathrm{ind}(S) = -1$. On the other hand, if N is a normal operator which is also Fredholm, then its index is zero. In case $S = N + K$ belongs to $\mathcal{N} + C$, then we would have

$$\mathrm{ind}(S) = \mathrm{ind}(N + K) = \mathrm{ind}(N) = 0,$$

which is impossible. This shows that $S \notin \mathcal{N} + C$. Secondly, if N is a normal operator with essential spectrum equal to the unit circle (as ensured by Example 1.3.8), then $\sigma_e(N) = \sigma_e(S)$

by Example 1.5.6. If these two operators were essentially equivalent, then for some unitary U, we would have

$$\text{ind}(N) = \text{ind}(N + K) = \text{ind}(U^*SU) = \text{ind}(S),$$

which is again not possible. ∎

The preceding example shows that the essential spectrum is not the only invariant for essential equivalence of essentially normal operators. The remarkable theorem of Brown, Douglas, and Fillmore says that the essential spectrum together with certain index data is a complete set of invariants for essential equivalence.

Proof of the Weyl–von Neumann–Berg–Sikonia theorem

The essential idea in the proof of the Weyl–von Neumann–Berg–Sikonia theorem is to represent any self-adjoint operator as a compact perturbation of a diagonalizable operator with the spectrum not containing isolated eigenvalues of finite multiplicity, so that one can use Proposition 2.1.1. The general case relies on the fact that any normal operator is a continuous function of a self-adjoint operator. Here is the first step in this process (see Exercise 2.8.3 for a finer version).

Lemma 2.2.1

Let $A \in \mathcal{L}(\mathcal{H})$, $f \in \mathcal{H}$, and $\epsilon > 0$. If A is a self-adjoint operator, then there is a finite rank projection P and a self-adjoint, finite rank operator K such that $f \in \text{ran}\,P$, $\|K\| < \epsilon$, and $\text{ran}\,P$ is reducing for $A + K$.

Proof   Assume that A is a self-adjoint operator. Let $a, b \in \mathbb{R}$ be such that $\sigma(A) \subseteq [a,b]$. Let $E(\cdot)$ denote the spectral measure of A and let n be a positive integer such that $(b-a)/n < \epsilon$. Let P denote the orthogonal projection onto the space $\mathcal{M}$ spanned by $f_1, \ldots, f_n$, where $f_1 = E([a, a + (b-a)/n] \cap \sigma(A))f$ and

$$f_k = E((a + (b-a)(k-1)/n, a + (b-a)k/n] \cap \sigma(A))f, \quad k = 2, \ldots, n.$$

If some $f_k = 0$, we discard it and hence, without loss of generality, we assume that all f_ks are nonzero. Note that $\{f_1/\|f_1\|, \ldots, f_n/\|f_n\|\}$ forms an orthonormal basis for $\mathcal{M}$. Let P denote the orthogonal projection of $\mathcal{H}$ onto $\mathcal{M}$. Thus,

$$\|P^\perp AP\| = \max_{k=1}^{n} \|P^\perp APf_k\| = \max_{k=1}^{n} \|P^\perp (A - \lambda_k)Pf_k\| \leqslant (b-a)/2n,$$

where $P^\perp = I - P$ and λ_k is the mid-point of $[a + (b-a)(k-1)/n, a + (b-a)k/n]$ for $k = 1, \ldots, n$. Finally, note that for the choice $K = -P^\perp AP - PAP^\perp$, $A + K = PAP + P^\perp AP^\perp$ reduces the range of P. □

In the next step, we show that any self-adjoint operator is a compact perturbation of a diagonalizable self-adjoint operator with perturbation as small as possible.

Lemma 2.2.2

Let $\epsilon > 0$ and let $A \in \mathcal{L}(\mathcal{H})$ be an operator on a separable Hilbert space $\mathcal{H}$. If A is self-adjoint, then there is a diagonalizable self-adjoint operator D such that $A - D$ is compact with norm at most ϵ.

Proof   Assume that A is a self-adjoint operator. Let $\{f_n\}_{n\geqslant 1}$ be a countable dense subset of $\mathcal{H}$. We apply the preceding lemma to A and f_1 to get a finite rank projection P_1 and a self-adjoint finite rank operator K_1 such that $\|K_1\| \leqslant \epsilon$, $f_1 \in \operatorname{ran} P_1$ and $\operatorname{ran} P_1$ is reducing for $A + K_1$. Next, we apply the last lemma to $(A + K_1)P_1^\perp$ and $P_1^\perp f_2$ to get a finite rank projection P_2 and a self-adjoint finite rank operator K_2 such that $\|K_2\| \leqslant \epsilon/2$, $P_1^\perp f_2 \in \operatorname{ran} P_2$, and $\operatorname{ran} P_2$ is reducing for $(A + K_1)P_1^\perp + K_2$. Extend K_2 to $\mathcal{H}$ by letting $K_2 = 0$ on $P_1\mathcal{H}$. Since $P_2 \leqslant P_1^\perp$, $A + K_1 + K_2$ reduces $\operatorname{ran} P_2$. Continuing this, by induction, we get sequences $\{P_n\}_{n\geqslant 1}$ of orthogonal projections, $\{f_n\}_{n\geqslant 1}$ of vectors in $\mathcal{H}$, and $\{K_n\}_{n\geqslant 1}$ of self-adjoint finite rank operators such that $P_j P = 0$, $j \neq n$, $f_n \in \operatorname{ran} P_1 + \cdots + \operatorname{ran} P_n$, $(P_1 + \cdots + P_n)\mathcal{H}$ is reducing for $A + K_1 + \cdots + K_n$, $K_{n+1}(P_1 + \cdots + P_n) = 0$ and $\|K_n\| \leqslant \epsilon/2^n$, $n \geqslant 1$. It is now easy to see that $\sum_n P_n = I$, $K = \sum_n K_n$ is compact with norm at most ϵ, and $A + K$ is diagonalizable. $\square$

Remark 2.2.3   It is easy to see, using Exercise 2.8.3, that the diagonalizable self-adjoint operator D of the lemma can be chosen so that $A - D$ is in the Schatten p-class with $\|A - D\|_p < \epsilon$, $p \in (1, \infty)$. The case of $p = 2$ of this fact is due to von Neumann [101], whereas the general case is due to Kuroda [88]. Kuroda's theorem fails for the trace-class perturbations, that is, the case of $p = 1$ (see [36, Page 212]).

The following proposition shows that in Lemma 2.2.2, the spectrum and the essential spectrum of the diagonalizable operator D may be taken to be the same.

Proposition 2.2.1

Let $\epsilon > 0$ and let $A \in \mathcal{L}(\mathcal{H})$ be a self-adjoint operator. Assume that $A = D + K$ for a diagonalizable self-adjoint operator $D \in \mathcal{L}(\mathcal{H})$ and a compact operator $K \in \mathcal{L}(\mathcal{H})$ such that $\|K\| < \epsilon$. Then, there is a diagonalizable self-adjoint operator D_2 such that $A - D_2$ is compact and $\sigma(D_2) = \sigma_e(D_2)$.

Proof   Write $\sigma(D) = \sigma_e(D) \cup \{\mu_n\}_{n\geqslant 1}$ (see (1.3.16)), where $\{\mu_n\}_{n\geqslant 1}$ is a set of isolated eigenvalues of D of finite multiplicity (repeated as often as their multiplicity) with $\{h_n\}_{n\geqslant 1} \subseteq \mathcal{H}$ denoting the respective eigenvectors of D. For each $j \geqslant 1$, let $\lambda_j \in \sigma_e(D)(= \sigma_e(A))$ be such that $|\mu_j - \lambda_j| = d(\mu_j, \sigma_e(D)) > 0$. Thus, we have the decomposition

$$D = D_2 + K_2, \quad D_2 = D + \sum_{j=1}^\infty (\lambda_j - \mu_j) h_j \otimes h_j, \quad K_2 = \sum_{j=1}^\infty (\mu_j - \lambda_j) h_j \otimes h_j.$$

We claim that $\sigma(D) \subseteq \{\lambda : d(\lambda, \sigma(A)) < \epsilon\}$. By the spectral theorem, $d(\lambda, \sigma(A)) \geqslant \epsilon$ implies that $A - \lambda \geqslant \epsilon$, and hence, $D - \lambda \geqslant \epsilon - K$. Since K is a self-adjoint operator with $\|K\| < \epsilon$, $D - \lambda$ is invertible, and the claim stands verified. This also implies that $|\mu_j - \lambda_j| < 2\epsilon$ since $\mu_j, \lambda_j \in \sigma(D)$. Thus, $\|K_2\| < 2\epsilon$. Note that $\{\mu_j - \lambda_j\}_{j\geqslant 1}$ converges to 0 as any limit point of $\{\mu_j\}_{j\geqslant 1}$ belongs to $\sigma_e(D)$, and thus, K_2 is compact. Finally, note that $\sigma(D_2) = \sigma_e(D_2)$ (cf. Corollary 1.3.7). $\square$

For later use, we isolate separately a consequence of the last result.

Corollary 2.2.1

Let $T \in \mathcal{L}(\mathcal{H})$. If $\pi(T)$ is an orthogonal projection, then T is a compact perturbation of an orthogonal projection in $\mathcal{L}(\mathcal{H})$.

Proof Assume that $\pi(T)$ is an orthogonal projection. Since $\pi(T)$ is self-adjoint, $\pi(\mathfrak{R}(T)) = \pi(T)$. Hence, without loss of generality, we may assume that T is self-adjoint. By Proposition 2.2.1 and Lemma 2.2.2, there exists a self-adjoint operator D such that $T - D$ is compact and $\sigma(D) = \sigma_e(D)$. It follows that

$$\sigma(D) = \sigma_e(D) = \sigma_e(T) = \sigma(\pi(T)) \subseteq \{0, 1\}.$$

By the spectral theorem, D is an orthogonal projection. Further, T is a compact perturbation of D. $\square$

Propositions 2.2.1 and 2.1.1 immediately yield the following result due to Weyl and von Neumann:

Theorem 2.2.4

Two self-adjoint operators T_1 and T_2 in $\mathcal{L}(\mathcal{H})$ are essentially equivalent if and only if $\sigma_e(T_1) = \sigma_e(T_2)$.

Proof The necessity part is already recorded in Remark 2.1.2. To see the sufficiency part, assume that $\sigma_e(T_1) = \sigma_e(T_2)$. By Proposition 2.2.1, there is a diagonalizable operator D_i such that $T_i - D_i$ is compact and $\sigma(D_i) = \sigma_e(D_i)$ for $i = 1, 2$. By hypothesis,

$$\sigma(D_1) = \sigma_e(D_1) = \sigma_e(T_1) = \sigma_e(T_2) = \sigma_e(D_2) = \sigma(D_2).$$

By Lemma 2.1.1, there exists a unitary map U such that $D_1 - U^* D_2 U$ is compact. On the other hand, since $(T_1 - U^* T_2 U) - (D_1 - U^* D_2 U)$ is compact, T_1 and T_2 are essentially equivalent. $\square$

To extend Theorem 2.2.4 to normal operators, we need a lemma of Halmos, which is of independent interest (refer to [73]).

Lemma 2.2.3 (Halmos' Lemma)

If $N \in \mathcal{L}(\mathcal{H})$ is a normal operator, then there exists a self-adjoint operator $S \in \mathcal{L}(\mathcal{H})$ such that $C^*(N) \subseteq C^*(S)$. In particular, there exists a continuous function $f : \sigma(S) \to \mathbb{C}$ such that $N = f(S)$.

Proof Let N be a normal operator and let $E(\cdot)$ denote its spectral measure. For each $n \geqslant 1$, consider a Borel partition $\{\Delta_{n,j} : j = 1, \ldots, k_n\}$ of $\sigma(N)$ of diameter less than $1/n$. Define $P_{n,j} = E(\Delta_{n,j})$ and consider the abelian C^*-algebra C generated by

$$\mathscr{P} := \{P_{n,j} : j = 1, \ldots, k_n, \ n \in \mathbb{N}\}.$$

Let X denote its maximal ideal space of C and let $\rho : C \to C(X)$ be the Gelfand map as ensured by Corollary C.1.2. By the spectral theorem, $C^*(N) \subseteq C$. Hence, it suffices to check that C is generated by a self-adjoint operator.

Let $\{Q_m\}_{m\in\mathbb{N}}$ be an enumeration of $\mathscr{P}$. Note that

$$\rho(Q_m)^2 = \rho(Q_m), \quad \overline{\rho(Q_m)} = \rho(Q_m), \quad m \in \mathbb{N}.$$

It follows that $f = \sum_{m=0}^{\infty} 3^{-m}(2\rho(Q_m) - 1)$ belongs to $C_\mathbb{R}(X) \subseteq C(X)$. Note that $\{\rho(Q_m)\}_{m\in\mathbb{N}}$ generates $C(X)$, and hence, it separates the points of X. Let $x \neq y$ in X be given. Then there is a smallest integer $k \geqslant 0$ such that $\rho(Q_k)(x) \neq \rho(Q_k)(y)$. Since for any $m \in \mathbb{N}$, $\rho(Q_m)(x) - \rho(Q_m)(y) \in \{-1, 0, 1\}$, we have

$$|f(x) - f(y)| = 2\left|\sum_{m=k}^{\infty} 3^{-m}(\rho(Q_m)(x) - \rho(Q_m)(y))\right|$$

$$\geqslant 23^{-k} - 2\sum_{m=k+1}^{\infty} 3^{-m} = 3^{-k}.$$

We now apply the Stone–Weierstrass theorem to check that $C^*(f) = C(X)$ (see Theorem B.1.1). It follows that C is generated by the self-adjoint operator $\rho^{-1}(f)$. This completes the proof. □

The following is a counter-part of Proposition 2.2.1 for normal operators.

Lemma 2.2.4

If $N \in \mathcal{L}(\mathcal{H})$ is a normal operator, then there is a diagonalizable operator $\tilde{D} \in \mathcal{L}(\mathcal{H})$ such that $\sigma(\tilde{D}) = \sigma_e(\tilde{D})$ and $N - \tilde{D}$ is compact.

Proof Assume that $N \in \mathcal{L}(\mathcal{H})$ is a normal operator. By Halmos' lemma, there exists $f \in C(\sigma(A))$ and a self-adjoint operator $S \in \mathcal{L}(\mathcal{H})$ such that $f(S) = N$. By Proposition 2.2.1, there exists a diagonalizable operator D and compact operator K such that $S = D + K$ and $\sigma(D) = \sigma_e(D)$. Note first that

$$\sigma(D) = \sigma_e(D) = \sigma_e(S) \subseteq \sigma(S). \tag{2.2.2}$$

Let $\{p_n\}_{n\in\mathbb{N}}$ be a sequence of polynomials such that $\|p_n - f\|_{\infty,\sigma(S)} \to 0$ as $n \to \infty$. By the spectral theorem, $\{p_n(S)\}_{n\in\mathbb{N}}$ is a convergent sequence in $\mathcal{L}(\mathcal{H})$. By (2.2.2), $\{p_n(D)\}_{n\in\mathbb{N}}$ is a norm convergent sequence in $\mathcal{L}(\mathcal{H})$. Thus, the sequence $\{p_n(S) - p_n(D)\}_{n\in\mathbb{N}}$ converges to a compact operator $K = f(S) - f(D)$. It follows that $N = f(S) = \tilde{D} + K$ with $\tilde{D} = f(D)$. □

One may now in the same manner as the proof of Theorem 2.2.4 complete the proof of the Weyl–von Neumann–Berg–Sikonia theorem.

Theorem 2.2.5 (Weyl–von Neumann–Berg–Sikonia theorem)

Any two normal operators N_1 and N_2 in $\mathcal{L}(\mathcal{H})$ are essentially equivalent if and only if $\sigma_e(N_1) = \sigma_e(N_2)$.

We point out that the previous theorem extends to normal operators acting on two separable Hilbert spaces. Indeed, if $N_i \in \mathcal{L}(\mathcal{H}_i)$, $i = 1, 2$ and $U : \mathcal{H}_1 \to \mathcal{H}_2$ is any unitary, then N_1 and N_2 are essentially equivalent if and only if N_1 and U^*N_2U, both acting on $\mathcal{H}_1$, are essentially equivalent. Finally, note that Theorem 2.2.5 fails if the underlying Hilbert space is nonseparable (see Exercise 2.8.12).

2.3 Extensions and Essentially Unitary Operators

Although the main objective in [26] was the classification of the essentially normal operators, Brown, Douglas, and Fillmore actually considered a related problem more general in nature. Before we explain this, we need a formal definition of the extensions of C^*-algebras (we have already encountered this notion in Corollary 1.6.1).

For a map $\varphi : \mathcal{A} \to \mathcal{B}$ between C^*-algebras $\mathcal{A}$ and $\mathcal{B}$, let $\operatorname{Im}\varphi$ denote the image $\{\varphi(f) : f \in \mathcal{A}\}$ and $\ker\varphi$ denote the kernel $\{f \in \mathcal{A} : \varphi(f) = 0\}$ of φ.

Definition 2.3.1

If $\mathcal{A}$ and C are C^*-algebras, then an *extension of C by $\mathcal{A}$* is a short exact sequence

$$0 \longrightarrow \mathcal{A} \xrightarrow{\varphi} \mathcal{B} \xrightarrow{\psi} C \longrightarrow 0,$$

for some C^*-algebra $\mathcal{B}$. In other words, there exists a C^*-algebra $\mathcal{B}$ and $*$-homomorphisms $\varphi : \mathcal{A} \to \mathcal{B}$ and $\psi : \mathcal{B} \to C$ such that $\ker\phi = \{0\}$, $\operatorname{Im}\varphi = \ker\psi$ and $\operatorname{Im}\psi = C$. We let the pair $(\mathcal{B}, \psi)$ denote this extension and say that $\mathcal{B}$ *is an extension of C by $\mathcal{A}$*.

With this notion, first, observe that if S_T is the C^*-algebra generated by the essentially normal operator T, the compact operators $C(\mathcal{H})$ and the identity operator I on the Hilbert space $\mathcal{H}$, then $S_T/C(\mathcal{H})$ is isomorphic to the C^*-algebra generated by $\pi(I)$ and $\pi(T)$ in the Calkin algebra $Q(\mathcal{H})$. In particular, $S_T/C(\mathcal{H})$ is a commutative C^*-algebra with the maximal ideal space $\sigma_e(T)$, and we have

$$\begin{array}{ccc} S_T & \longrightarrow & C(\sigma_e(T)) \\ \downarrow{\scriptstyle\pi} & & \uparrow{\scriptstyle\Gamma_{S_T/C(\mathcal{H})}} \\ S_T/C(\mathcal{H}) & \hookrightarrow & S_T/C(\mathcal{H}) \subseteq Q(\mathcal{H}) \end{array}$$

where $\Gamma_{S_T/C(\mathcal{H})}$ is the Gelfand map as ensured by Corollary C.1.2. Further, we have an *extension of $C(\mathcal{H})$ by $C(\sigma_e(T))$*, that is

$$0 \longrightarrow C(\mathcal{H}) \hookrightarrow S_T \xrightarrow{\varphi_T} C(\sigma_e(T)) \longrightarrow 0$$

is exact, where $\varphi_T = \Gamma_{S_T/C(\mathcal{H})} \circ \pi$. Conversely, if S is any C^*-algebra of operators on the Hilbert space $\mathcal{H}$ containing compact operators and X is any compact subset of the complex plane $\mathbb{C}$ such that

$$0 \longrightarrow C(\mathcal{H}) \hookrightarrow S \xrightarrow{\varphi} C(X) \longrightarrow 0$$

is exact, then for any T in S, $\varphi(TT^* - T^*T) = 0$ and it follows that T is essentially normal. Fix any T in S such that $\varphi(T) = \mathrm{id}|_X$. Let S_T be the C^*-algebra generated by the operator T, the compact operators and the identity on $\mathcal{H}$. Now, $\varphi(S_T)$ is a C^*-subalgebra of $C(X)$ containing the identity function and therefore, by the Stone–Weierstrass theorem, it must be all of $C(X)$. If S is any operator in S, then there is always an operator S' in S_T such that $\varphi(S) = \varphi(S')$, so that $S - S'$ is compact and hence, S is in S_T. This also shows that $S = S_T$. This discussion is summarized as follows.

Lemma 2.3.1

If S_T denotes the unital C^*-algebra generated by the essentially normal operator $T \in \mathcal{L}(\mathcal{H})$, the C^*-algebra $C(\mathcal{H})$ of compact operators and the identity operator on $\mathcal{H}$, then

$$S_T \xrightarrow{\pi} S_T/C(\mathcal{H}) \xrightarrow{\Gamma_{S_T/C(\mathcal{H})}} C(\sigma_e(T)),$$

where π is the (restriction) of the Calkin map and $\Gamma_{S_T/C(\mathcal{H})}$ is the Gelfand map. Moreover, we have an extension, that is,

$$0 \longrightarrow C(\mathcal{H}) \hookrightarrow S_T \xrightarrow{\varphi_T} C(\sigma_e(T)) \longrightarrow 0 \tag{2.3.3}$$

is exact, where φ_T is governed by $\varphi_T(p(T,T^*)) = p(z,\bar{z})$ for every polynomial p in two complex variables z and w.

Conversely, if S is a unital C^*-algebra of operators in $\mathcal{L}(\mathcal{H})$ containing all compact operators, that is $C(\mathcal{H}) \subseteq S \subseteq \mathcal{L}(\mathcal{H})$, and X is any compact subset of the complex plane $\mathbb{C}$ such that

$$0 \longrightarrow C(\mathcal{H}) \hookrightarrow S \xrightarrow{\varphi} C(X) \longrightarrow 0 \tag{2.3.4}$$

is exact, then there exists an essentially normal operator T in $\mathcal{L}(\mathcal{H})$ with essential spectrum X such that (2.3.3) is exact.

Thus, the short exact sequence, as given by (2.3.4), is an extension (S, φ) of $C(\mathcal{H})$ by $C(X)$. We have shown earlier that there is a natural correspondence between essentially normal operators T with essential spectrum, a compact set $X \subseteq \mathbb{C}$, and extensions (S_T, φ_T) of $C(\mathcal{H})$ by $C(X)$. We now relate unitary equivalence modulo the compacts of essentially normal operators to extensions. If one has two essentially equivalent essentially normal operators T_1 and T_2 (with corresponding extensions (S_1, φ_1) and (S_2, φ_2)), then $U^*T_2U = T_1 + K$ for some unitary operator U and compact operator K, and hence, $U^*S_2U = S_1$ (by continuity of the map $T \mapsto U^*TU$) and $\varphi_2(T) = \varphi_1(U^*TU)$ for all T in S_2. This suggests the following definition.

Definition 2.3.2

Two extensions (S_1, φ_1) and (S_2, φ_2) defined by the C^*-subalgebras S_1 and S_2 of $\mathcal{L}(\mathcal{H})$ are *equivalent* if there exists a unitary operator U such that $U^*S_2U = S_1$ and $\varphi_2(T) = \varphi_1(U^*TU)$.

We formally record that the classification problem of essential normal operators modulo the compacts is equivalent to classifying extensions (S, φ).

Theorem 2.3.3

If two essentially normal operators T_1 and T_2 in $\mathcal{L}(\mathcal{H})$ are essentially equivalent, then the corresponding extensions of $C(\mathcal{H})$ by $C(X)$ are equivalent, where $\sigma_e(T_i) = X$ for $i = 1, 2$. Conversely, if two extensions (S_1, φ_1) and (S_2, φ_2) are equivalent, then the corresponding essentially normal operators T_1 and T_2 in $\mathcal{L}(\mathcal{H})$ are essentially equivalent, where $T_i \in \phi_i^{-1}(id|_X)$ for $i = 1, 2$.

Proof We have already seen the first half. To prove the converse, note that

$$\varphi_1(U^*T_2U) = \varphi_2(T_2) = \mathrm{id}|_X = \varphi_1(T_1),$$

and hence, $U^*T_2U - T_1$ is compact. □

The classification problem for essentially normal operators and for extensions of $C(\mathcal{H})$ by $C(X)$ are identical for any compact subset X of $\mathbb{C}$. The extension point of view of course has many advantages. The extension problem makes sense for any C^*-algebra not necessarily just the algebra of continuous functions $C(X)$ leading to surprising connections with many other areas of mathematics. We have discussed some of these later developments in the "Epilogue" to these notes. Moreover, in this reformulation, it is possible to use tools from topology and homological algebra.

For any compact metrizable space X, let $\mathrm{Ext}(X)$ denote the set of equivalence classes of the extensions (S, φ) of $C(\mathcal{H})$ by $C(X)$. If X is a compact subset of the complex plane $\mathbb{C}$, then by Theorem 2.3.3, $\mathrm{Ext}(X)$ may be identified with the set of equivalence classes $[T]$ of essentially normal operators T with $\sigma_e(T) = X$. In what follows, we will use both these realizations of $\mathrm{Ext}(X)$.

Proposition 2.3.1

Let X and Y be compact metrizable spaces. If X and Y are homeomorphic, then there is a bijection from $\mathrm{Ext}(X)$ to $\mathrm{Ext}(Y)$.

Proof Let $p : X \to Y$ be a homeomorphism. Then, the map $p^* : C(Y) \to C(X)$ defined by $f \mapsto f \circ p$ is an isomorphism. If (S, φ) is an extension of $C(\mathcal{H})$ by $C(Y)$, then $(S, p^*\varphi)$ is an extension of $C(\mathcal{H})$ by $C(X)$. If (S_1, φ_1) and (S_2, φ_2) are two equivalent extensions of $C(\mathcal{H})$ by $C(Y)$ and U is the unitary operator implementing this unitary equivalence, then, $p^*\varphi_1(U^*TU) = p^*\varphi_2(T)$, and hence, the two extensions $(S_1, p^*\varphi_1)$ and $(S_2, p^*\varphi_2)$ are equivalent. This shows that the map $\psi : (S, \varphi) \to (S, p^*\varphi)$ from $\mathrm{Ext}(X)$ to $\mathrm{Ext}(Y)$ is well-defined. Note that ψ is invertible with inverse given by $(S, \varphi) \to (S, (p^{-1})^*\varphi)$. □

If X is a subset of the real line, then $\mathrm{Ext}(X)$ is trivial.

Example 2.3.4 Let Δ be a compact subset of the real line. Let us compute $\mathrm{Ext}(\Delta)$. By Theorem 2.3.3, $\mathrm{Ext}(\Delta)$ is the set of equivalence classes of essentially normal operators S with $\sigma_e(S) = \Delta$. For $j = 1, 2$, let S_j be an essentially normal operator with essential spectrum Δ. If π is the Calkin map, then $\pi(S_j)$ is a normal operator with $\sigma(\pi(S_j)) \subseteq \mathbb{R}$. By the spectral theorem, $\pi(S_j)$ is a self-adjoint operator, and hence, $S_j = \mathfrak{R}(S_j) + K$ for some compact operator K, where $\mathfrak{R}(S_j)$ is the real part of S_j. It follows that

$$\sigma_e(\mathfrak{R}(S_1)) = \sigma_e(S_1) = \sigma_e(S_2) = \sigma_e(\mathfrak{R}(S_2)).$$

Hence, by Weyl–von Neumann–Berg theorem, $\mathfrak{R}(S_1)$ and $\mathfrak{R}(S_2)$ are essentially equivalent (see Theorem 2.2.4). It follows that S_1 and S_2 are essentially equivalent. Thus, $\mathrm{Ext}(\Delta)$ contains only one equivalence class of extensions, that is, $\mathrm{Ext}(\Delta) = \{0\}$, where 0 denotes a self-adjoint operator with essential spectrum Δ as ensured by Example 1.3.8. ■

If T is an essentially normal operator with essential spectrum homeomorphic to a subset of the real line, then by Example 2.3.4 and Proposition 2.3.1, $\text{Ext}(\sigma_e(T)) = \{0\}$, and hence, T is a compact perturbation of a self-adjoint operator. Here is an interesting consequence of this fact, first obtained by Olsen [105] using different means (cf. [26, Remark 6.3]).

Proposition 2.3.2

Let $C, D \in \mathcal{L}(\mathcal{H})$. If CD is compact, then there exists an orthogonal projection $P \in \mathcal{L}(\mathcal{H})$ such that both CP and $(I - P)D$ are compact operators.

Proof Assume that CD is compact. In view of the polar decomposition (see Theorem 1.1.3 and Corollary 1.1.1), after replacing C by $|C|$ and D by $|D^*|$ respectively, we may assume that C and D are positive. Further, since C^*CDD^* and DD^*C^*C are compact, in view of the square-root lemma (see the discussion following Theorem C.1.12), we may also assume that CD and DC are compact. It follows that $C + iD$ is essentially normal. Further, by the spectral mapping property (see Corollary C.1.5), the essential spectrum of $C + iD$ is homeomorphic to the union of subsets of $[0, \infty)$. Without loss of generality, we may assume that $\sigma_e(C + iD)$ is a compact subset of $\mathbb{R}$. Then, by the discussion prior to the proposition, $C + iD$ is a compact perturbation of a self-adjoint operator S. Let P be any rank one orthogonal projection, and note that CP is compact. Finally, one may compare the imaginary part of $(I - P)(C + iD) = (I - P)(S + K)$ to conclude that $(I - P)D$ is also compact. $\qquad\qquad \square$

Let us determine what happens to essentially normal operators with essential spectrum homeomorphic to the unit circle $\mathbb{T}$. The next theorem shows that $\text{Ext}(\mathbb{T})$ can be identified with the set $\mathbb{Z}$ of integers.

Theorem 2.3.5

If $T \in \mathcal{L}(\mathcal{H})$ is essentially unitary, then T is a compact perturbation of a unitary operator, a shift of multiplicity n or the adjoint of the shift of multiplicity n, according as $\text{ind}(T) = 0$, $\text{ind}(T) = -n < 0$ or $\text{ind}(T) = n > 0$.

Proof Assume that $T \in \mathcal{L}(\mathcal{H})$ is essentially unitary. Since $T^*T - I$ is compact and $C = |T| + I$ is invertible, after multiplying by C^{-1}, we see that $|T| - I$ is also compact. If $T = W|T|$ is the polar decomposition for T with W denoting a partial isometry (see Theorem 1.1.3), then $T = W + K$ for the compact operator $K = W(|T| - I)$. Thus, T is essentially equivalent to W. Since T is Fredholm, so is W.

To obtain the desired result, we need the fact that $U \oplus S$ is unitarily equivalent to a compact perturbation of S, where S is the shift operator (see Example 1.5.2) and U is a unitary. Assuming this result to be called, Absorption Lemma, for the moment, the proof of the theorem can be completed as follows.

Case I $\text{ind}(T) = n \leqslant 0$:

Note that $\text{ind}(W) = \text{ind}(T) = n \leqslant 0$ and $\dim \ker W \leqslant \dim(\text{ran } W)^\perp$. If U denotes the isometric embedding of $\ker W$ into $(\text{ran } W)^\perp$, then $L = U \oplus 0$ from $\ker W \oplus (\ker W)^\perp$ into $(\text{ran } W)^\perp \oplus \text{ran } W$ is a finite rank partial isometry with initial space $\ker W$ and final space contained in $(\text{ran } W)^\perp$. It is easy to see that the operator $V = W + L$ is an isometry. Thus, V is essentially equivalent to

W. We can now apply the Wold–von Neumann decomposition to the isometry V to obtain an unitary operator U and a unilateral shift S of some multiplicity such that $V = U \oplus S$. Since

$$\text{ind}(T) = \text{ind}(V) = \text{ind}(U) + \text{ind}(S) = \text{ind}(S),$$

it follows that $\text{ind}(S) = n$, which in turn implies that S is a shift of multiplicity n. Thus, T is essentially equivalent to $U \oplus S$. By the Absorption Lemma, T is essentially equivalent to S.

Case II $ind(T) > 0$:

For an operator T with ind $T > 0$, apply the preceding case to the adjoint T^* of T. This completes the proof. $\qquad\qquad\qquad\qquad\qquad\qquad\qquad\qquad\qquad\qquad\qquad\qquad\qquad\qquad$ □

Remark 2.3.6 The set $\text{Ext}(T)$ can be identified with $\mathbb{Z}$ (see Exercise 2.8.27). The foregoing result shows that for any compact subset σ of $\mathbb{T}$, $\text{Ext}(\sigma)$ can be identified with a subset of $\mathbb{Z}$. It turns out that $\text{Ext}(\sigma)$ is trivial whenever σ is a proper subset of $\mathbb{T}$ (see Corollary 4.5.2).

2.4 Absorption Lemma

In this section, we will prove the Absorption Lemma, which will complete the proof of Theorem 2.3.5. This lemma will also be used later in proving $\text{Ext}(\mathbb{T}) = \mathbb{Z}$. Keeping later developments in mind, we prove a little more than the Absorption Lemma.

For $n \in \mathbb{N} \cup \{\infty\}$, let $N = (N_1, \dots, N_n)$ denote a family of commuting normal operators $N_1, \dots, N_n$ and consider the commutative C^*-algebra $C^*(N)$ generated by N. The *joint spectrum* $\sigma(N)$ of N is defined as the collection

$$\sigma(N) = \{(\phi(N_k))_{k=0}^{n} : \phi \text{ is a complex } *\text{-homomorphism of } C^*(N)\}$$

(see the discussion following Remark C.1.5).

The following lemma shows that an "approximate joint eigenvector" for a family of essentially normal operators can be constructed within the complement of arbitrary finite dimensional subspace.

Lemma 2.4.1

For $n \in \mathbb{N} \cup \{\infty\}$, let $T = (T_1, \dots, T_n)$ be a family of operators in $\mathcal{L}(\mathcal{H})$ such that $\pi(T) = (\pi(T_1), \dots, \pi(T_n))$ is a commuting family of normal elements on $\mathcal{Q}(\mathcal{H})$ and let λ in $\mathbb{C}^n$ be in the joint spectrum $\sigma(\pi(T))$ of $\pi(T)$. Given $\epsilon > 0$ and a finite dimensional subspace $\mathcal{M}$ of $\mathcal{H}$, there exists a nonzero vector φ in $\mathcal{M}^{\perp}$ such that

$$\|(T_m - \lambda_m)\varphi\| < \epsilon, \quad m = 1, \dots, n.$$

Proof Let $C^*(\pi(T)) \subseteq \mathcal{Q}(\mathcal{H})$ denote the commutative C^*-algebra generated by $\pi(T_1), \dots, \pi(T_n)$. Set $\alpha_m = \|T_m - \lambda_m\|^2 + 1$, $m = 1, \dots, n$,

$$S = \sum_{m=1}^{n} \frac{(T_m - \lambda_m)^*(T_m - \lambda_m)}{2^m \alpha_m},$$

and note that $S \in \mathcal{L}(\mathcal{H})$ is a positive operator. Let $\Gamma : C^*(\pi(T)) \to C(\sigma(\pi(T)))$ be the Gelfand map and note that $\Gamma(\pi(S))$ vanishes at λ. Indeed, $\Gamma(\pi(S)) = \sum_{m=1}^{n} \frac{|z_m - \lambda_m|^2}{2^m \alpha_m}$. It follows that $0 \in \sigma(\pi(S)) = \sigma_e(S)$. Moreover,

$$S = \chi_{\sigma(S)\backslash[0,\epsilon)}(S)S + \chi_{[0,\epsilon)}(S)S. \tag{2.4.5}$$

Since $\chi_{\sigma(S)\backslash[0,\epsilon)}(S)S$ is invertible, we conclude that $\chi_{[0,\epsilon)}(S)$ is a projection of infinite rank.

Let $\mathcal{M}$ be a finite dimensional subspace of $\mathcal{H}$. If $(\mathrm{ran}\chi_{[0,\epsilon)}(S)) \cap (\mathcal{M}^\perp \backslash \{0\}) = \emptyset$, then the projection $P_\mathcal{M}$ would map the infinite dimensional space $\mathrm{ran}\chi_{[0,\epsilon)}(S)$ injectively into the finite dimensional space $\mathcal{M}$ (see Lemma 1.5.2). This contradiction guarantees the existence of $\varphi \in \mathcal{M}^\perp \cap \mathrm{ran}\chi_{[0,\epsilon)}(S)$ such that $\|\varphi\| = 1$. To complete the proof, note that by (2.4.5),

$$
\begin{aligned}
\sum_{m=1}^{n} \|(T_m - \lambda_m)\varphi\|^2 &= \langle S\varphi, \varphi \rangle \\
&= \langle \chi_{\sigma(S)\backslash[0,\epsilon)}(S)S\varphi, \varphi \rangle + \langle \chi_{[0,\epsilon)}(S)S\varphi, \varphi \rangle \\
&= \langle \chi_{[0,\epsilon)}(S)S\varphi, \varphi \rangle < \epsilon,
\end{aligned}
$$

where we used the assumption that $\varphi \in \mathrm{ran}\chi_{[0,\epsilon)}(S)$ and the Borel functional calculus of S. $\qquad\square$

The preceding lemma enables us to decompose simultaneously modulo compacts a family of essentially normal operators.

Theorem 2.4.1

Suppose that $\lambda^{(r)} = \{\lambda_m^{(r)} \in \sigma_e(T_m) : m \geqslant 1\}$ is given for each integer $r \geqslant 1$ and that $\{T_m\}_{m=1}^{\infty}$ is a family of essentially normal operators in $\mathcal{L}(\mathcal{H})$. Then there exists an orthonormal sequence $\{\psi_r\}_{r=1}^{\infty}$ in $\mathcal{H}$ such that for every integer $m \geqslant 1$,

$$T_m = \begin{bmatrix} D_m & 0 \\ 0 & R_m \end{bmatrix} + K_m \quad \text{for some } K_m \in C(\mathcal{H}).$$

The decomposition of T_m is with respect to the subspaces $\mathcal{M}$ and $\mathcal{M}^\perp$ with $\mathcal{M}$ is equal to the closed linear span $\bigvee_{r=1}^{\infty}\{\psi_r\}$ of $\{\psi_r\}_{r=1}^{\infty}$. Moreover, D_m and R_m are bounded linear operators on $\mathcal{M}$ and $\mathcal{M}^\perp$ respectively. Further, D_m in $\mathcal{L}(\mathcal{M})$ is a diagonal operator given by

$$D_m \psi_r = \lambda_m^{(r)} \psi_r, \quad r \geqslant 1. \tag{2.4.6}$$

Proof First consider the case of self-adjoint operators $\{\pi(T_m)\}_{m=1}^{\infty}$. We construct by induction on r an orthonormal sequence $\{\psi_s\}_{s=1}^{\infty}$ such that

$$\|T_m \psi_r - \lambda_m^{(r)} \psi_r\| < 1/2^r, \quad m \leqslant r.$$

Indeed, if $\psi_1, \ldots, \psi_{r-1}$ are pairwise orthogonal and satisfy the preceding inequality, then apply Lemma 2.4.1 with $n = r$, $\lambda = (\lambda_1^{(r)}, \ldots, \lambda_n^{(r)})$, $\mathcal{M}_r = \mathrm{span}\{\psi_1, \ldots, \psi_{r-1}\}$ and $\epsilon = (1/2)^{r-1}$ to obtain ψ_r as desired. Fix an integer $m \geqslant 1$. Now, decompose each T_m with respect to $\mathcal{M}$ and $\mathcal{M}^\perp$ as

$$T_m = \begin{bmatrix} X_m & Y_m^* \\ Y_m & R_m \end{bmatrix},$$

and note that

$$T_m - D_m \oplus R_m = \begin{bmatrix} X_m - D_m & Y_m^* \\ Y_m & 0 \end{bmatrix},$$

where D_m is given by (2.4.6). It follows that for any integer $r \geqslant 1$,

$$
\begin{aligned}
\|(X_m - D_m)\psi_r\|^2 + \|Y_m\psi_r\|^2 &= \|(T_m - D_m)\psi_r\|^2 \\
&= \|T_m\psi_r - \lambda_m^{(r)}\psi_r\|^2 \\
&< 1/4^r.
\end{aligned}
$$

This shows that $X_m - D_m$ and Y_m are both Hilbert–Schmidt operators, and hence, by Corollary 1.4.2, $T_m - D_m \oplus R_m$ is compact.

To complete the proof in the general case, we apply the foregoing technique to the commuting family $\{\Re(\pi(T_m)), \Im(\pi(T_m))\}_{m=1}^{\infty}$ of real and imaginary parts of each $\pi(T_m)$. □

The preceding decomposition theorem together with the Weyl–von Neumann theorem yields the Absorption Lemma.

Lemma 2.4.2 (Absorption Lemma)

Let $T \in \mathcal{L}(\mathcal{H})$ be an essentially normal operator and let $N \in \mathcal{L}(\mathcal{H})$ be a normal operator. If the essential spectrum of N is contained in the essential spectrum of T, then $T \oplus N$ and $N \oplus T$ are essentially equivalent to T.

Proof Assume that $\sigma_e(N) \subseteq \sigma_e(T)$. Consider a sequence $\{\lambda^{(r)}\}_{r=1}^{\infty}$ dense in $\sigma_e(T)$ with isolated points being counted infinitely often. By Theorem 2.4.1,

$$T = D \oplus R + K$$

for some bounded linear operator R, a compact operator K, and the diagonal operator D with diagonal entries $\lambda^{(r)}$. The operators T and D have the same essential spectrum, and hence,

$$\sigma_e(T) = \sigma_e(D) = \sigma_e(D \oplus N).$$

By the Weyl–von Neumann theorem (see also the discussion following Theorem 2.2.5), $D \oplus N$ is essentially equivalent to D, and therefore, $T \oplus N$ is essentially equivalent to T. Similarly, one can check that $N \oplus T$ is essentially equivalent to T. This completes the proof. □

Note that the essential spectrum of the finite direct sum of bounded linear operators is the union of their essential spectra (see Exercise 1.7.38). Since the essential spectrum is invariant under essential equivalence, the following is immediate from the Absorption Lemma.

Theorem 2.4.2

If $T \in \mathcal{L}(\mathcal{H})$ is an essentially normal operator and $N \in \mathcal{L}(\mathcal{H})$ is a normal operator, then the following statements are equivalent:

(1) $T \oplus N$ is essentially equivalent to T.

(2) $N \oplus T$ is essentially equivalent to T.

(3) The essential spectrum of N is contained in the essential spectrum of T.

Here is a simple situation in which the last result is applicable.

Corollary 2.4.1

For a positive integer n, let S_n denote the unilateral shift of multiplicity n. For positive integers m, n and unitary operators U and V, $U \oplus S_n$ is essentially equivalent to $V \oplus S_m$ if and only if $m = n$.

Proof By Corollary 1.5.3 and Example 1.5.6, S_n is essentially normal with essential spectrum equal to the unit circle $\mathbb{T}$. Note that $\mathrm{ind}(U \oplus S_n) = n$ and $\mathrm{ind}(V \oplus S_m) = m$, and hence, by Remark 2.1.2, we obtain the necessity part. Further, by the Absorption Lemma, $U \oplus S_n$ and $V \oplus S_n$ are essentially equivalent to S_n, and the sufficiency part follows. □

2.5 Weakly and Strongly Equivalent Extensions

We have already seen that the problem of classifying essentially normal operators with essential spectrum X is equivalent to that of classifying extensions of $C(\mathcal{H})$ by $C(X)$. We now introduce yet another way of looking at the same problem. Let (S, φ) be an extension and pick a τ making the following diagram commutative:

$$
\begin{array}{ccccccccc}
0 & \longrightarrow & C(\mathcal{H}) & \hookrightarrow & S & \xrightarrow{\varphi} & C(X) & \longrightarrow & 0 \\
& & \| & & \downarrow & & \downarrow \tau & & \\
0 & \longrightarrow & C(\mathcal{H}) & \hookrightarrow & \mathcal{L}(\mathcal{H}) & \xrightarrow{\pi} & Q(\mathcal{H}) & \longrightarrow & 0
\end{array}
\qquad (2.5.7)
$$

Consequently, on the one hand, the map $\tau : C(X) \to Q(\mathcal{H})$ is determined by the equality $\tau\varphi(A) = \pi(A)$, $A \in S$, is a monomorphism. In fact, $\pi(A) = 0$ if and only if A is compact if and only if $\phi(A) = 0$. On the other hand, given a unital $*$-monomorphism $\tau : C(X) \to Q(\mathcal{H})$, one may set $S = \pi^{-1}(\mathrm{Im}\,\tau)$ and $\varphi = \tau^{-1} \circ \pi$, where $\mathrm{Im}\,\tau$ denotes the image $\{\tau(f) : f \in C(X)\}$ of τ. Note that S is a unital C^*-algebra containing the ideal $C(\mathcal{H})$ of compact operators. Moreover, $\ker \phi$ is precisely $C(\mathcal{H})$. Thus, the pair (S, φ) obtained in this manner is an extension of $C(\mathcal{H})$ by $C(X)$.

Given an essentially normal operator T, we obtain the associated extension (S_T, φ_T), which in turn gives rise to the unital $*$-monomorphism $\tau : C(X) \to Q(\mathcal{H})$ given by

$$
\tau(\phi_T(A)) = \pi(A), \quad A \in S_T.
$$

What is the relationship of τ to the operator T? To answer this, note that $\varphi_T(p(T)) = p$ for any polynomial p, and therefore,

$$
\tau(p) = \pi(p(T)) = p(\pi(T)).
$$

Thus, the map τ extends the polynomial functional calculus for the operator $\pi(T)$. Further, for any $f \in C(X)$, $\tau(f)$ lies in the image of the Calkin map. Thus, there exists an operator $T^f \in \mathcal{L}(\mathcal{H})$ (depending of course on f) such that $\tau(f) = \pi(T^f)$. However, we also have $\tau(p) = p(\pi(T))$, where $T \in \mathcal{L}(\mathcal{H})$ is any essentially normal operator such that $\tau(\mathrm{id}|_X) = \pi(T)$. By the continuity of τ and functional calculus of $\pi(T)$, we obtain $\tau(f) = f(\pi(T))$ for any $f \in C(X)$. Thus, τ is nothing but the continuous functional calculus of the normal operator $\pi(T)$. It is thus natural to refer to the assignment $f \mapsto T^f$ as the *generalized functional calculus* of T, which is well-defined modulo compact. Needless to say, in case T is normal, it recovers the continuous functional calculus of T, that is, T^f can be chosen to be $f(T)$.

If we start with a unital $*$-monomorphism $\tau : C(X) \to Q(\mathcal{H})$, then by letting T be any operator T such that $\pi(T) = \tau(\mathrm{id}|_X)$, we obtain an essentially normal operator T with essential spectrum equal to X. Conversely, if T is essentially normal with essential spectrum X, then $\tau(f) = f(\pi(T))$ defines a $*$-monomorphism on $C(X)$. For later purposes, we record the following lemma (see Exercise 2.8.5).

Lemma 2.5.1

Let $T \in \mathcal{L}(\mathcal{H})$ be an essentially normal operator and let $\tau : C(X) \to Q(\mathcal{H})$ be a $*$-homomorphism given by $\tau(f) = f(\pi(T))$. If $\sigma_e(T) = X$, then τ is a $*$-monomorphism. In this case, τ is isometric on $C(X)$.

Proof   The continuous functional calculus for a normal operator is isometric (see Corollary C.1.2). This yields the desired conclusions.   □

How do we define equivalence for the unital $*$-monomorphisms $\tau_i : C(X) \to Q(\mathcal{H}_i)$, $i = 1, 2$? If we start with two essentially equivalent essentially normal operators $T_i \in \mathcal{L}(\mathcal{H}_i)$, $i = 1, 2$ (implemented by a unitary $U : \mathcal{H}_1 \to \mathcal{H}_2$ and compact operator K), with corresponding unital $*$-monomorphisms τ_1, τ_2, then for any f in $C(X)$,

$$\tau_1(f) = f(\pi(T_1)) = f(\pi(U^* T_2 U + K)) = f(\pi(U^* T_2 U)) = (\alpha_U \tau_2)(f),$$

where the unital $*$-automorphism $\alpha_U : Q(\mathcal{H}_2) \to Q(\mathcal{H}_1)$ is given by

$$\alpha_U(\pi(A)) = \pi(U^* A U), \quad A \in \mathcal{L}(\mathcal{H}_2). \tag{2.5.8}$$

The previous discussion leads us to the following.

Definition 2.5.1

Let $\tau_i : C(X) \to Q(\mathcal{H}_i)$, $i = 1, 2$ be unital $*$-monomorphisms. We say that τ_1 and τ_2 are *equivalent* if there exists a unitary $U : \mathcal{H}_1 \to \mathcal{H}_2$ such that $\tau_1 = \alpha_U \circ \tau_2$.

We record the following elementary fact for ready reference.

Lemma 2.5.2

Let $\tau : C(X) \to Q(\mathcal{H})$ be a $*$-monomorphism. Any unitary transformation $U : \mathcal{H} \to \mathcal{K}$ induces an isomorphism $\alpha_U : Q(\mathcal{H}) \to Q(\mathcal{K})$ as given by (2.5.8). Moreover, the $*$-monomorphism $\alpha_U \tau : C(X) \to Q(\mathcal{K})$ is strongly equivalent to τ.

Proof Note that α_u is an isomorphism with inverse $\alpha_{U^{-1}}$. The remaining part is now immediate. □

In the discussion prior to Definition 2.5.1, we have seen that equivalent extensions (or equivalent ∗-monomorphisms) give rise to equivalent essentially normal operators, and vice versa. Thus, the classification problem for essentially normal operators is again identical to that of classifying the unital ∗-monomorphisms $\tau : C(X) \to Q(\mathcal{H})$. We will for the rest of this discussion, work only with these objects and occasionally use essentially normal operators to provide motivation for some definitions. The equivalence class $[\tau]$ of a ∗-monomorphisim τ will be called an *extension*. The set of all such equivalence classes for a fixed compact metrizable space X is again denoted by Ext(X) without any ambiguity. Sometimes, we write $[\tau_X]$ in place of $[\tau]$ to emphasize that it is an element of Ext(X).

Since the main problem is to study normal elements in the Calkin algebra, it would seem that the correct notion of equivalence is to some extent weaker than the equivalence we have defined earlier.

Definition 2.5.2

Two extensions $[\tau_1]$ and $[\tau_1]$ are said to be *weakly equivalent* if there is an essentially unitary operator T such that

$$\pi(T)\tau_1(f)\pi(T)^* = \tau_2(f) \text{ for all } f \in C(X).$$

Remark **2.5.3** The extensions $[\tau_1]$ and $[\tau_1]$ are weakly equivalent if and only if there exists an essentially unitary operator T such that $\pi(TT_1^f T^*) = \pi(T_2^f)$ for every $f \in C(X)$, where $\tau_i(f) = \pi(T_i^f)$, $i = 1, 2$.

Perhaps, weak equivalence is the more natural equivalence in this setting. We will however show that this weaker notion of equivalence is actually equivalent to the equivalence defined in Definition 2.5.1. This fact is no longer true if we consider unital ∗-monomorphisms of non-abelian C^*-algebras (see [27, Remark 3.6]).

Proposition 2.5.1

Weakly equivalent extensions are equivalent.

Proof For Hilbert spaces $\mathcal{H}_1$ and $\mathcal{H}_2$, let $\tau_1, \tau_2 : C(X) \to Q(\mathcal{H}_k)$ be weakly equivalent ∗-monomorphisms. In view of Lemma 2.5.2, we may assume $\mathcal{H}_1 = \mathcal{H} = \mathcal{H}_2$ without loss of generality. If $\tau, \tau' : C(X) \to Q(\mathcal{H})$ are weakly equivalent ∗-monomorphisms, then there exists an essentially unitary operator S such that

$$(\alpha_S \tau)(f) = \pi(S)\tau(f)\pi(S^*) = \tau'(f).$$

If $\pi(S')$ is any other unitary element in $Q(\mathcal{H})$ commuting with the image Im τ of τ, then

$$
\begin{aligned}
(\alpha_{ss'}\tau)(f) &= \pi(S)\pi(S')\tau(f)\pi(S'^*)\pi(S^*) \\
&= \pi(S)\tau(f)\pi(S^*) = \tau'(f).
\end{aligned}
$$

If S' can be chosen such that SS' is a compact perturbation of an unitary, then τ would be strongly equivalent to τ'. The fact that $SS' = U + K$ for some unitary U and compact K in turn would follow from Theorem 2.3.5 provided we verify that $\mathrm{ind}(SS') = 0$, that is, $\mathrm{ind}(S') = -\mathrm{ind}(S)$. We now establish the existence of the essentially unitary operator S' with index equal to $-\mathrm{ind}(S)$, which further satisfies the relation $\pi(S')\tau(f) = \tau(f)\pi(S')$ for every $f \in C(X)$.

Let $\{f_m\}_{m \geqslant 1}$ be a dense sequence in $C(X)$ and $\{\lambda_m\}_{m \geqslant 1}$ be such that λ_m belongs to the spectrum of $\tau(f_m)$ for every $m \geqslant 1$. Fix operator T^{f_m} such that $\pi(T^{f_m}) = \tau(f_m)$ for every $m \geqslant 1$. Apply Theorem 2.4.1 with $\lambda^{(r)} = \{\lambda_m\}_{m \geqslant 1}$ for all $r \geqslant 1$ to obtain an orthonormal sequence $\{\psi_r\}_{r \geqslant 1}$ such that

$$T^{f_m} = \lambda_m I \oplus R_m + K_m,$$

where K_m is compact and the decomposition of T^{f_m} is with respect to $\mathcal{H} = \mathcal{M} \oplus \mathcal{M}^{\perp}$ with $\mathcal{M} = \vee_{r \geqslant 1}\{\psi_r\}$. Let U_+ be the shift operator on $\mathcal{M}$ given by $U_+\psi_r = \psi_{r+1}$, and define

$$U_+^{(n)} = \begin{cases} U_+^n & n \geqslant 0, \\ U_+^{*|n|} & n < 0. \end{cases}$$

Further, define the operator

$$S_n' = \begin{cases} U_+^{(n)} & \text{on} \quad \mathcal{M}, \\ \mathrm{Id} & \text{on} \quad \mathcal{M}^{\perp}, \end{cases}$$

and note that S_n' is essentially unitary with $\mathrm{ind}(S_n') = n$. To verify that $\pi(S_n')$ commutes with $\mathrm{Im}\,\tau$, observe that

$$[S_n', T^{f_m}] = S_n'T^{f_m} - T^{f_m}S_n' \text{ is a compact operator.}$$

Thus,

$$\pi(S_n')\tau(f_m) = \pi(S_n'T^{f_m}) = \pi(T^{f_m}S_n') = \tau(f_m)\pi(S_n').$$

Since $\pi(S_n')$ commutes with dense subset of $\mathrm{Im}\,\tau$, it follows that $\pi(S_n')$ commutes with all of $\mathrm{Im}\,\tau$ and the proof is complete.   □

The previous proposition can be rephrased in completely operator-theoretic terms.

Corollary 2.5.1

Let $S, T \in \mathcal{L}(\mathcal{H})$ be essentially normal. If there exists an essentially unitary $V \in \mathcal{L}(\mathcal{H})$ such that $V^*SV - T$ is compact, then there exists an unitary $U \in \mathcal{L}(\mathcal{H})$ such that $U^*SU - T$ is compact.

2.6   Existence and Uniqueness of Trivial Class

The main goal in this section is to show that for any compact metric space X, $\mathrm{Ext}(X)$ is an abelian semigroup with an identity. The fact that $\mathrm{Ext}(X)$ is a group will be established much later (see Corollary 4.3.1).

First note that if T_1 and T_2 are two essentially normal operators with essential spectrum X, then $T_1 \oplus T_2$ is also essentially normal with essential spectrum equal to X (see Exercises 1.7.38 and 2.8.15) and the class $[T_1 \oplus T_2]$ of $T_1 \oplus T_2$ depends only on those of T_1 and T_2. Thus, for $X \subseteq \mathbb{C}$, we may define addition in $\mathrm{Ext}(X)$ by

$$[T_1] + [T_2] = [T_1 \oplus T_2].$$

If τ_1 and τ_2 are the unital *-monomorphisms corresponding to the operators T_1 and T_2, then the sum $\tau_1 + \tau_2$ should be given by the functional calculus for $T_1 \oplus T_2$:

$$(\tau_1 + \tau_2)(f) = f(\pi(T_1 \oplus T_2)).$$

Now $\tau_1 + \tau_2$ is a mapping from $C(X)$ into $Q(\mathcal{H}_1) \oplus Q(\mathcal{H}_2)$. However, we can think of $Q(\mathcal{H}_1) \oplus Q(\mathcal{H}_2)$ as a subalgebra of $Q(\mathcal{H}_1 \oplus \mathcal{H}_2)$. Indeed, if ρ is the injective *-homomorphism determined by the diagram

$$
\begin{array}{ccc}
\mathcal{L}(\mathcal{H}_1) \oplus \mathcal{L}(\mathcal{H}_2) & \lhook\joinrel\longrightarrow & \mathcal{L}(\mathcal{H}_1 \oplus \mathcal{H}_2) \\
\pi \oplus \pi \downarrow & & \downarrow \pi \\
\mathrm{Im}\,\tau_1 \oplus \mathrm{Im}\,\tau_2 \subseteq Q(\mathcal{H}_1) \oplus Q(\mathcal{H}_2) & \overset{\rho}{\dashrightarrow} & Q(\mathcal{H}_1 \oplus \mathcal{H}_2),
\end{array}
\qquad (2.6.9)
$$

then we have

$$
\begin{aligned}
(\tau_1 + \tau_2)(f) &= f(\pi \oplus \pi(T_1 \oplus T_2)) \\
&= \rho(f(\pi(T_1)) \oplus f(\pi(T_2))) \\
&= \rho(\tau_1(f) \oplus \tau_2(f)).
\end{aligned}
$$

We formally define sum of τ_1 and τ_2 as follows:

Definition 2.6.1

For any compact metrizable space X, we define the sum $\tau_1 + \tau_2$ by the formula

$$(\tau_1 + \tau_2)(f) = \rho(\tau_1(f) \oplus \tau_2(f)), \quad f \in C(X), \qquad (2.6.10)$$

where $\rho : Q(\mathcal{H}_1) \oplus Q(\mathcal{H}_2) \to Q(\mathcal{H}_1 \oplus \mathcal{H}_2)$ is the injective *-homomorphism determined by the diagram (2.6.9).

Lemma 2.6.1

The sum $\tau_1 + \tau_2$ defined by (2.6.10) makes $\mathrm{Ext}(X)$ into an abelian semigroup.

Proof Let us check that $\tau_1 + \tau_2$ is a *-monomorphism if $\tau_1, \tau_2 \in \mathrm{Ext}(X)$. By the preceding discussion, $(\tau_1 + \tau_2)(f) = 0$ if and only if $f(\pi(T_1) \oplus \pi(T_2)) = 0$ if and only if $f(\pi(T_1)) = 0$ and $f(\pi(T_2)) = 0$. By the spectral mapping property, $f|_{\sigma_e(T_1)} = 0$ and $f|_{\sigma_e(T_2)} = 0$. However, $\sigma_e(T_1) = X = \sigma_e(T_2)$,; hence, f is identically zero. Since τ_1 and τ_2 are unital, so is $\tau = \tau_1 + \tau_2$. We leave it to the reader to check that the class $[\tau]$ depends only on the class of τ_1 and τ_2, and is therefore well defined as an element of $\mathrm{Ext}(X)$. $\square$

What would be the identity element in Ext(X)? Again, we examine at the level of an essentially normal operator to answer this question. If an essentially normal operator N with essential spectrum $\sigma_e(N) = X$ is in the class $\mathcal{N} + \mathcal{C}$, then the Absorption Lemma implies that for any essentially normal operator T with $\sigma_e(T) = X$, the operator $T \oplus N$ is essentially equivalent to T. Thus, for $X \subseteq \mathbb{C}$ and any operator N in $\mathcal{N} + \mathcal{C}$, we have

$$[T] = [T \oplus N] = [T] + [N].$$

This amounts to saying that the class $[N]$, N in $\mathcal{N} + \mathcal{C}$ acts as the identity in Ext(X); for this reason, operators in $\mathcal{N} + \mathcal{C}$ will be called *trivial*. For $X \subseteq \mathbb{C}$, the Weyl–von Neumann theorem states that the class of any such operator must be uniquely determined. As we have pointed out earlier, we can compactly perturb normal operator N to obtain another normal operator N' such that $\sigma(N') = X = \sigma_e(N')$ (see Lemma 2.2.4). If $\mathcal{N}_X$ is the associated unital $*$-monomorphism, that is, $\mathcal{N}_X : C(X) \to \mathcal{Q}(\mathcal{H})$ is given by $\mathcal{N}_X(f) = f(\pi(N'))$, then the diagram

$$
\begin{array}{ccc}
 & & \mathcal{L}(\mathcal{H}) \\
 & \overset{\mathcal{N}_0}{\nearrow} & \downarrow{\scriptstyle \pi} \\
C(X) & \underset{\mathcal{N}_X}{\longrightarrow} & \mathcal{Q}(\mathcal{H})
\end{array}
$$

is commutative, where $\mathcal{N}_0$ is defined by $\mathcal{N}_0(f) = f(N')$. This leads us to the following definition.

Definition 2.6.2

For a compact metrizable space X, a unital $*$-monomorphism $\mathcal{N}_X : C(X) \to \mathcal{Q}(\mathcal{H})$ is *trivial* if we can find a unital $*$-monomorphism $\mathcal{N}_0 : C(X) \to \mathcal{L}(\mathcal{H})$ such that $\mathcal{N}_X = \pi \circ \mathcal{N}_0$. We say that $\mathcal{N}_0$ trivializes $\mathcal{N}_X$ and that $\mathcal{N}_X$ lifts to $\mathcal{L}(\mathcal{H})$.

***Remark* 2.6.3**   Let (X, d) be a metric space. Let $\mathcal{N}_0 : C(X) \to \mathcal{L}(\mathcal{H})$ be a unital $*$-homomorphism. Then there exists a unique spectral measure $\mathbb{E} : \mathcal{B}(X) \to$ of N_0 defined on all Borel subsets of X such that for every $x, y \in \mathcal{H}$, $\langle \mathbb{E}(\cdot)x, y \rangle$ is a regular measure and

$$\mathcal{N}_0(f) = \int_X f \, d\mathbb{E}, \quad f \in C(X).$$

This can be deduced from Theorem B.1.14 (see [35, Theorem 1.14] for a proof). We claim that Im $\mathcal{N}_0 \subseteq \mathcal{Z}_0$, where $\mathcal{Z}_0$ is a C^*-algebra generated by a countable family of commuting projections. In view of the separability of $C(X)$ (see Theorem B.1.6), it suffices to check that for every continuous function f on X, $\mathcal{N}_0(f)$ is generated by a countable family of commuting projections. To see this, let $\{U_n\}_{n \geqslant 0}$ be a basis of open sets in X and $\epsilon > 0$. By the uniform continuity of f, there exists $\delta > 0$ such that $|f(x) - f(x')| < \epsilon$ whenever $d(x, x') < \delta$ for any $x, x' \in X$. Now if $\{U_{n_k}\}_{k \geqslant 0}$ is a basis of open sets in X of diameter less than δ, then by the compactness of X, there exists a finite pairwise disjoint covering $\{\tilde{U}_{k_1}, \dots, \tilde{U}_{k_N}\}$ of Borel measurable subsets of X of diameter less than δ such that

$$|f(x) - f(x')| < \epsilon, \quad x, x' \in \tilde{U}_n, \ n = k_1, \dots, k_N.$$

Fix x_n in $\tilde{U}_n$ for each n, and note that for any $x \in X$,

$$f(x) - \sum_{i=1}^{N} f(x_{k_i})\chi_{\tilde{U}_{k_i}}(x) = \sum_{i=1}^{N} (f(x) - f(x_{k_i}))\chi_{\tilde{U}_{k_i}}(x).$$

It follows that $\|f - \sum_{i=1}^{N} f(x_{k_i})\chi_{\tilde{U}_{k_i}}\|_\infty < \epsilon$, completing the verification of the claim (cf. [104, Proof of Theorem 3]).

In this section, we show that the class $[\tau]$ of a trivial extension in $\mathrm{Ext}(X)$ is uniquely determined. This is a generalization of the Weyl–von Neumann theorem. First, we need a variant of Lemma 2.2.3. Recall that a topological space X is *totally disconnected* if the only non-empty connected subsets of X are singletons (see Appendix A for examples of totally disconnected spaces and Exercise 2.8.7 for a characterization of totally disconnected compact Hausdorff spaces).

An alternate proof of the following lemma is outlined in Exercise 2.8.10 (cf. [104, Theorem 2]).

Lemma 2.6.2 (Arveson Lemma)

If $\mathcal{L}$ is an abelian C^*-algebra generated by a countable family of orthogonal projections, then the maximal ideal space $\mathcal{M}$ of $\mathcal{L}$ is totally disconnected. Moreover, $\mathcal{L}$ has a single self-adjoint generator.

Proof Let $\{e_n\}_{n \geqslant 1}$ be the set of orthogonal projections generating the C^*-algebra $\mathcal{L}$. Then there exists a collection $\{U_n\}_{n \geqslant 1}$ of non-empty clopen sets such that $\Gamma_{\mathcal{L}}(e_n) = \chi_{U_n}$, where $\Gamma_{\mathcal{L}} : \mathcal{L} \to C(\mathcal{M})$ is the Gelfand map. Consider the map

$$\gamma : \mathcal{M} \to \{0,1\}^{\mathbb{N}}, \quad \gamma(x) = (\chi_{U_n}(x)),$$

where $\{0,1\}^{\mathbb{N}}$ carries the product topology. Since each χ_{U_n} is continuous, so is γ. As $\{e_n\}_{n \geqslant 1}$ generates $\mathcal{L}$, $\{U_n\}_{n \geqslant 1}$ separates points of $\mathcal{M}$. It follows that γ is one to one. Thus, γ is a homeomorphism onto a compact subset of $\{0,1\}^{\mathbb{N}}$. We next note that the map

$$\varphi : \{0,1\}^{\mathbb{N}} \to [0,1], \quad \varphi(\{a_n\}) = \sum_{n=1}^{\infty} \frac{2a_{n-1}}{3^n}$$

is a one to one map from $\{0,1\}^{\mathbb{N}}$ onto the Cantor set (see proof of Theorem A.1.5). Moreover, since the series $\sum_{n=1}^{\infty} \frac{2a_{n-1}}{3^n}$ converges absolutely, φ is easily seen to be continuous. Thus, the map $h = \varphi \circ \gamma : \mathcal{M} \to [0,1]$ is a one to one map in $C(\mathcal{M})$, and hence, by Corollary B.1.3, the C^*-algebra $C^*(h)$ generated by h is equal to $C(\mathcal{M})$. Note that the maximal ideal space of $C^*(h)$ equals the image of h. As the image of h (being the Cantor set) is totally disconnected, so is the maximal ideal space $\mathcal{M}$. $\qquad\square$

Remark 2.6.4 Note that $\mathcal{M}$ has a basis of clopen sets.

Before we present a generalization of the Weyl–von Neumann theorem, we introduce some useful operations used frequently in what follows. For compact Hausdorff spaces X, Y and a continuous map $p : X \to Y$, we define the *pullback* $p^* : C(Y) \to C(X)$ by

$$p^*(f) = f \circ p, \quad f \in C(Y). \tag{2.6.11}$$

For a $*$-homomorphism $\tau : C(X) \to Q(\mathcal{H})$, we introduce the $*$-homomorphism $p_* \tau : C(Y) \to Q(\mathcal{H})$ by

$$p_* \tau(f) = \tau(f \circ p), \quad f \in C(Y). \tag{2.6.12}$$

Note that $p_* \tau = \tau p^*$.

The following theorem establishes the existence and uniqueness of the trivial element.

Theorem 2.6.5

If X is a compact metric space, then there exists a trivial extension N_X in Ext(X). Further, any two trivial extensions are equivalent.

Proof   Let $\{x_n\}_{n \geqslant 0}$ be a dense set in X, where each isolated point x_n is counted infinitely often. Let $\mathcal{H} = \ell^2(\mathbb{N})$ and define $N_X : C(X) \to Q(\mathcal{H})$ by

$$N_X(f) = \pi(\mathrm{diag}(f(x_n))), \quad f \in C(X),$$

where $\mathrm{diag}(f(x_n))$ denotes the diagonal operator on $\ell^2(\mathbb{N})$ with diagonal entries $\{f(x_n)\}_{n \geqslant 0}$. The map N_X is obviously a $*$-homomorphism that factors through π. If $N_X(f) = 0$, then $\mathrm{diag}(f(x_n))$ is compact. Therefore, $f(x_n) \to 0$, and hence, by the choice of $\{x_n\}_{n \geqslant 0}$, f is identically 0. Thus, N_X is a $*$-monomorphism.

We next show that any two $*$-monomorphisms of this type are equivalent. Let N_X' be a $*$-monomorphisms corresponding to another such sequence $\{y_n\}_{n \geqslant 0}$. By Proposition 2.1.1, there exists a permutation ν of $\mathbb{N}$ such that $d(x_n, y_{\sigma(n)}) \to 0$, where d is the given metric on X. By the uniform continuity of f, $f(x_n) - f(y_{\sigma(n)}) \to 0$ as $n \to \infty$. Define a unitary operator $U : \ell^2(\mathbb{N}) \to \ell^2(\mathbb{N})$ by setting

$$U(e_n) = e_{\sigma(n)}, \quad n \in \mathbb{N},$$

where $\{e_n\}_{n \geqslant 0}$ denotes the standard orthonormal basis for $\ell^2(\mathbb{N})$. Note that for any $m \in \mathbb{N}$,

$$
\begin{aligned}
U \mathrm{diag}(f(x_n)) e_m &= f(x_m) e_{\sigma(m)}, \\
\mathrm{diag}(f(y_n)) U e_m &= f(y_{\sigma(m)}) e_{\sigma(m)},
\end{aligned}
$$

which implies that

$$U \mathrm{diag}(f(x_n)) - \mathrm{diag}(f(y_n)) U = \mathrm{diag}(f(x_n) - f(y_{\sigma(n)})) = K$$

for some compact operator $K \in \mathcal{L}(\ell^2(\mathbb{N}))$. Thus, N_X and N_X' are equivalent.

Finally, we show that any trivial map is equivalent to the one we have described earlier. If $N : C(X) \to Q(\mathcal{H})$ is trivial, then there is a trivializing map $N_0 : C(X) \to \mathcal{L}(\mathcal{H})$ such that $\pi \circ N_0 = N$. Let $\mathbb{E}$ be the spectral measure of N_0 supported on X such that

$$N_0(f) = \int_X f \, d\mathbb{E}, \quad f \in C(X).$$

By Remark 2.6.3, $\mathrm{Im}\, N_0$ is contained in a C^*-algebra $\mathcal{Z}_0$ generated by countably many commuting orthogonal projections. Further, by Lemma 2.6.2, $\mathcal{Z}_0$ has a single self-adjoint generator H. Let $\Gamma_{\mathcal{Z}_0} : \mathcal{Z}_0 \to C(\tilde{X}_0)$ be the Gelfand map, where $\tilde{X}_0$ is the spectrum $\sigma(H)$ of H. It follows that the map $\Gamma_{\mathcal{Z}_0} N_0 : C(X) \to C(\tilde{X}_0)$ is injective. Therefore, it is induced by the continuous surjection $p_0 : \tilde{X}_0 \to X$, where

$$
\begin{array}{ccc}
p_0^* : C(X) & \dashrightarrow & C(\tilde{X}_0) \\
N_0 \downarrow & & \uparrow \Gamma_{\mathcal{Z}_0} \\
\mathrm{Im}\, N_0 & \hookrightarrow & \mathcal{Z}_0 \subseteq \mathcal{L}(\mathcal{H})
\end{array}
$$

and p_0^* is the pushback map (see (2.6.11)). Thus, $\Gamma_{\mathcal{Z}_0} N_0 = p_0^* : C(X) \to C(\tilde{X}_0)$ satisfies

$$N_0(f) = \Gamma_{\mathcal{Z}_0}^{-1} p_0^*(f) = \Gamma_{\mathcal{Z}_0}^{-1}(f \circ p_0) = f \circ p_0(H).$$

However, by the Weyl–von Neumann–Berg theorem, there exists a diagonal operator D with $\sigma_e(D) = \sigma_e(H)$ such that $H - D$ is compact. We may further assume that $\sigma(D) = \tilde{X}_0 = \sigma_e(D)$. As $\pi(f \circ p_0(D)) = (f \circ p_0)(\pi(D))$, it follows that

$$N_0(f) = f \circ p_0(H) = f \circ p_0(D + K) = f \circ p_0(D) + K_f$$

for some compact operator K_f. If $D = \mathrm{diag}(\lambda_n)$, then $\{\lambda_n\}_{n \geqslant 0}$ is dense in $\tilde{X}_0$ and consequently, $\{x_n = p_0(\lambda_n)\}_{n \geqslant 0}$ is dense in X. Moreover,

$$N_X(f) = \pi \circ N_0(f) = \pi(f \circ p_0(D)) = \pi(\mathrm{diag}(f(x_n)))$$

for all f in $C(X)$. Therefore, N_X arises as earlier from the sequence $\{x_n\}_{n \geqslant 0}$. This completes the proof. $\qquad\qquad\square$

Remark 2.6.6 Note that $[N_X] + [N_X] = [N_X]$.

The proof of the following corollaries are contained in that of the preceding theorem.

Corollary 2.6.1

Any normal operator $N \in \mathcal{L}(\mathcal{H})$ is a compact perturbation of a diagonal operator.

The following is Remark 2.6.3 (see [104, Theorem 3] for a generalization).

Corollary 2.6.2

If X is a compact metric space, then the image of a trivial map $N_0 : C(X) \to \mathcal{L}(\mathcal{H})$ is contained in a C^*-algebra generated by countably many commuting orthogonal projections.

The converse of the previous corollary will be proved in the next section (see Theorem 2.7.3).

2.7   Identity Element for Ext(X)

We have seen in Theorem 2.6.5 that the class of a trivial map

$$
\begin{array}{ccc}
 & \mathcal{L}(\mathcal{H}) & \\
{\scriptstyle N_0}\nearrow & \downarrow{\scriptstyle \pi} & \\
C(X) \xrightarrow{\;N_X\;} & \mathcal{Q}(\mathcal{H}) &
\end{array}
$$

is uniquely determined. We now show that the class $[N_X]$ acts as the identity in the abelian semigroup Ext(X).

Theorem 2.7.1

Let X be a compact metric space. If $[\tau]$ is any extension in Ext(X) and $N_X : C(X) \to \mathcal{Q}(\mathcal{H})$ is a trivial element, then

$$[\tau] + [N_X] = [\tau] = [N_X] + [\tau].$$

Proof   Let $\{x_r\}_{r\geq 0}$ be a countable dense set in X. If $\{f_m\}_{m\geq 1}$ is dense in $C(X)$ and T^{f_m} in $\mathcal{L}(\mathcal{H})$ is chosen such that $\pi(T^{f_m}) = \tau(f_m)$, then $\{T^{f_m}\}_{m\geq 1}$ is essentially normal, that is, $[T^{f_m},(T^{f_n})^*] \in C(\mathcal{H})$ for all $m,n \geq 1$. Let

$$\lambda^{(r)} = \{\lambda_m^{(r)} \in \sigma_e(T^{f_m}) \mid m \geq 1\}, \quad r \geq 1,$$

where $\lambda_m^{(r)} = f_m(x_r)$. Let $\mathcal{M} = \bigvee_{r\geq 1}\{\psi_r\}$ be as in Theorem 2.4.1 such that

$$T^{f_m} = \begin{bmatrix} D_m & 0 \\ 0 & R_m \end{bmatrix} + K_m, \quad K_m \in C(\mathcal{H}). \tag{2.7.13}$$

Note that by Lemma 2.5.1, if $f_n \to f$ in $C(X)$, then $\pi(T^{f_n}) \to \pi(T^f)$. As (2.7.13) holds for a dense set, it follows that for any f in $C(X)$ and T^f in $\mathcal{L}(\mathcal{H})$ satisfying $\tau(f) = \pi(T^f)$, we have

$$T^f = \begin{bmatrix} D_f & 0 \\ 0 & R_f \end{bmatrix} + K_f, \quad K_f \in C(\mathcal{H}), \ f \in C(X).$$

The fact that D_f and R_f are determined upto a compact operator implies that the maps

$$\tau_1(f) = \pi(D_f) \text{ and } \tau_2(f) = \pi(R_f), \quad f \in C(X)$$

are well defined. The off diagonal entries in T^f are compact. Therefore, both τ_1 and τ_2 are homomorphisms. Furthermore, in obtaining the decomposition (2.7.13) by using the $\lambda^{(r)}$ twice in succession, that is, by applying Theorem 2.4.1 to $\{\lambda^{(r)} : r \geqslant 1\} \cup \{\lambda^{(r)} : r \geqslant 1\}$, the operator R_m can be replaced by $D_m \oplus R'_m$. In particular,

$$\sigma_e(R_m) = \sigma_e(T^{f_m}) = \sigma_e(D_m)$$

(see Corollary C.1.6). Therefore,

$$\|\pi(R_m)\| = \|\pi(T^{f_m})\| = \|f_m\|_\infty,$$

and similarly

$$\|\pi(D_m)\| = \|\pi(T^{f_m})\| = \|f_m\|_\infty.$$

By continuity of the Calkin map π, we obtain

$$\|\pi(D_f)\| = \|f\|_\infty = \|\pi(R_f)\|.$$

Thus, both the maps τ_1 and τ_2 are $*$-monomorphisms.

If $\mathcal{N}_0 : X \to \mathcal{L}(\mathcal{H})$ is the $*$-monomorphism given by $\mathcal{N}_0(f) = \mathrm{diag}(f(x_r))$, then τ_1 and $\pi \circ \mathcal{N}_0$ agree on a dense set. Therefore, $\tau_1 = \pi \circ \mathcal{N}_0$, which is a trivial map. By the uniqueness part of Theorem 2.6.5, we must have $[\tau_1] = [\mathcal{N}_X]$. As $\tau = \tau_1 + \tau_2$ by construction, it follows from Remark 2.6.6, commutativity and associativity of Ext(X) that $[\tau] + [\mathcal{N}_X] = [\tau]$. $\qquad\square$

What we have seen here and a little more can be stated in the language of category theory. We refer the reader to [116] for a discussion *on category theory*. However, we reproduce some of the terminology so that the reader can fully appreciate the following result. A *functor* is a correspondence between two categories C_1 and C_2 that maps objects X in C_1 (e.g., compact metrizable spaces, abelian semigroups) to objects $\tilde{X}$ in C_2 and morphisms f between two objects in C_1 (continuous functions, homomorphisms) to morphisms $\tilde{f}$ between two objects in C_2. A functor is called *covariant* if $f : X \to Y$ is mapped to $\tilde{f} : \tilde{X} \to \tilde{Y}$, that is, if it preserves the directions of arrows.

Corollary 2.7.1

The correspondence $X \mapsto Ext(X)$ is a covariant functor from compact metrizable spaces and continuous maps to abelian semigroups and homomorphisms.

Proof   Given a continuous function $p : X \to Y$ and an extension $\tau : C(X) \to Q(\mathcal{H})$, consider the $*$-homomorphism $p_*\tau$ given by

$$p_*\tau(f) = \tau(f \circ p), \quad f \in C(Y).$$

The map $p_*\tau : C(Y) \to Q(\mathcal{H})$ is injective if p is surjective. In fact,

$$\ker p_*\tau = \{f \in C(Y) : f|_{p(X)} = 0\}.$$

To eliminate this kernel and obtain a $*$-monomorphism, define

$$(p_*\tau)(f) = \tau p^*(f) + \mathcal{N}_Y(f), \tag{2.7.14}$$

where $\mathcal{N}_Y : C(Y) \to Q(\mathcal{H})$ is a trivial element chosen appropriately. This can be done as follows. Note that $\tau(f \circ p) = f(\pi(T^p))$ for some essentially normal operator T^p with essential spectrum $p(X) \subseteq Y$. If we take a normal operator N' such that $\sigma_e(N') = Y = \sigma(N')$, then $T^p \oplus N'$ is essentially normal with essential spectrum Y. It is now easy to see that the trivial element $\mathcal{N}_Y(f) = f(\pi(N'))$ induced by N' makes $(p_*\tau)(f) = \tau(p \circ f) + \mathcal{N}_Y(f)$ a $*$-monomorphism.

Thus, although $p_*\tau$ given by (2.7.14) is not well defined, it determines a well defined map $p_* : \mathrm{Ext}(X) \to \mathrm{Ext}(Y)$ by $p_*([\tau]) = [p_*\tau]$, where we have used a fixed but arbitrary trivial map in defining $p_*\tau$. As $p_*\tau$ and $p_*\tau'$ are equivalent if τ and τ' are equivalent, it follows that the map $p_* : \mathrm{Ext}(X) \to \mathrm{Ext}(Y)$ is well defined. Moreover, if p is surjective, then τp^* is a $*$-monomorphism. By Theorem 2.7.1, $\tau \circ p^*$ and $\tau \circ p^* + \mathcal{N}_Y$ determine the same class in $\mathrm{Ext}(Y)$. Clearly, p_* preserves the semigroup structure, $p_*[\mathcal{N}_X] = [\mathcal{N}_Y]$, and for any continuous map $q : Y \to Z$, $(q \circ p)_* = q_* p_*$. Thus, Ext is a covariant functor. This completes the proof. $\square$

Remark 2.7.2   Given that $p : X \to Y$ is surjective, if $\tau : C(X) \to Q(\mathcal{H})$ is a $*$-monomorphism, then so is $p_*\tau : C(Y) \to Q(\mathcal{H})$.

It was shown in Corollary 2.6.2 that if τ is a trivial map, then the image $\mathrm{Im}\,\tau$ of τ is contained in an abelian C^*-algebra generated by orthogonal projections. The following theorem establishing the converse leads naturally to the concept of splitting.

Theorem 2.7.3

Let X be a compact metric space. If $\tau : C(X) \to Q(\mathcal{H})$ is a $*$-monomorphism with the image $\mathrm{Im}\,\tau$ of τ contained in an abelian C^*-algebra $\mathcal{Z}$ generated by orthogonal projections, then τ is trivial.

Proof   Recall that $C(X)$ is norm separable (see Theorem B.1.6). It follows that so is the image $\mathrm{Im}\,\tau$ of τ. Hence, without loss of generality, we may assume that $\mathrm{Im}\,\tau$ is contained in an abelian C^*-algebra $\mathcal{Z}$ generated by countably many orthogonal projections. Now, by Lemma 2.6.2, the abelian C^*-algebra $\mathcal{Z}$ is $*$-isomorphic to $C(\tilde{X})$, where $\tilde{X}$ is a compact subset of $\mathbb{R}$. If $\Gamma_{\mathcal{Z}} : \mathcal{Z} \to C(\tilde{X})$ is the Gelfand map, then $\Gamma_{\mathcal{Z}} \circ \tau : C(X) \to C(\tilde{X})$ being injective is induced by a subjective continuous map $p : \tilde{X} \to X$, so that we have the commutative diagram

$$
\begin{array}{ccc}
C(X) & \overset{p^*}{\dashrightarrow} & C(\tilde{X}) \\
\tau \downarrow & & \uparrow \Gamma_{\mathcal{Z}} \\
\mathrm{Im}\,\tau & \hookrightarrow & \mathcal{Z} \subseteq Q(\mathcal{H})
\end{array}
$$

Thus,

$$\Gamma_{\mathcal{Z}} \circ \tau = p^* \text{ or } \tau = \Gamma_{\mathcal{Z}}^{-1} \circ p^* = p_*(\Gamma_{\mathcal{Z}}^{-1}).$$

However, $\tilde{X}$ is a subset of $\mathbb{R}$, and hence, Γ_{Z}^{-1} is trivial (see Example 2.3.4). Therefore, τ is trivial, as required. $\quad\Box$

We present here several interesting consequences of Theorem 2.7.3.

Corollary 2.7.2

Let X be a compact metric space and let $\tau : C(X) \rightarrow Q(\mathcal{H})$ be a $*$-monomorphism. If $Im\,\tau$ is contained in $\pi(C)$ for some abelian C^*-algebra C of $\mathcal{L}(\mathcal{H})$, then τ is trivial.

Proof By the spectral theorem, the abelian C^*-algebra C, and hence, $\pi(C)$ is generated by orthogonal projections. Now apply Theorem 2.7.3. $\quad\Box$

In general, the foregoing results fail for non-metrizable spaces. Indeed, there are no trivial extensions of compact operators by continuous functions on non-metrizable spaces on a separable Hilbert space (see Exercise 2.8.29).

Corollary 2.7.3

If X is a totally disconnected compact metric space, then $Ext(X)$ is trivial.

Proof If X is a totally disconnected compact metric space, then $C(X)$ is generated by characteristic functions in $C(X)$. It follows that the image of any $*$-monomorphism is contained in a C^*-algebra generated by orthogonal projections, and hence, by Theorem 2.7.3, $Ext(X)$ is trivial. $\quad\Box$

One may employ the last corollary to reveal the structure of essentially normal operators with finite essential spectra.

Example 2.7.4 Let $T \in \mathcal{L}(\mathcal{H})$ be an essentially normal operator with essential spectra equal to $\{\lambda_1, \ldots, \lambda_k\}$, $k \geq 1$. By Corollaries 2.7.3 and 2.6.1, T must be a compact perturbation of a diagonal operator D with spectrum and essential spectrum equal to $\{\lambda_1, \ldots, \lambda_k\}$. Further, by Theorem 1.3.7, D is a diagonal operator with diagonal entries being λ_j repeated infinitely often. Thus, T is unitarily equivalent to a compact perturbation of $\oplus_{j=1}^{k} \lambda_j I_{\mathcal{H}_j}$, where $\mathcal{H}_j$ is an infinite dimensional subspace of $\ell^2(\mathbb{N})$. $\quad\blacksquare$

We conclude this chapter with a generalization of Corollary 2.2.1.

Corollary 2.7.4

If $\{e_n\}_{n \geq 0}$ is a sequence of mutually orthogonal projections in $Q(\mathcal{H})$, then there exists a sequence $\{E_n\}_{n \geq 0}$ of mutually orthogonal projections in $\mathcal{L}(\mathcal{H})$ such that $\pi(E_n) = e_n$ for every integer $n \geq 0$.

Proof Consider the C^*-algebra $\mathcal{Z}$ generated by the sequence $\{e_n\}_{n \geq 0}$ of mutually orthogonal projections and let X be the maximal ideal space of $\mathcal{Z}$. If $\Gamma_{\mathcal{Z}}$ is the Gelfand map, then $\Gamma_{\mathcal{Z}}(e_n) = \chi_n$ for a sequence $\{\chi_n\}_{n \geq 0}$ of characteristic functions in $C(X)$. As shown in the proof of Theorem 2.7.3, $\Gamma_{\mathcal{Z}}^{-1}$ is equivalent to the trivial element N_X associated with a normal operator N. Thus, there exists a unitary U such that

$$e_n = \Gamma_{\mathcal{Z}}^{-1}(\chi_n) = \pi(U)^* N_X(\chi_n)\pi(U) = \pi(U^*\chi_n(N)U), \quad n \geq 0.$$

Note that the choice $E_n = U^* \chi_n(N) U$, $n \geqslant 0$ provides a sequence of orthogonal projections with desired properties. $\square$

2.8   Notes and Exercises

- The proof of the Weyl–von Neumann–Berg–Sikonia theorem is borrowed from [36], whereas Proposition 2.3.2 is taken from [105].

- The treatment of the remaining sections is based largely on the expositions [25, 26, 27, 104].

Exercises

Exercise 2.8.1   Let $\mathcal{H}$ be a separable Hilbert space and let D_1, D_2 be two diagonalizable operators in $\mathcal{L}(\mathcal{H})$. If $\sigma_e(D_1) = \sigma(D_1)$ and $\sigma_e(D_2) = \sigma(D_2)$, then show that there exists a unitary operator $U \in \mathcal{L}(\mathcal{H})$ such that $D_1 - U^* D_2 U$ is of trace-class.

Hint: Examine the proof of Proposition 2.1.1 and apply the full strength of Lemma 2.1.1.

Exercise 2.8.2   Show, by an example, that there are essentially equivalent normal operators $N_1, N_2 \in \mathcal{L}(\mathcal{H})$ such that $N_1 - U^* N_2 U$ does not belong to the Schatten p-class $\mathscr{I}_p$ for any $p \geqslant 1$ and for any unitary operator $U \in \mathcal{L}(\mathcal{H})$.

Hint: Take $N_1 = 0$ and N_2 to be a suitable diagonal operator.

Exercise 2.8.3   Let $A \in \mathcal{L}(\mathcal{H})$ be a self-adjoint operator, $f \in \mathcal{H}$ and $\epsilon > 0$. Show that for any $p \in (1, \infty)$, there is a finite rank projection P and a self-adjoint, finite rank operator K such that $f \in \operatorname{ran} P$, $\|K\|_p < \epsilon$, and $\operatorname{ran} P$ is reducing for $A + K$.

Hint: By the proof of Lemma 2.2.1, K is of rank at most n with $\|K\| \leqslant (b-a)/n$, Now use Exercise 1.7.18 (this argument does not work for $p = 1$).

Exercise 2.8.4 ([97])   If U is a unitary operator on $\mathcal{H}$, then show that there are diagonal unitary operators D_1, D_2 relative to some orthonormal basis in $\mathcal{H}$ such that $U = D_1 D_2$.

Hint: By Lemma 2.2.4, $U = D + K$ for a unitary diagonal operator D. Let $D_1 = D$ and $D_2 = I + D^* K$.

Exercise 2.8.5   Show that the conclusion of Lemma 2.5.1 no longer holds if $\sigma_e(T)$ is a proper subset of X.

Hint: Let $X = \mathbb{T}$ and let $T \in \mathcal{L}(\mathcal{H})$ be a unitary operator such that $\sigma_e(T)$ is a proper subset of $\mathbb{T}$. Use Urysohn's Lemma to construct a continuous nonzero function $f \in C(\mathbb{T})$ such that $f|_{\sigma_e(T)} = 0$. Verify that $\tau : C(X) \to Q(\mathcal{H})$ given by the functional calculus of T satisfies $\tau(f) = 0$.

Exercise 2.8.6 ([89]) Show that every essentially normal operator with essential spectrum lying on a simple arc is unitarily equivalent to a compact perturbation of a normal operator.

Hint: Use Proposition 2.3.1 and Example 2.3.4.

Exercise 2.8.7 Let X be a compact Hausdorff space. Show that X is totally disconnected if and only if $C(X)$ is generated by idempotents. Show further that if $C(X)$ is generated by the idempotents $\{\chi_m\}_{m \geqslant 0}$, then $C(X)$ is singly generated with generator in $C_{\mathbb{R}}(X)$ given by $\sum_{m=0}^{\infty} 3^{-m}(2\chi_m - 1)$.

Hint: Assume that X is totally disconnected. Consider the algebra $\mathscr{A}$ generated by the idempotents in $C(X)$. If $x_1, x_2 \in X$ are distinct, then there exists disjoint clopen sets X_1, X_2 such that $x_1 \in X_1$, $x_2 \in X_2$ and $X = X_1 \sqcup X_2$. Apply Corollary B.1.1 to conclude that $\overline{A} = C(X)$. Conversely, if $C(X)$ is generated by idempoents and $x_1, x_2 \in X$ are distinct, then there exists an idempotent $f \in C(X)$ such that $f(x_1) \neq f(x_2)$. However, $f = \chi_{X_1}$ for some clopen subset X_1 of X. which yields the disconnection $X_1 \sqcup (X \backslash X_1)$ of X. For the remaining part, follow the steps of the proof of Lemma 2.2.3 (with $\rho(Q_m)$ replaced by χ_m).

Exercise 2.8.8 Show that every essentially normal operator with spectrum equal to the Cantor set is unitarily equivalent to a compact perturbation of a normal operator.

Hint: Use Corollary 2.7.3.

Exercise 2.8.9 ([104]) Show that every separable abelian C^*-algebra in $\mathcal{L}(\mathcal{H})$ is singly generated.

Hint: Note that every element in a separable abelian C^*-algebra is normal and hence, by spectral theorem can be approximated by countably many orthogonal projections. Now apply Lemma 2.6.2.

Exercise 2.8.10 Deduce Lemma 2.6.2 from Exercise 2.8.7. Further, deduce the sufficiency part of Exercise 2.8.7 from Lemma 2.6.2.

Hint: Let $\{e_n\}_{n \geqslant 1}$ be the set of orthogonal projections generating the C^*-algebra $\mathcal{L}$. If $\Gamma_{\mathcal{L}} : \mathcal{L} \to C(\mathcal{M})$ is the Gelfand map with $\mathcal{M}$ denoting the maximal ideal space of $\mathcal{L}$, then $C(\mathcal{M})$ is generated by idempotents. Now apply Exercise 2.8.7.

Exercise 2.8.11 ([75]) Let $\{e_n\}_{n \in \mathbb{Z}}$ be the standard orthonormal basis for $\ell^2(\mathbb{Z})$. Define $U : \ell^2(\mathbb{Z}) \to \ell^2(\mathbb{Z})$ by $Ue_n = e_{n+1}$, $n \in \mathbb{Z}$, which can be extended as a bounded linear operator on $\ell^2(\mathbb{Z})$. Verify the following statements:

(i) The closed linear span of $\{e_n : n \geqslant 0\}$ is reducing for $U - e_0 \otimes e_{-1}$ (see (1.2.2)).

(ii) $U - e_0 \otimes e_{-1}$ is unitarily equivalent to $U_+^* \otimes U_+$ on $\ell^2(\mathbb{N}) \otimes \ell^2(\mathbb{N})$, where U_+ is the shift operator on $\ell^2(\mathbb{N})$ (see Example 1.3.2).

Conclude that the spectrum is not invariant under compact perturbation.

Hint: The spectrum of U equals the unit circle whereas that of $U_+^* \otimes U_+$ is the closed unit disc.

Exercise 2.8.12 ([72]) Verify the following assertions:

(i) There exists a bounded linear self-adjoint operator acting on a nonseparable Hilbert space with no eigenvalues.

(ii) Suppose that there exists a self-adjoint operator $A \in \mathcal{L}(\mathcal{H})$ and a diagonal operator $D \in \mathcal{L}(\mathcal{H})$ such that $A - D$ is compact. If $\mathcal{H}$ is nonseparable, then A must have an eigenvalue.

Hint: For the first part, try the uncountable orthogonal direct sum of any Hilbert space (possibly separable) with this property. For the second part, one may argue as in Exercise 1.7.29 to show that there exists a separable subspace of $\mathcal{H}$ that reduces both $A - D$ and D with the orthogonal complement contained in $\ker(A - D)$.

Exercise 2.8.13 Show that any Toeplitz operator on $H^2(\mathbb{T})$ with a continuous symbol is essentially normal.
Hint: Apply Theorem 1.6.5.

Exercise 2.8.14 ([3, 4, 5]) Let S be the Szegö shift on the Hardy space $H^2(\mathbb{D})$. For $\sigma > 0$ and $\theta \in \mathbb{R}$, consider the bounded linear operator $B_{\sigma,\theta} : H^2(\mathbb{D}) \oplus \mathbb{C} \to H^2(\mathbb{D}) \oplus \mathbb{C}$ given by

$$T = \begin{bmatrix} S & \sigma(1 \oplus 1) \\ 0 & e^{i\theta} \end{bmatrix} \text{ on } H^2(\mathbb{D}) \otimes \mathbb{C}.$$

Show that $B_{\sigma,\theta}$ is essentially normal with $\sigma_e(B_{\sigma,\theta}) = \mathbb{T}$.

Exercise 2.8.15 Show that the finite orthogonal direct sum of essentially normal operators is essentially normal. This fails for infinite orthogonal direct sums.
Hint: For the second part, consider infinite copies of the shift operator of multiplicity 1.

Exercise 2.8.16 Give an example of a non-normal operator $T \in \mathcal{L}(\mathcal{H})$ with $\sigma_e(T) = \overline{\mathbb{D}}$.

Exercise 2.8.17 For $n \in \mathbb{N} \cup \{\infty\}$, let $T = (T_1, \ldots, T_n)$ be a family of normal operators in $\mathcal{L}(\mathcal{H})$. Show that λ in $\mathbb{C}^n$ is in the joint spectrum of T if and only if the bounded linear operator $\sum_{m=1}^{n} \frac{(T_m - \lambda_m)^*(T_m - \lambda_m)}{2^m(\|T_m - \lambda_m\|^2 + 1)}$ in $\mathcal{L}(\mathcal{H})$ is not invertible.

Exercise 2.8.18 ([77]) Let $\{e_n\}_{n \in \mathbb{N}}$ be the standard orthonormal basis for $\ell^2(\mathbb{N})$. Let $\mathcal{H}$ be any separable Hilbert space with an orthonormal basis $\{f_n\}_{n \in \mathbb{N}}$. Define a linear transformation $U : \mathcal{H} \to \ell^2(\mathbb{N})$ by $U f_n = e_n$, $n \in \mathbb{N}$, which can be extended as a unitary transformation on $\mathcal{H}$. For $T \in \mathcal{L}(\mathcal{H})$, define $\tilde{T} : \ell^2(\mathbb{N}) \to \ell^2(\mathbb{N})$ by

$$\tilde{T} f_n = U T e_n, \quad n \in \mathbb{N},$$

which can be extended as a bounded linear operator on $\mathcal{H}$. Verify the following assertions:

(i) T is essentially normal if and only if so is $\tilde{T}$.

(ii) $[\tilde{T}] \in \text{Ext}(\sigma_e(T))$.

Conclude that $\text{Ext}(X)$ may be defined as the collection of equivalence classes of essentially normal operators with essential spectrum X on a fixed separable Hilbert space.

Exercise 2.8.19 Let $T \in \mathcal{L}(\mathcal{H})$ be such that $\pi(T)$ is a positive operator. Then, T is a compact perturbation of a positive operator P in $\mathcal{L}(\mathcal{H})$. Moreover, if $\pi(T) \leqslant I$, then one can choose P such that $P \leqslant I$.

Hint: Recall that for any self-adjoint operator P, we have $0 \leqslant P \leqslant I$ if and only if $\sigma(P) \subseteq [0,1]$ (see Theorem C.1.4). Both parts can now be obtained by adopting the proof of Corollary 2.2.1 to the present situation.

Exercise 2.8.20 ([27]) If $C \in \mathcal{L}(\mathcal{H})$ is a positive operator and $a, b \in Q(\mathcal{H})$ are positive elements such that $\pi(C) = a + b$ and $\epsilon > 0$, then show that there exist positive operators $A, B \in Q(\mathcal{H})$ such that

$$\pi(A) = a, \quad \pi(B) = b, \quad \|C - A - B\| < \epsilon.$$

Hint: Let $d := (a + b + \epsilon/2)^{-1/2}$. If $f := dad$, then $0 \leqslant f \leqslant I$, and hence, there exists $F \in \mathcal{L}(\mathcal{H})$ such that $0 \leqslant F \leqslant I$ and $\pi(F) = f$ (see Exercise 2.8.19). Set $G := (C + \epsilon/2)^{1/2}$, $A := GFG$ and $B' := G(I - F)G$. Note that $A, B' \geqslant 0$, $A + B' = C + \epsilon/2$ and $\pi(A) = a$. It follows that $\pi(B') = b + \epsilon/2$. Define $B := \phi(B')$, where $\phi(t) = \max\{t - \epsilon/2, 0\}$, $t \in \sigma(B')$. Then, $\pi(B) = b$, $\|B' - B\| \leqslant \epsilon/2$, and $\|C - A - B\| < \epsilon$.

Exercise 2.8.21 For a positive integer n, let S_n denote the unilateral shift of multiplicity n and let U, V denote unitary operators on $\ell^2(\mathbb{N})$. Show that $U \oplus S_n$ is essentially equivalent to $S_m \oplus V$ if and only if $m = n$.

Hint: Examine the proof of Corollary 2.4.1.

Exercise 2.8.22 Show that two unitary operators on a separable Hilbert space are essentially equivalent if and only if their essential spectra are the same.

Hint: Apply Theorem 2.3.5.

Exercise 2.8.23 Show that the *Toeplitz extension* $\tau : C(\mathbb{T}) \to Q(\mathcal{H})$ induced by (1.6.27) is non-trivial.

Hint: Either appeal to the usual index argument or consult Theorem 2.3.5.

Exercise 2.8.24 Let X be a compact metric space. Let $\mathcal{N}_0 : C(X) \to \mathcal{L}(\mathcal{H})$ be a *-monomorphism such that $P = \mathcal{N}_0(1)$ is an orthogonal projection. For any $x_0 \in X$, define $\tilde{\mathcal{N}}_0 : C(X) \to \mathcal{L}(\mathcal{H})$ by

$$\tilde{\mathcal{N}}_0(f) = \mathcal{N}_0(f) + f(x_0)(I - P), \quad f \in C(X).$$

Show that $\tilde{\mathcal{N}}_0$ is a unital *-monomorphism. Further, if $\mathcal{N}_0$ trivializes $\mathcal{N}_X$ (and that $\mathcal{N}_X$ lifts to $\mathcal{L}(\mathcal{H})$), then show that $\tilde{\mathcal{N}}_0$ trivializes $\mathcal{N}_X$.

Exercise 2.8.25 Show that replacing the unitary by an invertible operator in the notion of essential equivalence of operators gives exactly the same notion of equivalence.

Hint: See Corollary C.1.10.

Exercise 2.8.26 ([40]) If $C(X)$ is generated by a single real-valued continuous function, then show that $\mathrm{Ext}(X)$ is trivial. Deduce the Weyl–von Neumann theorem from this statement.

Hint: For the first part, apply Corollary 2.7.2 and for the second part, appeal to the Stone–Weierstrass theorem.

Exercise 2.8.27 Define $\gamma : \mathrm{Ext}(\mathbb{T}) \to \mathbb{Z}$ by $\gamma([T]) = \mathrm{ind}(T)$. Show that γ defines a semi-group isomorphism.

Hint: Infer this from Theorem 2.3.5.

Exercise 2.8.28 Let γ be a simple closed curve in $\mathbb{C}$ and let $\phi : \mathbb{T} \to \gamma$ be an orientation-preserving homeomorphism. If $T \in \mathcal{L}(\mathcal{H})$ is an essentially normal operator with essential spectrum $\sigma_e(T) = \gamma$ and $\mathrm{ind}(T - \lambda I) = n$ inside γ, then show that T is unitarily equivalent to a compact perturbation of M_ϕ or $T_{\phi \,\circ\, z^{-n}}$ according as $n = 0$ or $n \neq 0$.

Hint: Use Corollary 2.7.1 and Exercise 2.8.27.

Exercise 2.8.29 Let $\beta\mathbb{N}$ denote the Stone–Čech compactification of the set of nonnegative integers. Show that there is no trivial extension of $C(\mathcal{H})$ by $C(\beta\mathbb{N}\backslash\mathbb{N})$ on a separable Hilbert space $\mathcal{H}$.

Hint: Note that $\beta\mathbb{N}\backslash\mathbb{N}$ contains a continuum of pairwise disjoint clopen sets. If there is a trivial extension, then there are uncountably many mutually orthogonal projections in $\mathcal{L}(\mathcal{H})$, and hence, $\mathcal{H}$ is not separable.

3

Splitting and the Mayer–Vietoris Sequence

In this chapter, we address the question of decomposing the semigroup Ext(X) provided a decomposition of X is given. In particular, we introduce the notion of splitting of an extension, and note that a closed disjoint cover $\{A, B\}$ of X yields direct sum decomposition of Ext(X) into Ext(A) and Ext(B). As the first major step in the ultimate splitting lemma (to be proved in the next chapter), we establish the first splitting lemma, which states that such a decomposition holds for a closed cover $\{A, B\}$ if $A \cap B$ is a singleton.

3.1 Splitting

Let X be a compact Hausdorff space. Given a $*$-monomorphism $\tau : C(X) \to Q(\mathcal{H})$ and $f \in C(X)$, write T^f for any operator in $\mathcal{L}(\mathcal{H})$ such that $\pi(T^f) = \tau(f)$. It will always be understood that T^f is determined only up to simultaneous unitary equivalence modulo the compacts. Recall that Im τ stands for the image $\{\tau(f) : f \in C(X)\}$ of τ. If T is in $\mathcal{L}(\mathcal{H})$ and E is an orthogonal projection in $\mathcal{L}(\mathcal{H})$, then write T_E for the operator $ET|_{E\mathcal{H}}$ in $\mathcal{L}(E\mathcal{H})$.

Lemma 3.1.1

Let e be an orthogonal projection in the Calkln algebra $Q(\mathcal{H})$ and $\tau_e : C(X) \to Q(\mathcal{H})$ be a $*$-monomorphism such that $\tau_e(1) = e$. If E is an orthogonal projection in $\mathcal{L}(\mathcal{H})$ such that $\pi(E) = e$ (see Corollary 2.2.1), then we have the following statements:

(1) (Existence) There exists a unital $*$-monomorphism $\tau_{e,E} : C(X) \to Q(E\mathcal{H})$ such that

$$\tau_{e,E}(f) = \pi(T_E^f), \text{ where } T_E^f = ET^f|_{E\mathcal{H}} \text{ and } \pi(T^f) = \tau_e(f).$$

(2) (Uniqueness) If F is another orthogonal projection such that $\pi(F) = e$, then $\tau_{e,E}$ is equivalent to $\tau_{e,F}$.

Proof (1) Note that $\tau_e(1) = e$ implies that $\tau_e(f) = \tau_e(1 \cdot f \cdot 1) = e\tau_e(f)e$, that is, $\pi(T^f - ET^f E) = 0$. Thus, the map $\tau_{e,E}(f) = \pi(T_E^f)$ is well defined. Similarly, one can see that the projection e commutes with $\operatorname{Im} \tau_e$, and hence,

$$\pi(T^f E - ET^f) = 0, \quad \pi(ET^f - ET^f E) = 0, \quad \pi(T^f E - ET^f E) = 0. \tag{3.1.1}$$

If we decompose the operator T^f with respect to E and $I - E$, then by (3.1.1), the off diagonal entries are compact. Thus, the map $\tau_{e,E}(f) = \pi(T_E^f)$ is $*$-homomorphism.

(2) Let F be an orthogonal projection in $\mathcal{L}(\mathcal{H})$ such that $\pi(F) = e$ and note that $E - F$ is a compact operator. Let U and V be isometries in $\mathcal{L}(\mathcal{H})$ such that $UU^* = E$, $VV^* = F$ and EU, FV are unitaries (see Corollary 1.5.1). Define $\tilde{\tau}_{e,E} : C(X) \to Q(E(\mathcal{H}))$ and $\tilde{\tau}_{e,F} : C(X) \to Q(F(\mathcal{H}))$ by

$$\tilde{\tau}_{e,E}(f) = \pi(U^* T_E^f U) \quad \text{and} \quad \tilde{\tau}_{e,F}(f) = \pi(V^* T_F^f V).$$

Since EU and FV are unitaries,

$$[\tilde{\tau}_{e,E}] = [\tau_{e,E}] \text{ and } [\tilde{\tau}_{e,F}] = [\tau_{e,F}].$$

As weakly equivalent extensions are equivalent (see Proposition 2.5.1), it suffices to show that $\tilde{\tau}_{e,E}$ is weakly equivalent to $\tilde{\tau}_{e,F}$. Observe that

$$
\begin{aligned}
\pi(V^* U)\tilde{\tau}_{e,E}(f)\pi(U^* V) &= \pi(V^* U)\pi(U^* T_E^f U)\pi(U^* V) \\
&= \pi(V^* UU^* T_E^f UU^* V) \\
&= \pi(V^* ET_E^f EV) = \pi(V^* ET^f EV) \\
&= \pi(V^* FT^f FV) = \tilde{\tau}_{e,F}(f).
\end{aligned}
$$

In the last but one equality, we have used the fact that E and F differ by a compact operator. Finally note that

$$U^* VV^* U = U^* FU = U^*(E + \text{compact})U = I + \text{compact},$$

and similarly, $V^* UU^* V = I + \text{compact}$. Thus, the operator $V^* U$ is essentially unitary, and the proof of the lemma is complete. $\square$

Let $\tau : C(X) \to Q(\mathcal{H})$ be a $*$-monomorphism. If $\mathcal{Z}$ is a separable abelian C^*-subalgebra of the Calkin algebra $Q(\mathcal{H})$ such that $\operatorname{Im} \tau \subseteq \mathcal{Z}$, and $\Gamma_{\mathcal{Z}} : \mathcal{Z} \to C(\tilde{X})$ is the Gelfand map, then we have

$$
\begin{array}{ccc}
C(X) & \xrightarrow{\ p^*\ } & C(\tilde{X}) \\
\tau \downarrow & & \uparrow \Gamma_{\mathcal{Z}} \\
\operatorname{Im}\tau & \lhook\joinrel\longrightarrow & \mathcal{Z} \subseteq Q(\mathcal{H})
\end{array}
$$

where $\Gamma_Z \circ \tau$ is an injection of $C(X)$ into $C(\tilde{X})$ and it is induced by a continuous surjection $p : \tilde{X} \to X$, that is, for any $f \in C(X)$,

$$\Gamma_Z \circ \tau(f) = p^*(f) = f \circ p,$$
$$\tau(f) = \Gamma_Z^{-1} p^*(f),$$
$$\tau = p_*(\Gamma_Z^{-1}),$$

(see (2.6.11) and (2.6.12)).

Let e be an orthogonal projection in $Q(\mathcal{H})$ commuting with $\operatorname{Im}\tau$ and consider the separable C^*-algebra $Z = C^*[\operatorname{Im}\tau, e]$ generated by the image of τ and e. Then it is possible to split the extension $[\tau]$ with respect to certain subsets of X. In the following discussion, we make this precise. As e is a projection in Z and $\Gamma_Z : Z \to C(\tilde{X})$ is the Gelfand map, it follows that there exists a clopen subset $\tilde{X}_1$ of $\tilde{X}$ such that $\Gamma_Z(e)$ is equal to the characteristic function $\chi_{\tilde{X}_1}$ of the set $\tilde{X}_1$. Let $\tilde{X}_2 = \tilde{X} \backslash \tilde{X}_1$. Thus, $\tilde{X} = \tilde{X}_1 \sqcup \tilde{X}_2$ is the disjoint union of the two clopen sets $\tilde{X}_1$ and $\tilde{X}_2$. We claim that the map $p : \tilde{X} \to X$ is one to one on $\tilde{X}_1$ and also on $\tilde{X}_2$, which will show that $p|_{\tilde{X}_i} : \tilde{X}_i \to p(\tilde{X}_i)$ is a homeomorphism on these sets for $i = 1, 2$. In fact, by the Stone–Weierstrass Theorem (see Theorem B.1.1), $\chi_{\tilde{X}_1}$ together with $\Gamma_Z(\operatorname{Im}\tau)$ must separate points of $\tilde{X}$. However, the function $\chi_{\tilde{X}_1}$ cannot distinguish any point of $\tilde{X}_1$, and therefore, all the points in $\tilde{X}_1$ must be separated by $\Gamma_Z(\operatorname{Im}\tau)$. However, if $p(x) = p(y)$ for any two points x, y in $\tilde{X}_1$, then

$$\Gamma_Z\tau(f)(x) = f(p(x)) = f(p(y)) = \Gamma_Z\tau(f)(y)$$

for every $f \in C(X)$, which implies that $x = y$. The fact that p is one to one on $\tilde{X}_2$ follows similarly and the claim stands verified. Note that p^* is an injective map by construction and therefore, the map $p : \tilde{X} \to X$ must be surjective. In particular, $p(\tilde{X}_1) \cup p(\tilde{X}_2) = X$. We thus identify $\tilde{X}_1$ with the closed subset $X_1 = p(\tilde{X}_1)$ of X and similarly, $\tilde{X}_2$ is identified with the closed subset $X_2 = p(\tilde{X}_2)$ of X such that $X_1 \cup X_2 = X$. Now, if $\tau_e^d : C(X) \to Q(\mathcal{H})$ is the $*$-homomorphism defined by $\tau_e^d(f) = e\tau(f)$, then τ_e^d need not be injective. Moreover, it is not unital as $\tau_e^d(1) = e$. It however induces a $*$-monomorphism $\tau_e : C(X_1) \to Q(\mathcal{H})$. We now collect what we have said so far plus a little more in the following lemma.

Lemma 3.1.2

Let $\tau : C(X) \to Q(\mathcal{H})$ be a $*$-monomorphism and let e be an orthogonal projection in $Q(\mathcal{H})$ commuting with the image $\operatorname{Im}\tau$ of τ. Consider the C^*-algebra $Z = C^*[\operatorname{Im}\tau, e]$ generated by $\operatorname{Im}\tau$ and e. If $\tilde{X}$ is the maximal ideal space of Z and $\Gamma_Z : Z \to C(\tilde{X})$ is the Gelfand map, then there exists a continuous surjection p from $\tilde{X} = \tilde{X}_1 \sqcup \tilde{X}_2$ into X such that $p|_{\tilde{X}_i} : \tilde{X}_i \to X_i$ is a homeomorphism, where $X_i = p(\tilde{X}_i)$ for $i = 1, 2$. Further, we have the following:

(1) If $\tau_e^d : C(X) \to Q(\mathcal{H})$ is given by $\tau_e^d(f) = e\tau(f)$, $f \in C(X)$, then

$$\ker\tau_e^d = \{f \in C(X) : f|_{X_1} = 0\}.$$

In particular, $\psi_e : C(X)/\ker\tau_e^d \to C(X_1)$ given by $\psi_e([f]) = f|_{X_1}$ is an isomorphism.

(2) $\tau_e : C(X_1) \to Q(E\mathcal{H})$ given below is a $*$-monomorphism:

$$\tau_e(f_1) = \tau_e^d(f), \text{ where } \psi_e([f]) = f_1.$$

(3) If $\tau_{1-e}^d : C(X) \to Q(\mathcal{H})$ is given by $\tau_{1-e}^d(f) = (1-e)\tau(f)$, $f \in C(X)$, then

$$\ker \tau_{1-e}^d = \{f \in C(X) : f|_{X_2} = 0\}.$$

In particular, $\psi_{1-e} : C(X)/\ker\tau_{1-e}^d \to C(X_2)$ given by $\psi_{1-e}([f]) = f|_{X_2}$ is an isomorphism.

(4) $\tau_{1-e} : C(X_2) \to Q((I-E)\mathcal{H})$ given below is a $*$-monomorphism:

$$\tau_e(f_2) = \tau_{1-e}^d(f), \text{ where } \psi_e([f]) = f_2.$$

Proof   We have already proved the first half of the lemma in the preceding discussion. To prove the second half, we verify the statements (1) and (2). As $\Gamma_Z(e) = \chi_{\tilde{X}_1}$,

$$\tau_e^d(f) = 0 \iff e\tau(f) = 0 \iff \Gamma_Z(e\tau(f)) = 0 \iff \text{We ve}$$
$$\chi_{\tilde{X}_1}p^*(f) = 0 \iff (f \circ p)\chi_{\tilde{X}_1} = 0 \iff f|_{X_1} = 0.$$

It follows that $\ker\tau_e^d = \{f \in C(X) : f|_{X_1} = 0\}$. This yields (1). Thus,

$$C(X_1) \overset{\psi_e^{-1}}{\cong} C(X)/(\ker\tau_e^d) \overset{\phi_e}{\cong} \tau_e^d(C(X)),$$

where ϕ_e is an isomorphism given by $\phi_e([f]) = \tau_e^d(f)$. As $\tau_e = \phi_e \circ \psi_e^{-1}$. This verifies (2). The statements (3) and (4) can be verified in an analogous manner and the proof of the lemma is complete.   □

Let τ be as defined in Lemma 3.1.2. In view of Lemma 3.1.2, we may now apply Lemma 3.1.1 to the $*$-monomorphism τ and the orthogonal projection $e = \tau(1)$ to get a unital $*$-monomorphism $\tau_{e,E} : C(X_1) \to Q(E\mathcal{H})$. This depends only on the class e and not on the representative E. Similarly, by considering the $*$-monomorphism τ_{1-e} and the orthogonal projection $1 - e$, we obtain the unital $*$-monomorphism $\tau_{1-e,I-E} : C(X_2) \to Q((I-E)\mathcal{H})$.

3.2   Disjoint Sum of Extensions

Let X_1 and X_2 be topological spaces. The customary notation $X_1 \sqcup X_2$ is employed for the *disjoint union of the spaces X_1 and X_2*, which we defined as the space

$$X_1 \sqcup X_2 = \{(x_i, i) : x_i \in X_i \text{ for } i = 1, 2\}.$$

For each $k = 1, 2$, let $j_{X_k,X} : X_k \to X_1 \sqcup X_2$ be the canonical injection defined by $j_{X_k,X}(x) = (x, k)$. The *disjoint union topology* on $X_1 \sqcup X_2$ is defined as the finest topology on X for which the canonical injections $j_{X_1,X}$ and $j_{X_2,X}$ are continuous. Thus, a subset U of X is open in the disjoint union topology if and only if $j_{X_k,X}^{-1}(U)$ is open for $k = 1, 2$. Note that $X_1 \sqcup X_2$ is a compact Hausdorff space if X_1 and X_2 are compact Hausdorff spaces.

Definition 3.2.1

Let X_1, X_2 be compact Hausdorff spaces. Given extensions $[\tau_{X_k}] : C(X_k) \to Q(\mathcal{H}_1)$ for $k = 1, 2$, define the map $\tau_{X_1} \sqcup \tau_{X_2} : C(X_1 \sqcup X_2) \to Q(\mathcal{H}_1) \oplus Q(\mathcal{H}_2) \subseteq Q(\mathcal{H}_1 \oplus \mathcal{H}_2)$ (see (2.6.9)) by

$$\tau_{X_1} \sqcup \tau_{X_2}(f) = \tau_{X_1}(f \circ j_{X_1,X}) \oplus \tau_{X_2}(f \circ j_{X_2,X}), \quad f \in C(X_1 \sqcup X_2).$$

Remark 3.2.2   Clearly, $\tau_{X_1} \sqcup \tau_{X_2}$ is a $*$-monomorphism. Moreover, $[\tau_{X_1} \sqcup \tau_{X_2}]$ depends only on the classes $[\tau_{X_1}]$ and $[\tau_{X_2}]$.

Let X_1, X_2 be compact Hausdorff spaces. Define the continuous surjection $p : X_1 \sqcup X_2 \to X = X_1 \cup X_2$ by $p(x, k) = x$ for $k = 1, 2$. Recall that $p^*(f) = f \circ p$ and $p_*([\tau]) = [\eta]$, where

$$\eta(f) = \tau p^*(f) + N_X(f), \quad f \in C(X)$$

for some suitable trivial element N_X of $\mathrm{Ext}(X)$. For a subset A of X, let $i_{A,X}$ denote the inclusion map.

Lemma 3.2.1

Let X, X_1, X_2 and p be as in the preceding discussion. Let e be an orthogonal projection in $Q(\mathcal{H})$ and let $\tau : C(X) \to Q(\mathcal{H})$ be a $*$-monomorphism. If $\mathrm{Im}\,\tau$ commutes with e, then

$$p_*[\tau_1 \sqcup \tau_2] = [\tau] = (i_{X_1,X})_*[\tau_1] + (i_{X_2,X})_*[\tau_2],$$

where $\tau_1 = \tau_e$, $\tau_2 = \tau_{1-e}$ (see Lemma 3.1.2), and $i_{X_k,X}$ is the inclusion map from X_k into X for $k = 1, 2$.

Proof   Assume that $\mathrm{Im}\,\tau$ commutes with e. The equality

$$p_*[\tau_1 \sqcup \tau_2] = (i_{X_1,X})_*[\tau_1] + (i_{X_2,X})_*[\tau_2]$$

follows from the definition. Indeed, for f in $C(X)$,

$$
\begin{aligned}
p_*(\tau_1 \sqcup \tau_2)(f) &= (\tau_1 \sqcup \tau_2)(f \circ p) = \tau_1(f \circ p \circ j_{X_1,X}) \oplus \tau_2(f \circ p \circ j_{X_2,X}) \\
&= \tau_1(f|_{X_1}) \oplus \tau_2(f|_{X_2}) = ((i_{X_1,X})_* \tau_1 + (i_{X_2,X})_* \tau_2)(f).
\end{aligned}
$$

On the other hand, for f in $C(X)$,

$$
\begin{aligned}
(\tau_1(i_{X_1,X})^* + \tau_2(i_{X_2,X})^*)(f) &= \tau_1(f \circ i_{X_1,X}) \oplus \tau_2(f \circ i_{X_2,X}) \\
&= \tau_1(f|_{X_1}) \oplus \tau_2(f|_{X_2}) \\
&= e\tau(f) \oplus (1-e)\tau(f) \\
&= \tau(f),
\end{aligned}
$$

which completes the proof of the lemma. $\square$

All this can be said, in the language of matrix decomposition, which is rather elegant. To see this, recall that $\tau_1(f|_{X_1}) = \pi(ET^f|_{E\mathcal{H}})$ and $\tau_2(f|_{X_2}) = \pi((I-E)T^f|_{(I-E)\mathcal{H}})$, where $\pi(T^f) = \tau(f)$.

Moreover, note that in the matrix decomposition of T^f with respect to $E\mathcal{H}$ and $(I - E)\mathcal{H}$, the off diagonal entries are compact (see (3.1.1)). It follows that for f in $C(X)$,

$$
\begin{aligned}
p_*(\tau_1 \sqcup \tau_2)(f) &= \tau_1(f|_{X_1}) \oplus \tau_2(f|_{X_2}) \\
&= \pi(ET^f|_{E\mathcal{H}}) \oplus \pi((I-E)T^f|_{(I-E)\mathcal{H}}) \\
&= \pi(ET^f|_{E\mathcal{H}} \oplus (I-E)T^f|_{(I-E)\mathcal{H}}) \\
&= \pi(T^f) = \tau(f).
\end{aligned}
\tag{3.2.2}
$$

What we have done is to simultaneously obtain an orthogonal direct sum decomposition of the essentially normal operators T^f modulo the compacts.

For any two extensions $[\tau_{X_1}]$ and $[\tau_{X_2}]$ in $\mathrm{Ext}(X_1)$ and $\mathrm{Ext}(X_2)$ respectively, define the two maps $\beta : \mathrm{Ext}(X_1) \oplus \mathrm{Ext}(X_2) \to \mathrm{Ext}(X)$ by

$$
\beta([\tau_{X_1}], [\tau_{X_2}]) = (i_{X_1, X})_*[\tau_{X_1}] + (i_{X_2, X})_*[\tau_{X_2}]
\tag{3.2.3}
$$

and $\lambda : \mathrm{Ext}(X_1) \oplus \mathrm{Ext}(X_2) \to \mathrm{Ext}(X_1 \sqcup X_2)$ by

$$
\lambda([\tau_{X_1}], [\tau_{X_2}]) = [\tau_{X_1} \sqcup \tau_{X_2}].
\tag{3.2.4}
$$

Sometimes, we denote $[\tau_{X_1} \sqcup \tau_{X_2}]$ by $[\tau_{X_1}] \sqcup [\tau_{X_2}]$. We will see later that the class of $\tau_{X_1} \sqcup \tau_{X_2}$ depends only on the class of τ_{X_1} and τ_{X_2}, and so does λ. What we have almost shown in the preceding lemma is that $p_* \lambda = \beta$. Now we make the following definition. Recall first that a *closed cover* of X is a pair $\{X_1, X_2\}$ of closed non-empty subsets of X such that $X = X_1 \cup X_2$.

Definition 3.2.3

Let X be a compact Hausdorff space. An extension $[\tau_X]$ is said to *split with respect to a closed cover* $\{X_1, X_2\}$ *of* X if it is in the range of β or equivalently in the range of $p_* \lambda$.

The following proposition relates the Ext groups of disjoint union and its components.

Proposition 3.2.1

If $\{X_1, X_2\}$ *is a closed cover of a compact Hausdorff space X, then the natural map $p : X_1 \sqcup X_2 \to X$ is a continuous surjection and the operation $\sqcup$ induces the isomorphism $\lambda : \mathrm{Ext}(X_1) \oplus \mathrm{Ext}(X_2) \to \mathrm{Ext}(X_1 \sqcup X_2)$ given by (3.2.4).*

Proof If $[\tau]$ is an extension in $\mathrm{Ext}(X_1 \sqcup X_2)$, then as in Lemma 3.1.2 (with orthogonal projections $e = \tau(\chi_{X_1})$ and $1 - e = \tau(\chi_{X_2})$), we obtain two extensions $[\tau_1] = [\tau_e]$ and $[\tau_2] = [\tau_{1-e}]$, which depend only on the projection e. Moreover, $p_*[\tau_1 \sqcup \tau_2] = [\tau]$ as observed in (3.2.2). Define $\mu : \mathrm{Ext}(X_1 \sqcup X_2) \to \mathrm{Ext}(X_1) \oplus \mathrm{Ext}(X_2)$ by $\mu([\tau]) = ([\tau_1], [\tau_2])$. It is clear that $\lambda \circ \mu = \mathrm{id}$ on $\mathrm{Ext}(X_1 \sqcup X_2)$.

To show that $\mu \circ \lambda = \mathrm{id}$ on $\mathrm{Ext}(X_1) \oplus \mathrm{Ext}(X_2)$, let E be the orthogonal projection of $\mathcal{H}_l \oplus \mathcal{H}_2$ onto $\mathcal{H}_1$ in $\mathcal{L}(\mathcal{H}_l \oplus \mathcal{H}_2)$. In view of Lemma 3.1.2, we may use this orthogonal projection to define the map μ. Finally, note that $\mu \circ \lambda$ is the identity map.   □

Remark 3.2.4 If X is the union of two disjoint closed sets X_1 and X_2, then X is actually equal to $X_1 \sqcup X_2$ and the map p_* is just the identity map. As λ is just seen to be an isomorphism, it follows that $p_* \lambda$ is an isomorphism. In other words, every extension in such a space splits.

Here is a particular instance in which Proposition 3.2.1 is applicable.

Corollary 3.2.1

Let X be a compact Hausdorff space. If X_0 is a finite subset of the set of isolated points of X, then Ext(X) is isomorphic to Ext(X\\X_0).

Proof   Note that $\{X_0, X \backslash X_0\}$ is a disjoint closed cover of X. By Proposition 3.2.1, $\mathrm{Ext}(X_0) \oplus \mathrm{Ext}(X \backslash X_0)$ is isomorphic to $\mathrm{Ext}(X)$. However, X_0, being a finite set, is homeomorphic to a subset of the real line. By the Weyl–von Neumann–Berg theorem, $\mathrm{Ext}(X_0)$ is trivial, and we conclude that $\mathrm{Ext}(X)$ is isomorphic to $\mathrm{Ext}(X \backslash X_0)$.   □

The conclusion of the forgoing corollary is true even if X_0 is not necessarily finite (for example, if $X = \overline{\mathbb{D}} \cup \{1 + 1/n : n \geqslant 1\}$). In the next section, we will see that τ_X may split even if X has a closed cover $\{X_1, X_2\}$ with $X_1 \cap X_2$ non-empty.

3.3   First Splitting Lemma

In this section, we prove the first splitting lemma, which is the first step in the iterated splitting argument.

Proposition 3.3.1 (First Splitting Lemma)

Let X and Y be compact metrizable spaces and $q : X \to Y$ be a continuous surjection. Let $[\tau]$ in Ext(X) be any extension. If $q_*[\tau]$ is trivial in Ext(Y), then the trivializing map $\mathcal{N}_0 : C(Y) \to \mathcal{L}(\mathcal{H})$ is induced by a spectral measure $\mathbb{E}$ on Y, that is,

$$\mathcal{N}_0(f) = \int_Y f \, d\mathbb{E}. \tag{3.3.5}$$

If C is any closed subset of Y such that $q|_{q^{-1}(\partial C)}$ is one-to-one, then the orthogonal projection $\pi(\mathbb{E}(C))$ commutes with $\mathrm{Im}\,\tau$, where ∂C denotes the boundary of C.

Proof   Assume that $q_*[\tau] = [\mathcal{N}_Y]$. The existence of spectral measure $\mathbb{E}$ on Y satisfying (3.3.5) follows from the spectral theorem. To see the remaining part, let C be a closed subset of Y such that $q|_{q^{-1}(\partial C)}$ is one-to-one. If f is any continuous function on X, then $f \circ q^{-1}$ is well defined and continuous on ∂C. Hence, by the Tietze extension theorem (see Theorem A.1.2), $f \circ q^{-1}$ extends to a continuous function f_1 on all of Y. Further, the function $g = f - f_1 \circ q$ is continuous on X and vanishes on $q^{-1}(\partial C)$. Then, for any $\epsilon > 0$, there exists a continuous function G on X vanishing in a neighborhood of $q^{-1}(\partial C)$ such that $\|g - G\|_\infty < \epsilon$ (see Corollary A.1.1). Any such function G is a sum of two continuous functions f_2 and f_3 such that

$$\mathrm{supp}\, f_2 \subseteq q^{-1}(\mathrm{int}\, C) \quad \text{and} \quad \mathrm{supp}\, f_3 \subseteq q^{-1}(Y \backslash C),$$

where supp and int denote support of a function and interior of a set respectively. Indeed, G is given by

$$G = f_2 \chi_{q^{-1}(\mathrm{int}\, C)} + f_3 \chi_{q^{-1}(Y \backslash C)}. \tag{3.3.6}$$

Hence, any function f in $C(X)$ can be approximated by a function of the form $f_1 \circ q + f_2 + f_3$. Therefore, it is enough to check that the projection $e = \pi(\mathbb{E}(C))$ commutes with each of $\tau(f_1 \circ q), \tau(f_2)$ and $\tau(f_3)$. Note that

$$\tau(f_1 \circ q) = (q_*\tau)(f_1) = \mathcal{N}_Y(f_1) = \pi \circ \mathcal{N}_0(f_1) = \pi\left(\int f_1 \, d\mathbb{E}\right). \tag{3.3.7}$$

As $\mathbb{E}(C) = \int \chi_C \, d\mathbb{E}$ commutes with $\int f_1 \, d\mathbb{E}$, we obtain

$$\pi(\mathbb{E}(C))\tau(f_1 \circ q) = \tau(f_1 \circ q)\pi(\mathbb{E}(C)).$$

Since the set $K = \operatorname{supp} f_2$ is a (compact) subset of X, by Urysohn's lemma (see Theorem A.1.1), one can find a continuous function h on Y, which takes value 1 on $q(K)$ and $\operatorname{supp} h$ is contained in $\operatorname{int} C$. Arguing as in (3.3.7), we obtain

$$\pi(\mathbb{E}(C))\tau(h \circ q) = \tau(h \circ q) = \tau(h \circ q)\pi(\mathbb{E}(C)). \tag{3.3.8}$$

Since $(h \circ q)f_2 = f_2 = f_2(h \circ q)$, we obtain

$$\pi(\mathbb{E}(C))\tau(f_2) = \pi(\mathbb{E}(C))\tau(h \circ q)\tau(f_2) \overset{(3.3.8)}{=} \tau(h \circ q)\tau(f_2) = \tau(f_2).$$

Similarly, one can see that $\tau(f_2)\pi(\mathbb{E}(C)) = \tau(f_2)$. The proof that $\tau(f_3)$ commutes with $\pi(\mathbb{E}(C))$ is identical, and the proof of the theorem is complete. □

Remark 3.3.1   As the trivializing map depends only on Y, $\pi(\mathbb{E}(C))$ is independent of the choice of q.

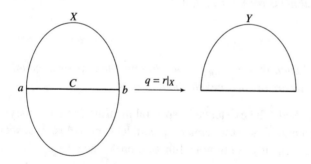

Figure 3.1 *Theta curve*

The first splitting lemma is illustrated with an example.

Example 3.3.2   Let X be the theta curve with middle edge C contained in the real axis $\mathbb{R}$ (see Figure 3.1) and let r be the reflection along $\mathbb{R}$, that is,

$$r(z) = \begin{cases} z & \text{if } \Im(z) \geqslant 0, \\ \bar{z} & \text{otherwise.} \end{cases}$$

Let $Y = r(X)$ and $q : X \to Y$ be the restriction $r|_X$ of r to X. Note that ∂C contains two distinct points, say a and b. Clearly, $q|_{q^{-1}(\partial C)} = q|_{\{a,b\}}$ is one-to-one. Hence, by Proposition 3.3.1, there exists a spectral measure $\mathbb{E}$ on Y such that the orthogonal projection $\pi(\mathbb{E}(C))$ commutes with $\operatorname{Im}\tau$ for any $[\tau] \in \operatorname{Ext}(X)$. ∎

The following is the first step towards the iterated splitting.

Corollary 3.3.1 (First Splitting Theorem)

Let X be a compact metric space and let A, B be closed subsets of X. If $X = A \cup B$ with $A \cap B = \{x_0\}$ for some $x_0 \in X$, then the map $\beta : \operatorname{Ext}(A) \oplus \operatorname{Ext}(B) \to \operatorname{Ext}(X)$, as defined in (3.2.3), is an isomorphism.

Proof Without loss of generality, we may assume that none of A and B are equal to $\{x_0\}$ (see Corollary 3.2.1). Define a continuous function $q : X \to [-1, 1]$ by

$$
q(x) = \begin{cases} d(x_0, x)/(d(x_0, x) + d(x, a)), & x \in A \\ -d(x_0, x)/(d(x_0, x) + d(x, b)), & x \in B \end{cases}
$$

where a in $A \setminus \{x_0\}$ is arbitrary and b in $B \setminus \{x_0\}$ is arbitrary. Let U be the component containing x_0. We may assume that U has more than a point (see Corollary 3.2.1). Let $C = q(X) \cap [0, 1]$. As $q < 0$ on B, we get $C = q(A)$. Now if we choose any a belonging to U, then $q(A)$ contains a connected set containing 0 and 1, and hence, $C = q(A)$ is equal to $[0, 1]$. Hence, we obtain the injectivity of q on $q^{-1}(\partial C)$ (since $\partial C = \{0, 1\}$).

If $[\tau]$ is any extension in $\operatorname{Ext}(X)$, then $q_*([\tau])$ is trivial. Indeed, by the Weyl–von Neumann–Berg theorem, $[q_*\tau] \in \operatorname{Ext}([-1, 1]) = \{0\}$. Let $Y = [-1, 1]$ and $C = [0, 1]$. As q is injective on $q^{-1}(\partial C) = \{x_0, a\}$, we may conclude from Proposition 3.3.1 that $\pi(\mathbb{E}(C))$ commutes with $\operatorname{Im}\tau$. An application of Lemma 3.2.1 with $e = \pi(\mathbb{E}(C))$ shows that $[\tau]$ splits into $[\tau_1]$ and $[\tau_2]$ with respect to some closed cover $\{X_1, X_2\}$ of X, that is,

$$
[\tau] = (i_{X_1,X})_*[\tau_1] + (i_{X_2,X})_*[\tau_2],
$$

where $i_{X_k,X} : X_k \to X$, $k = 1, 2$ is the inclusion map.

Let f_2 be the continuous function on X with $\operatorname{supp} f_2 \subseteq q^{-1}(\operatorname{int} C)$, as appearing in the proof of Proposition 3.3.1. Then,

$$
\tau(f_2) = \pi(\mathbb{E}(C))\tau(f_2),
$$

that is, the function f_2 is in $\ker \tau_{1-e}^d$. By Lemma 3.1.2(3), we have $f_2|_{X_2} = 0$. Moreover, $\operatorname{supp} f_2 \subseteq q^{-1}(\operatorname{int} C) = q^{-1}((0, 1)) = A \setminus \{x_0, a\}$. Thus, $f_2 = 0$ on $B \cup \{a\}$. We contend that $X_2 \subseteq B$. If possible, then there exists $x \in X_2$ such that $x \notin B$. As $x_0 \in B$, $x \neq x_0$. Moreover, we can choose $a \in A$ such that $x \neq a$ (see Remark 3.3.1). Let G be a continuous function vanishing in a neighborhood of $q^{-1}(\partial C) = \{x_0, a\}$ as given in (3.3.6) and note that

$$
f_2 = \chi_{q^{-1}(\operatorname{int} C)} G = \chi_{A \setminus \{x_0, a\}} G.
$$

Certainly, we can choose G such that $G(x) \neq 0$. Then, $f_2(x) \neq 0$, which is not possible as $x \in X_2$ and $f|_{X_2} = 0$. This proves that X_2 must be contained in B.

Similarly, if f_3 is a continuous function with $\operatorname{supp} f_3 \subseteq q^{-1}(Y \backslash \operatorname{int} C)$, as in the proof of Proposition 3.3.1, then f_3 is in $\ker \tau_e^d$ and $X_1 \subseteq A$. Since $i_{X_1,X} = i_{A,X} \circ i_{X_1,A}$ and $i_{X_2,X} = i_{B,X} \circ i_{X_2,B}$, it follows that

$$
\begin{aligned}
[\tau] &= (i_{A,X})_* ((i_{X_1,A})_*[\tau_1]) + (i_{B,X})_* ((i_{X_2,B})_*[\tau_2]) \\
&= (i_{A,X})_*[\tau_1'] + (i_{B,X})_*[\tau_2']
\end{aligned}
$$

for some $\tau_1' \in \operatorname{Ext}(A)$ and $\tau_2' \in \operatorname{Ext}(B)$. Therefore, β is surjective.

To check that β is injective, consider the map $r : X \to A$ given by

$$
r(x) = \begin{cases} x & x \in A, \\ x_0 & x \in B. \end{cases}
$$

Then r is continuous as A, B are closed and $A \cap B = \{x_0\}$. Thus, r is a retraction of X onto A. Similarly, let $s : X \to B$ be a retraction of X onto B. As β is surjective, one can define $(r_*, s_*) : \operatorname{Ext}(X) \to \operatorname{Ext}(A) \oplus \operatorname{Ext}(B)$ by

$$
\begin{aligned}
&((r_*, s_*)\beta)([\tau_1], [\tau_2]) \\
&= (r_*(i_{A,X})_*[\tau_1] + r_*(i_{B,X})_*[\tau_1], \ s_*(i_{A,X})_*[\tau_2] + s_*(i_{B,X})_*[\tau_2]).
\end{aligned}
$$

However, since $r \circ i_{A,X} = \operatorname{id}_A$ and $r \circ i_{B,X}$ is constant,

$$
r_*(i_{A,X})_*[\tau_1] = (r \circ i_{A,X})_*[\tau_1] = [\tau_1], \quad r_*(i_{B,X})_+[\tau_2] \text{ is trivial.}
$$

Similarly, $s_*(i_{A,X})_*[\tau_1]$ is trivial and $s_*(i_{B,X})_*[\tau_2] = [\tau_2]$. Therefore,

$$
(r_*, s_*)\beta = \operatorname{id}\big|_{\operatorname{Ext}(A) \oplus \operatorname{Ext}(B)},
$$

and hence, β is injective. $\qquad\qquad\qquad\qquad\qquad\qquad\qquad\qquad\qquad\qquad\qquad\qquad$ □

We illustrate the first splitting theorem with a simple example.

***Example* 3.3.3**   Let $\mathbb{D}$ denote the closed unit disc $\{z \in \mathbb{C} : |z| < 1\}$ in $\mathbb{C}$ and let $X = \overline{\mathbb{D}} \cup \{1 + 1/n : n \geqslant 1\}$. Let $A = \overline{\mathbb{D}}$ and $B = \{1\} \cup \{1 + 1/n : n \geqslant 1\}$. Note that by Corollary 2.7.3, $\operatorname{Ext}(B)$ is trivial. One may conclude from the first splitting theorem that $\operatorname{Ext}(X)$ is isomorphic to $\operatorname{Ext}(\overline{\mathbb{D}})$.   ■

We will see later that $\operatorname{Ext}(\overline{\mathbb{D}})$ is trivial (cf. Exercise 3.7.6).

3.4 Surjectivity of $(i_{A,X})_*$

Let X be a compact metric space and A be a closed subset of X. Then, the quotient space X/A is the one point compactification of the complement $X \backslash A$ of A in X (equivalently, X/A is the space obtained by collapsing A to a point). In particular, X/A is a compact Hausdorff space (see Lemma A.1.1). Moreover, if X is a metric space, then so is X/A (see Lemma A.1.3). We have seen in the previous chapter that $\operatorname{Ext}(X)$ is trivial if X is totally disconnected (Corollary 2.7.3).

In this section, we examine $\text{Ext}(X)$ in case the quotient space X/A is totally disconnected. In particular, we show that $\text{Ext}(X)$ is isomorphic to $\text{Ext}(A)$ whenever X/A is totally disconnected.

We begin with a preliminary lemma.

Lemma 3.4.1

Suppose (X,d) is a compact metric space and A is a closed subset of X. If the quotient space X/A is totally disconnected, then we have the following assertions:

(1) *The complement $X \backslash A$ of A in X can be written as the disjoint union of clopen sets X_n, $n \geqslant 1$ such that $\text{diam}\, X_n \to 0$ as $n \to \infty$.*

(2) *If a_n in ∂A is chosen such that $d(a_n, X_n) = d(\partial A, X_n)$ for every integer $n \geqslant 1$, then the map $r : X \to A$ defined by $r|_A = \text{id}$ and $r(x) = a_n$ for all x in X_n is a retraction,*

(3) *Let $f \in C(X)$ be such that $f|_A = 0$ and for every integer $n \geqslant 1$, let χ_n denote the characteristic function of X_n. Then, the series $\sum_{n=1}^{\infty} f\chi_n$ converges to f uniformly on X.*

Proof   Assume that the quotient space X/A is totally disconnected. Recall that a totally disconnected metric space has a basis of clopen sets. Let $q : X \to X/A$ be the quotient map and $\{U_n : n \geqslant 1\}$ be a decreasing neighborhood (or local) basis of $q(A)$ (as an element of X/A) consisting of clopen sets U_n in X/A. Let U_n^c denote the complement of U_n in the quotient space X/A. The set $q^{-1}(U_n^c)$ is homeomorphic to U_n^c. As U_n^c is a clopen set in a totally disconnected space, it follows that U_n^c is itself totally disconnected and hence, so is $q^{-1}(U_n^c)$. Therefore, there exists a finite cover of $q^{-1}(U_n^c)$ by clopen sets $F_{n,k}, 1 \leqslant k \leq m_n$, which can be chosen to have the additional properties

(i) $\text{diam}\, F_{n,k} \leqslant 1/n$ for all k,

(ii) $F_{n,k} \cap F_{n,k'} = \emptyset$ for $k \neq k'$.

An enumeration $\{X_n\}_{n \geqslant 1}$ of $\{F_{n,k} : 1 \leqslant k \leq m_n, n \geqslant 1\}$ would satisfy (1).

Let us check that r is sequentially continuous. The only difficult case is the case in which $a \in \partial A$ and for a sequence $\{x_n\}_{n \geqslant 1} \in X \backslash A$ such that $d(x_n, a) \to 0$ as $n \to \infty$. We must check that $d(r(x_n), a) \to 0$ as $n \to \infty$. For $n \geqslant 1$, let $x_n \in X_{k_n}$ for some subset $\{k_n\}_{n \geqslant 1}$ of positive integers and note that $r(x_n) = a_{k_n}$. Since $d(a_n, X_n) = d(\partial A, X_n)$ and $d(X_n, A) \to 0$ as $n \to \infty$, for every integer $n \geqslant 1$,

$$
\begin{aligned}
d(a_{k_n}, a) &\leqslant d(a_{k_n}, X_{k_n}) + d(X_{k_n}, a) \\
&\leqslant d(\partial A, X_{k_n}) + d(x_n, a) \\
&\leqslant d(a, x_n) + d(x_n, a),
\end{aligned}
$$

which converges to 0 as $n \to \infty$. This shows that the map r in (2) is continuous.

To see (3), let $f_n = -\sum_{k=1}^{n} f\chi_k$ for integers $n \geqslant 1$, $f = -\sum_{k=1}^{\infty} f\chi_k$, and apply Lemma B.1.1 to positive and negative parts of f.   □

Let us illustrate the preceding lemma with one simple example.

***Example* 3.4.1 (Example 3.3.3 continued)**   Note that the quotient space X/A may be identified with $\{(n+1)/n : n \geqslant 1\}$. In this case, X_n turns out to be the singleton set $\{(n+1)/n\}$, and one can choose a_n of Lemma 3.4.1(2) to be 1 for every positive integer n.   ∎

If A is clopen and X/A is totally disconnected, then it may be concluded from Proposition 3.2.1 and Corollary 2.7.3 that $Ext(A)$ and $Ext(X)$ are isomorphic. Hence, we may assume that A is not clopen in the remaining part of this section.

First, we prove the easier half of the main result, namely, Theorem 3.4.2.

Proposition 3.4.1

Suppose that A is a closed subset of a compact metrizable space X such that X/A is totally disconnected. If $i_{A,X} : A \to X$ denotes the inclusion map, then the induced map $(i_{A,X})_* : Ext(A) \to Ext(X)$ is injective.

Proof   Let $X_1, \ldots, X_n, \ldots$ be the clopen sets such that $X \backslash A = \sqcup_{n \geqslant 1} X_n$ and $r : X \to A$ be the retraction, as ensured by Lemma 3.4.1. As $r \circ i_{A,X} = \mathrm{id}|_A$, it follows from Corollary 2.7.1 that $r_*(i_{A,X})_* = \mathrm{id}|_{Ext(A)}$. This clearly yields the injectivity of $(i_{A,X})_*$.   □

It turns out that $(i_{A,X})_*$ appearing in the last proposition is actually surjective (Theorem 3.4.2). This relies heavily on the following simultaneous decomposition theorem modulo compact for several commuting self-adjoint operators with respect to a given family of mutually orthogonal projections. Recall that for two orthogonal projections P and Q in $\mathcal{L}(\mathcal{H})$, $Q \leqslant P$ if $P - Q$ is a positive operator, and Q *is of codimension k in P* if $P - Q$ is an operator of rank k. If k is finite, then we say that Q *is of finite codimension in P*.

Lemma 3.4.2

Let $T \in \mathcal{L}(\mathcal{H})$ be self-adjoint and let $\{F_n\}_{n \geqslant 1}$ be a family of mutually orthogonal projections in $\mathcal{L}(\mathcal{H})$ such that $\sum_{n=1}^{\infty} F_n$ converges to some $F \in \mathcal{L}(\mathcal{H})$. If there exists a bounded sequence $\{\lambda_n\}_{n \geqslant 1}$ of real scalars such that $TF_n - \lambda_n F_n$ is compact for all integers $n \geqslant 1$, then there exist mutually orthogonal projections $F'_n \leqslant F_n$ such that F'_n is of finite codimension in F_n and

(1) $T = \oplus_{n=0}^{\infty} T_n + K$ with respect to the decomposition $\mathcal{H} = \oplus_{n=0}^{\infty} F'_n \mathcal{H}$, where $F'_0 = I - \sum_{n=1}^{\infty} F'_n$ and K is compact,

(2)   there exists a compact operator K_n such that $T_n = \lambda_n I + K_n$ for every $n \geqslant 1$

More generally, if $\{T(m)\}_{m \geqslant 1}$ is any family of self-adjoint operators in $\mathcal{L}(\mathcal{H})$ with a bounded sequence $\{\lambda_n(m)\}_{m,n \geqslant 1}$ of real numbers such that $T(m)F_n - \lambda_n(m)F_n$ is compact for all integers $m, n \geqslant 1$, then (1) and (2) hold simultaneously for all integers $m \geqslant 1$.

Proof   Assume that $TF_n - \lambda_n F_n$ is compact for every integer $n \geqslant 1$. As T and F_n are self-adjoint, $[T, F_n] = TF_n - F_n T$ and $F_n TF_n - \lambda_n F_n$ are also compact. Let $\{F_n^{(k)}\}$ be a sequence of mutually orthogonal projections such that $F_n^{(k)} \leqslant F_n$ of codimension k and $F_n^{(k)} \to 0$ strongly as $k \to \infty$. Note that for any integer $n \geqslant 1$,

$$\| [T, F_n^{(k)}] \| \;\; = \;\; \| TF_n^{(k)} - F_n^{(k)} T \|$$

$$
\begin{aligned}
&= \quad \|TF_nF_n^{(k)} - F_n^{(k)}F_nT\| \\
&= \quad \|(TF_nF_n^{(k)} - F_nTF_n^{(k)}) + (F_nTF_n^{(k)} - F_n^{(k)}TF_n) \\
&+ \quad (F_n^{(k)}TF_n - F_n^{(k)}F_nT)\| \\
&\leqslant \quad \|[T,F_n]F_n^{(k)}\| + \|F_n^{(k)}TF_n - F_nTF_n^{(k)}\| + \|F_n^{(k)}[T,F_n]\|.
\end{aligned}
$$

Moreover, for some compact operator K, we have

$$
\begin{aligned}
F_nTF_n^{(k)} - F_n^{(k)}TF_n &= \quad F_nTF_nF_n^{(k)} - F_n^{(k)}F_nTF_n \\
&= \quad [F_nTF_n, F_n^{(k)}] \\
&= \quad [\lambda_nF_n + K, F_n^{(k)}] \\
&= \quad [\lambda_nF_n, F_n^{(k)}] + [K, F_n^{(k)}].
\end{aligned}
$$

It follows that $\|[T, F_n^{(k)}]\| \to 0$ as $k \to \infty$. Therefore, there exist mutually orthogonal projections $F_n' \leqslant F_n$ of finite codimension such that

$$
\|[T, F_n']\| \leqslant 1/n^2 \quad \text{for all integers } n \geqslant 1. \tag{3.4.9}
$$

Consider the matrix representation of T with respect to the decomposition,

$$
\mathcal{H} = \bigoplus_{n \geqslant 0} F_n'\mathcal{H}, \quad F_0' = I - \sum_{n=1}^{\infty} F_n'.
$$

All entries above the diagonal in this matrix representation are compact. Indeed, since $F_k'F_n' = 0$ and $[T, F_n']$ is compact,

$$
F_k'TF_n' = F_k'[T, F_n']F_n' \quad \text{is compact for all integers } n > k \geqslant 0.
$$

Similarly, all entries below the diagonal are also compact. By (3.4.9), we have $\|F_k'TF_n\| \leqslant 1/n^2, k \neq n$, and hence, the operator formed by these entries is compact. Therefore,

$$
T = \bigoplus_{n=0}^{\infty} T_n + \text{compact},
$$

where $T_n = F_n'T|_{F_n'\mathcal{H}}$. Moreover,

$$
T_n - \lambda_n I_{F_n'\mathcal{H}} \quad \text{is compact for all integers } n \geqslant 1.
$$

This completes the verification of (1) and (2).

If $\{T(m)\}_{m \geqslant 1}$ is a family of self-adjoint operators satisfying the hypothesis of the lemma, then

$$
\|[T(m), F_n^{(k)}]\| \to 0 \quad \text{as } k \to \infty \text{ for all integers } m, n \geqslant 1.
$$

Hence, for each n, there exists a projection F_n' of finite codimension in F_n such that

$$\|[T(m), F_n']\| \leq 1/n^2 \text{ for all integers } m \leq n.$$

In the decomposition of $T(m)$, all entries above the diagonal are compact. The difference is that, except for a finite number of entries with $n < m$, we have

$$\|[F_k' T(m) F_n']\| \leq 1/n^2 \text{ for all integers } m \leq n.$$

However, the operator formed by the entries above the diagonal is still compact. Since $T(m)$ is self-adjoint, the operator formed by entries below the diagonal is also compact. The proof in this case is completed as before.   □

We need one more observation in the proof of the main result of this section.

Lemma 3.4.3

Suppose that T is a self-adjoint operator in $\mathcal{L}(\mathcal{H})$ of the form $\lambda I + K_1$, where λ is a nonzero real scalar and $K_1 \in \mathcal{L}(\mathcal{H})$ is a compact operator. If, for $\epsilon > 0$, χ_ϵ denotes the characteristic function of $[\lambda - \epsilon, \lambda + \epsilon]$, then the following statements are true:

(1) $\chi_\epsilon(T) = I - K_2$ for some finite rank operator $K_2 \in \mathcal{L}(\mathcal{H})$,

(2) $\chi_\epsilon(T)\mathcal{H}$ is a reducing subspace for T,

(3) $T = \lambda\chi_\epsilon(T) + K_3 + T(I - \chi_\epsilon(T))$, where $K_3 \in \mathcal{L}(\mathcal{H})$ is a compact operator such that $\|K_3\| \leq \epsilon$.

Proof The first part follows from the assumption $T = \lambda I + K_1$, Theorem 1.3.3 and the functional calculus of T, whereas the second part is immediate from fact that T commutes with $\chi_\epsilon(T)$. To see the last part, note that $T\chi_\epsilon(T) = \lambda\chi_\epsilon(T) + K_1\chi_\epsilon(T)$. It follows that

$$T - \lambda\chi_\epsilon(T) - T(I - \chi_\epsilon(T)) = T\chi_\epsilon(T) - \lambda\chi_\epsilon(T) = K_1\chi_\epsilon(T).$$

Thus, (3) holds with $K_3 = K_1\chi_\epsilon(T)$.   □

The main result of this section stated below, shows that $\mathrm{Ext}(X)$ and $\mathrm{Ext}(A)$ coincide provided the quotient X/A is totally disconnected.

Theorem 3.4.2

If A is a closed subset of the compact metrizable space X such that X/A is totally disconnected, then the map $i_* : \mathrm{Ext}(A) \to \mathrm{Ext}(X)$ induced by the inclusion map $i : A \to X$ is an isomorphism.

Proof Assume that X/A is totally disconnected. Let $X_1, \ldots, X_n, \ldots$ be clopen sets such that $X \backslash A = \sqcup_{n \geq 1} X_n$, $\{a_n\}_{n \geq 1} \subseteq \partial A$ and $r : X \to A$ be the retraction map, as ensured by Lemma 3.4.1. In view of Proposition 3.4.1, it suffices to show that $i_* r_* = \mathrm{id}|_{\mathrm{Ext}(X)}$. Fix a $*$-monomorphism $\tau : C(X) \to \mathcal{Q}(\mathcal{H})$. We claim that there exist mutually orthogonal projections $\{E_n\}_{n \geq 1}$ such that $\pi(E_n) = \tau(\chi_n)$, where χ_n is the characteristic function of X_n, $E_0 = I - \sum_{n=1}^{\infty} E_n$, and

(1) $\tau(g \circ r) = \tau(g \circ r)\pi(E_0) + \pi(\sum_{n=1}^{\infty} g(a_n)E_n)$ for all $g \in C(A)$,

(2) $\tau' : C(A) \to Q(E_0\mathcal{H})$ given by $\tau'(g) = \tau(g \circ r)\pi(E_0)$ is a *-monomorphism.

To establish (1), it is enough to find mutually orthogonal projections $\{F_n\}_{n \geqslant 1}$ such that (1) holds for a sequence $\{g_m\}_{m \geqslant 1}$ dense in the space $C_{\mathbb{R}}(A)$ of real-valued continuous functions on A. Choose a commuting family of self-adjoint operators H^{g_m} such that

$$\pi(H^{g_m}) = \tau(g_m \circ r), \quad m \geqslant 1. \tag{3.4.10}$$

Since $\tau(\chi_n)$ is a family of commuting orthogonal projections in $Q(\mathcal{H})$, by Corollary 2.7.4, there exist mutually orthogonal projections F_n such that $\pi(F_n) = \tau(\chi_n)$. As $r|_{X_n} = a_n$, we further obtain

$$\pi(H^{g_m}F_n - g_m(a_n)F_n) = \tau((g_m \circ r)\chi_n - g_m(a_n)\chi_n) = 0,$$

so that $H_m^g F_n - g_m(a_n)F_n$ is compact for all $m, n \geqslant 1$. Thus, Lemma 3.4.2 is applicable, and we get orthogonal projections $F_n' \leqslant F_n$ such that F_n is of finite codimension in F_n', and

$$H^{g_m} = \bigoplus_{n \geqslant 0} H_n^{g_m} + K_m, \quad m \geqslant 1,$$

where $H_n^{g_m} = F_n' H^{g_m}|_{F_n'\mathcal{H}}$ and K_m is a compact operator. Further,

$$H_n^{g_m} = g_m(a_n)I_n + K_{mn}, \quad m, n \geqslant 1, \tag{3.4.11}$$

where K_{mn} is a compact operator and I_n is the identity operator on $F_n'\mathcal{H}$. Now, apply Lemma 3.4.3 to obtain an orthogonal projection $F_n^{(m)}$ of finite codimension in F_n' such that

$$H_n^{g_m} = g_m(a_n)F_n^{(m)} + K_n^{(m)} + H_n^{g_m}(I - F_n^{(m)})$$

with $K_n^{(m)}$ compact and $\|K_n^{(m)}\| \leqslant 1/n^2$ for all integers $m, n \geqslant 1$. For each n, let F_n'' be the projection on the intersection of the ranges of $F_n^{(1)}, \ldots, F_n^{(n)}$. Then, F_n'' is of finite codimension in F_n (see Exercise 3.7.8(iii)) and $H_n^{g_m}$ admits the decomposition

$$H_n^{g_m} = H_m' \oplus g_m(a_n)F_n'' + K_n^{(m)}, \quad n \geqslant 1 \tag{3.4.12}$$

with respect to the decomposition $\mathcal{H} = (I - \sum_{n \geqslant 1} F_n'')\mathcal{H} \oplus \sum_{n \geqslant 1} F_n''\mathcal{H}$, where H_m' is a finite dimensional operator. Finally, for every integer $n \geqslant 1$, let E_n be any projection of codimension 1 in F_n'' and let $E_0 = I - \sum_{n \geqslant 1} E_n$. Then, $\pi(E_n) = \pi(F_n) = \tau(\chi_n)$ for every integer $n \geqslant 1$. Further, note that by (3.4.12) and the compactness of $\sum_{n=1}^{\infty} K_n^{(m)}$,

$$\tau(g_m \circ r)\pi(E_0) = \tau(g_m \circ r)\pi\left(I - \sum_{n=1}^{\infty} E_n\right)$$

$$= \tau(g_m \circ r) - \pi(H^{g_m})\pi\left(\sum_{n=1}^{\infty} E_n\right)$$

$$= \tau(g_m \circ r) - \pi\Big(\sum_{n=1}^{\infty} (H'_m \oplus g_m(a_n)) E_n\Big)$$

$$= \tau(g_m \circ r) - \pi\Big(\sum_{n=1}^{\infty} g_m(a_n) E_n\Big),$$

and hence, (1) is verified for all g_m in $C_{\mathbb{R}}(A)$.

We now verify (2). First, observe that $\tau(g \circ r)$ commutes with $\pi(E_0)$ for $g = g_m$ by (3.4.10) and (3.4.11), and hence, for all g. It follows that

$$\tau(f \circ r)\pi(E_0)\tau(g \circ r)\pi(E_0) = \tau(f \circ r)\tau(g \circ r)\pi(E_0)^2$$
$$= \tau((f \circ r)(g \circ r))\pi(E_0),$$

and therefore, τ' is a $*$-homomorphism. We claim that τ' is injective. Note that for any $k \in \ker \tau'$, by (1),

$$\tau(k \circ r) = \pi\Big(\sum_{n=1}^{\infty} k(a_n) E_n\Big) = \sum_{n=1}^{\infty} k(a_n)\pi(E_n), \tag{3.4.13}$$

where the series on right-hand side converges in the Calkin algebra. It follows that $|k(a_n)| \|\pi(E_n)\|_Q \to 0$ as $n \to \infty$ (see (B.1.2)). As $\pi(E_n) = \tau(\chi_n)$ is a non-zero projection (otherwise X_n would be empty), $\|\pi(E_n)\|_Q = 1$ (see Exercise 3.7.7) and hence, $k(a_n) \to 0$ as $n \to \infty$. It is easy to see that $f = \sum_{n=1}^{\infty} k(a_n)\chi_n$ and hence, $h = k \circ r - f$ is continuous. However, $h = 0$ outside A, so $h = 0$ on ∂A. As $f = 0$ on A, it follows that k vanishes on ∂A and $k(a_n) = 0$ for all n. By (3.4.13), we must have $\tau(k \circ r) = 0$. As τ is injective, $k \circ r = 0$, and hence, $k = 0$. This completes the proof of the claim.

In view of the unique decomposition $f = (f - f \circ r) + f \circ r$, we have

$$C(X) = \mathcal{Z}(A) \oplus r^* C(A), \tag{3.4.14}$$

where $\oplus$ is understood as the linear direct sum of the ideal $\mathcal{Z}(A)$ of functions in $C(X)$ vanishing on A and the subalgebra $r^* C(A) = \{g \circ r : g \in C(A)\}$. As X_n is totally disconnected, by Corollary 2.7.3, $\mathrm{Ext}(X_n)$ is trivial. Thus, there exists a trivializer $v_n : C(X_n) \to \mathcal{L}(E_n \mathcal{H})$ such that $\tau|_{C(X_n)} = \pi \circ v_n$. Let μ_1 be the $*$-monomorphism from $\mathcal{Z}(A)$ into $Q((I - E_0)\mathcal{H})$ given by $\mu_1(f) = \sum_{n=1}^{\infty} v_n(f|_{X_n})$. Note that

$$\pi \circ \mu_1(f) = \sum_{n=1}^{\infty} \tau(f\chi_n). \tag{3.4.15}$$

Define the map $\mu_0 : r^* C(A) \to \mathcal{L}((I - E_0)\mathcal{H})$ by

$$\mu_0(g \circ r) = \sum_{n=1}^{\infty} g(a_n) E_n, \quad g \in C(A). \tag{3.4.16}$$

Further, define $\mu : C(X) \to \mathcal{L}(I - E_0)\mathcal{H}$ by

$$\mu(f) = \mu_1(f - f \circ r) + \mu_0(f|_A \circ r), \quad f \in C(X).$$

Note that $\pi(0+\mu_1) = \tau|_{\mathcal{Z}(A)}$, where 0 is the zero map in $\mathcal{L}(E_0\mathcal{H})$. Clearly, the map μ is $*$-linear. In order that it be a homomorphism, it is necessary and sufficient that

$$\mu_1((g \circ r)f) = \mu_0(g \circ r)\mu_1(f), \quad f \in \mathcal{Z}(A), \; g \in C(A). \tag{3.4.17}$$

Indeed, since μ_0 and μ_1 are homomorphisms, by (3.4.14), μ is a homomorphism if and only if $\mu(fg) = \mu(f)\mu(g)$ for all $f \in \mathcal{Z}(A)$ and $g \in r^*C(A)$, which is easily seen to be equivalent to (3.4.17). To check (3.4.17), let f in $\mathcal{Z}(A)$, and note that by Lemma 3.4.1(3), the expansion $f = \sum_{n=1}^{\infty} f\chi_n$ converges in the sup norm, so by linearity and continuity of μ, it is enough to verify the relation (3.4.17) for functions f satisfying $f\chi_n = f$. Note that $(g \circ r)f = g(a_n)f$, and hence,

$$\mu_1((g \circ r)f) = g(a_n)\mu_1(f).$$

On the other hand, $\mu_1(f) = \mu_1(\chi_n f) = \mu_1(\chi_n)\mu_1(f) = E_n\mu_1(f)$. Hence,

$$\mu(g \circ r)\mu_1(f) = \mu_0(g \circ r)E_n\mu_1(f) \stackrel{(3.4.16)}{=} g(a_n)E_n\mu_1(f) = g(a_n)\mu_1(f).$$

This completes the verification of (3.4.17). After adding a trivial element on both sides, if necessary, we may assume that $\pi \circ \mu$ defines a trivial element (see the proof of Corollary 2.7.1). Thus, it suffices to show that τ is equivalent to $\tau \circ r^* \circ i^* + \pi \circ \mu$, that is, there exists a unitary V such that

$$\tau(f) = \pi(V)^*(\tau(f \circ r) \oplus \pi \circ \mu(f))\pi(V)$$

for every $f \in C(X)$. By (1), (3.4.15), and (3.4.16), we have

$$\tau(f) = 0 \oplus \pi \circ \mu_1(f), \quad f \in \mathcal{Z}(A),$$
$$\tau(g \circ r) = \tau(g \circ r)\pi(E_0) \oplus \pi \circ \mu_0(g \circ r), \quad g \in C(A).$$

It follows that for any $f \in C(X)$,

$$
\begin{aligned}
\tau(f) &= \tau(f - f \circ r) + \tau(f \circ r) \\
&= 0 \oplus \pi \circ \mu_1(f - f \circ r) + \tau(f \circ r)\pi(E_0) \oplus \pi \circ \mu_0(f \circ r) \\
&= \tau'(f) \oplus \pi \circ \mu(f).
\end{aligned}
$$

By (1) and (2), $[r_*\tau]$ is equivalent to $[\tau']$ (implemented by the unitary U), and hence, we obtain the desired equivalence implemented by the unitary $V = U \oplus I$, where I is the identity operator in $\mathcal{L}((I - E_0)\mathcal{H})$. □

Here is one concrete application of Theorem 3.4.2 to normal operators (see Exercise 3.7.10).

Corollary 3.4.1

If $N \in \mathcal{L}(\mathcal{H})$ is a normal operator, then $Ext(\sigma(N))$ is isomorphic to $Ext(\sigma_e(N))$.

Proof Let $N \in \mathcal{L}(\mathcal{H})$ be a normal operator. In view of Theorem 3.4.2, it suffices to verify that $\sigma(N)/\sigma_e(N)$ is totally disconnected. By Corollary 1.3.7, $\sigma(N)\backslash\sigma_e(N)$ is totally disconnected.

As the quotient space $\sigma(N)/\sigma_e(N)$ is obtained by identifying $\sigma_e(N)$ to a single point and attaching it to $\sigma(N)\backslash\sigma_e(N)$, $\sigma(N)/\sigma_e(N)$ is totally disconnected and the proof is over. $\qquad\square$

The conclusion of Theorem 3.4.2 may be rephrased by saying that the short sequence

$$\text{Ext}(A) \xrightarrow{i_*} \text{Ext}(X) \xrightarrow{0} \{0\} \tag{3.4.18}$$

is exact whenever X/A is totally disconnected. Even if X/A is not totally disconnected, one gets an exact sequence with the only change that the extreme right entry $\{0\}$ in the exact sequence (3.4.18) is replaced by $\text{Ext}(X/A)$. This will be shown in the next section.

3.5   Ext(A) → Ext(X) → Ext(X/A) is Exact

Let X be a compact metrizable space. Let A be a closed subset of X and let X/A be the quotient space. Consider the inclusion map $i : A \to X$ and the quotient map $q : X \to X/A$. Note that $q \circ i : A \to X/A$ is a constant map, and hence, $(q_* \circ i_*)([\tau_A]) = (q \circ i)_*([\tau_A])$ is always trivial for any $[\tau_A] \in \text{Ext}(A)$. Therefore, $\text{Im}\, i_* \subseteq \ker q_*$. This shows that the short sequence

$$\text{Ext}(A) \xrightarrow{i_*} \text{Ext}(X) \xrightarrow{q_*} \text{Ext}(X/A)$$

forms a *complex*. The natural question arises whether it is *exact*, that is, whether $\text{Im}\, i_* = \ker q_*$. The main result of this section provides an affirmative answer to this question. The desired inclusion $\ker q_* \subseteq \text{Im}\, i_*$ is a consequence of Proposition 3.5.1. The idea of the proof is to add enough orthogonal projections to $\text{Im}\,\tau$ and obtain a C^*-algebra $\mathcal{Z}$ such that part of its maximal ideal space is totally disconnected, so that Theorem 3.4.2 is applicable.

Proposition 3.5.1

Let X, Y be compact metrizable spaces, $q : X \to Y$ be a continuous surjection, B be a closed subset of Y such that $q|_{q^{-1}(Y\backslash B)}$ is injective and let $A = q^{-1}(B) \subseteq X$. Consider the commutative diagram

$$
\begin{array}{ccc}
A & \stackrel{i}{\hookrightarrow} & X \\
q' \downarrow & & q \downarrow \\
B & \stackrel{j}{\hookrightarrow} & Y
\end{array}
$$

where $q' = q|_A$ and $i : A \hookrightarrow X$, $j : B \hookrightarrow Y$ are the inclusion maps. Then,

$$\ker q_* \subseteq i_*(\ker q'_*).$$

Proof   Let $\tau : C(X) \to \mathcal{Q}(\mathcal{H})$ be any $*$-monomorphism such that $[\tau]$ is in $\ker q_*$, that is, $q_*([\tau]) = \mathcal{N}_Y$ is trivial. By Proposition 3.3.1, there exists a spectral measure $\mathbb{E}$ on Y such that

$$\tau(g \circ q) = \mathcal{N}_Y(g) = \pi\left(\int_Y g\,d\mathbb{E}\right), \quad g \in C(Y). \tag{3.5.19}$$

Let $\{U_n\}_{n \geqslant 1}$ be a basis of open sets U_n for $X \backslash A$ such that cl U_n is disjoint from A for all $n \geqslant 1$ and let $C_n = q(\overline{U}_n)$. Note that C_n in closed in Y. Let $\mathcal{Z}$ denote the C^*-algebra generated by Im τ and all orthogonal projections $e_n = \pi(E_n)$, where $E_n = \mathbb{E}(C_n)$. In view of $q^{-1}(C_n) \subseteq q^{-1}(Y \backslash B)$, Proposition 3.3.1 is applicable, and hence, $\mathcal{Z}$ is commutative. Thus, we have

$$
\begin{array}{ccc}
C(Y) \xrightarrow{q^*} & C(X) \xrightarrow{p^*} & C(\tilde{X}) \\
\tau \downarrow & & \uparrow \Gamma_{\mathcal{Z}} \\
\text{Im}\,\tau & \hookrightarrow & \mathcal{Z} \subseteq Q(\mathcal{H})
\end{array}
$$

where the map p^* is induced by a continuous surjection $p : \tilde{X} \to X$ and $\Gamma_{\mathcal{Z}}$ is the Gelfand map such that $\Gamma_{\mathcal{Z}} \circ \tau = p^*$, that is, $p_*([\Gamma_{\mathcal{Z}}^{-1}]) = [\tau]$.

Let $\tilde{A} = p^{-1}(A) \subseteq \tilde{X}$. We claim that

(1) $p : \tilde{A} \to A$ is a homeomorphism.

(2) $\tilde{X} \backslash \tilde{A}$ and $\tilde{X} / \tilde{A}$ are totally disconnected.

For any arbitrary but fixed $\tilde{x}$ in $\tilde{A}$, if $x = p(\tilde{x}) \notin \overline{U}_n$, then $y = q \circ p(\tilde{x}) \notin C_n$. By Urysohn's lemma (see Theorem A.1.1), there exists a function $g \in C(Y)$ with $g(y) = 1$ and $g = 0$ on C_n, and therefore, by (3.5.19),

$$
\tau(g \circ q)e_n = \mathcal{N}_Y(g)e_n = \pi\Big(\int_Y g \; d\mathbb{E} \Big)\pi(\mathbb{E}(C_n)) = \pi\Big(\int g \chi_{C_n} \; d\mathbb{E} \Big) = 0.
$$

On the other hand, $\Gamma_{\mathcal{Z}}^{-1}(\chi_n) = e_n$, where χ_n is the characteristic function of the clopen set $\tilde{C}_n = p^{-1}(C_n) \subseteq \tilde{X}$. It follows that

$$
0 = (\Gamma_{\mathcal{Z}}(\mathcal{N}_Y(g)e_n))(\tilde{x}) = ((g \circ q \circ p)\chi_n)(\tilde{x}) = g(y)\chi_n(\tilde{x}) = \chi_n(\tilde{x}),
$$

which implies that $\chi_n(\tilde{x}) = 0$. In particular, if $\tilde{x} \in \tilde{A}$, then $\chi_n(\tilde{x}) = 0$ for all n. As $\mathcal{Z}$ is generated by Im τ and the orthogonal projections e_n, it follows that Im $\Gamma_{\mathcal{Z}} \circ \tau = \text{Im}\,p^*$ must separate points of $\tilde{A}$, which completes the proof of (1).

Proof of (2) is similar. Let $\tilde{x}$ in $\tilde{X}$ be such that $p(\tilde{x}) \in U_n$, $y = q \circ p(\tilde{x}) \in \text{int}\,C_n$. We claim that $p^{-1}(U_n) \subseteq \tilde{C}_n$. Once again, by Urysohn's lemma, there exists a continuous function g on Y with $g(y) = 1$ and supp g contained in C_n. As $g = \chi_{C_n} g$, arguing as in the last paragraph, we have

$$
e_n \mathcal{N}_Y(g) = \pi\Big(\int_Y \chi_{C_n} g \; d\mathbb{E} \Big) = \pi\Big(\int_Y g \; d\mathbb{E} \Big) = \mathcal{N}_Y(g),
$$

which implies that $(e_n - 1)\mathcal{N}(g) = 0$, and hence,

$$
\Gamma_{\mathcal{Z}}((1 - e_n)\mathcal{N}(g))(\tilde{x}) = 0.
$$

However, $\Gamma_{\mathcal{Z}}(e_n) = \chi_n$ and $\Gamma_{\mathcal{Z}}(\mathcal{N}(g))(\tilde{x}) = g(y)$, and hence, $\chi_n(\tilde{x}) = 1$, that is, $\tilde{x} \in \tilde{C}_n$. Thus, $p(\tilde{x}) \in U_n$ implies that $p^{-1}(U_n) \subseteq \tilde{C}_n$, and the claim stands verified. Let $\tilde{x}_1, \tilde{x}_2$ be any two points in $\tilde{X}\backslash\tilde{A}$ such that $p(\tilde{x}_1) \neq p(\tilde{x}_2)$. As $\{U_n\}_{n\geqslant 1}$ forms a basis for $X\backslash A$, it follows that $p(\tilde{x}_1) \in U_n$ and $p(\tilde{x}_2) \notin \overline{U_n}$ for some n, and hence,

$$(\Gamma_{\mathcal{Z}}(e_n))(\tilde{x}_1) = 1 \quad \text{and} \quad (\Gamma_{\mathcal{Z}}(e_n))(\tilde{x}_2) = 0.$$

Thus, we have clopen sets $\tilde{C}_n$ such that $\tilde{x}_1 \in \tilde{C}_n$ and $\tilde{x}_2 \notin \tilde{C}_n$, and hence, $\tilde{x}_1$ and $\tilde{x}_2$ are distinguished by the clopen set $\tilde{C}_n$. If $p(\tilde{x}_1) = p(\tilde{x}_2)$, then for any any $*$-monomorphism $\tau : C(X) \to \mathcal{Q}(\mathcal{H})$,

$$\Gamma_{\mathcal{Z}}(\tau(f))(\tilde{x}_1) = f \circ p(\tilde{x}_1) = f \circ p(\tilde{x}_2) = \Gamma_{\mathcal{Z}}(\tau(f))(\tilde{x}_2), \quad f \in C(X)$$

and hence, $\operatorname{Im}\Gamma_{\mathcal{Z}} \circ \tau$ cannot distinguish these two points, so that either they are separated by a clopen set or they are equal. We have shown that $\tilde{X}\backslash\tilde{A}$ has a basis consisting of clopen sets $\tilde{C}_n$ and hence, it is totally disconnected. It follows that the quotient space $\tilde{X}/\tilde{A}$ is obtained by identifying $\tilde{A}$ to a single point is also totally disconnected. This completes the proof of (2).

Recall that if $\tilde{i} : \tilde{A} \to \tilde{X}$ is the inclusion map, then by Theorem 3.4.2, $\tilde{i}_*$ is a surjection. As the map $p : \tilde{X} \to X$ is a homeomorphism on $\tilde{A}$ (see (1)), it follows that $(\tilde{i} \circ (p|_{\tilde{A}})^{-1})_*$ is also surjective. Therefore, for any $[\tilde{\tau}] \in \operatorname{Ext}(\tilde{X})$,

$$[\tilde{\tau}] = (\tilde{i} \circ (p|_{\tilde{A}})^{-1})_*[\tau'],$$

for some $[\tau']$ in $\operatorname{Ext}(A)$. As $p_*([\tilde{\tau}]) = [\tau]$, $p \circ \tilde{i} \circ (p|_{\tilde{A}})^{-1} = i$, by letting $\tilde{\tau} = \Gamma_{\mathcal{Z}}^{-1}$, we obtain

$$[\tau] = p_*([\tilde{\tau}]) = (p_*\tilde{i}_*(p|_{\tilde{A}})_*^{-1})[\tau'] = i_*[\tau'].$$

Thus, it suffices to verify that $q'_*([\tau']) = 0$. If $\mathcal{Z}'$ is the commutative C^*-algebra generated by $\operatorname{Im}q_*[\tau] \subseteq \operatorname{Im}\tau$ and the orthogonal projections e_n, then $\mathcal{Z}'$ is isomorphic to $C(\tilde{Y})$. Thus, we obtain the commutative diagram

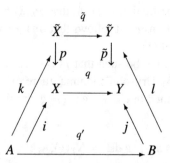

where $k = \tilde{i} \circ (p|_{\tilde{A}})^{-1}$, $\ell = \tilde{j} \circ (p|_{\tilde{B}})^{-1}$ with $\tilde{B} = \tilde{p}^{-1}(B)$ and $\tilde{j} : \tilde{B} \hookrightarrow \tilde{Y}$ being the obvious inclusion map. Note that $\tilde{q} \circ k = \ell \circ q'$. However, as $\mathcal{Z}'$ is contained in the C^*-algebra generated by the orthogonal projections in $\pi(\operatorname{Im}\mathbb{E})$, by Theorem 2.7.3, $\tilde{q}_*k_*([\tau'])$ is trivial. Hence, $(\ell \circ q')_*[\tau']$ is also trivial. We can again show that the map $\tilde{p}$ is a homeomorphism on $\tilde{p}^{-1}(B)$ and that

$\tilde{Y}\backslash\tilde{p}^{-1}(B)$ is totally disconnected, and apply Theorem 3.4.2 one more time to infer that $\tilde{j}_*$ is an isomorphism. As $\ell = \tilde{j} \circ (p|_{\tilde{B}})^{-1}$, it follows that l_* is injective, and hence, $q'_*[\tau']$ is trivial. $\quad\square$

We now state and prove the main theorem of this section.

Theorem 3.5.1

If A is a closed subset of the compact metrizable X, then

$$Ext(A) \overset{i_*}{\to} Ext(X) \overset{q_*}{\to} Ext(X/A)$$

is exact, where $i : A \to X$ is the inclusion map and $q : X \to X/A$ is the quotient map.

Proof   Assume that A is not singleton; if it is, the proof is evident. We have already seen the inclusion $Im\, i_* \subseteq \ker q_*$ in the beginning of this section. The other inclusion may be deduced from Proposition 3.5.1 applied to the quotient space $Y = X/A$ and $B = q(A)$ (the closed subset of X/A containing all points with more than one preimage under q in X). $\quad\square$

$$X = A \times B$$

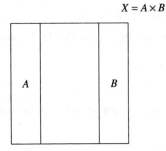

Figure 3.2 *Pictorial representation of the situation of Corollary 3.5.1*

We conclude this section with a consequence of Theorem 3.5.1.

Corollary 3.5.1

Let X_1 be a closed subset of $[0,1]$, X_2 be a compact metrizable space, and $X = X_1 \times X_2$ (with sup product metric). If

$$A = (X_1 \cap [0, 1/2]) \times X_2, \quad B = (X_1 \cap [1/2, 1]) \times X_2$$

(see Figure 3.2). then any $[\tau]$ in $Ext(X)$ splits with respect to A and B.

Proof   If $1/2 \notin X_1$, then $X_1 \times X_2 = A \sqcup B$, and hence, the desired conclusion follows from Proposition 3.2.1. Otherwise, define $r : X \to B$ by

$$r(x_1, x_2) = \begin{cases} (x_1, x_2) & \text{if } x_1 \geqslant 1/2 \\ (1/2, x_2) & \text{if } x_1 \leqslant 1/2. \end{cases}$$

Clearly, r is a continuous surjection. Let $i : A \hookrightarrow X$, $j : B \hookrightarrow X$ be inclusion maps, and let $q : X \to X/A$ be the quotient map. By Theorem 3.5.1, the complex

$$Ext(A) \overset{i_*}{\to} Ext(X) \overset{q_*}{\to} Ext(X/A)$$

is exact. As $q \circ j \circ r = q$, for any $[\tau] \in \mathrm{Ext}(X)$,

$$q_*([\tau]) = q_* j_* r_*([\tau]) \text{ or } q_*([\tau]) - j_* r_*([\tau]) = 0,$$

so that $[\tau] = i_*[\eta] + j_*(r_*([\tau]))$ for some $[\eta] \in \mathrm{Ext}(A)$.   □

3.6 Mayer–Vietoris Sequence

Let X be a compact metrizable space. Let X_1 and X_2 be closed subsets of X such that $X_1 \cup X_2 = X$ and let $A = X_1 \cap X_2$. Let $\mathrm{Ext}_1(A)$ denote the group of invertible elements in $\mathrm{Ext}(A)$ and let $i_{A,X_k} : A \to X_k$, $i_{X_k,X} : X_k \to X$ for $k = 1,2$ be inclusion maps. Define the map $\alpha : \mathrm{Ext}_1(A) \to \mathrm{Ext}(X_1) \oplus \mathrm{Ext}(X_2)$ by

$$\alpha([\tau_A]) = [(i_{A,X_1})_*([\tau_A]), (i_{A,X_2})_*(-[\tau_A])], \tag{3.6.20}$$

where $-[\tau_A]$ denotes the additive inverse of $[\tau_A]$. Define the map $\beta : \mathrm{Ext}(X_1) \oplus \mathrm{Ext}(X_2) \to \mathrm{Ext}(X)$ by

$$\beta([\tau_1], [\tau_2]) = (i_{X_1,X})_*([\tau_1]) + (i_{X_2,X})_*([\tau_2]). \tag{3.6.21}$$

Note that $\beta \circ \alpha[\tau_A]$ is a trivial extension for any $\tau_A \in \mathrm{Ext}_1(A)$. In particular, the short sequence

$$\mathrm{Ext}_1(A) \overset{\alpha}{\to} \mathrm{Ext}(X_1) \oplus \mathrm{Ext}(X_2) \overset{\beta}{\to} \mathrm{Ext}(X) \tag{3.6.22}$$

forms a complex.

The following analog of Mayer–Vietoris sequence enables us to relate $\mathrm{Ext}(X)$ with $\mathrm{Ext}(X_1)$, $\mathrm{Ext}(X_2)$ and $\mathrm{Ext}(X_1 \cap X_2)$.

Theorem 3.6.1 (Mayer–Vietoris)

The complex given in (3.6.22) is exact, that is, $\mathrm{Im}\,\alpha = \ker \beta$, where α and β are given by (3.6.20) and (3.6.21) respectively.

Proof   It suffices to verify that $\ker \beta \subseteq \mathrm{Im}\,\alpha$. To see that, suppose $\beta([\tau_1],[\tau_2]) = 0$ for some $[\tau_1] \in \mathrm{Ext}(X_1)$ and $[\tau_2] \in \mathrm{Ext}(X_2)$. Consider the commutative diagram

$$
\begin{array}{ccc}
X_1 \sqcup X_2 & \overset{q}{\longrightarrow} & X \\
j \uparrow & & \uparrow i \\
A \sqcup A & \overset{q'}{\longrightarrow} & A
\end{array}
$$

where $i : A \hookrightarrow X$, $j : A \sqcup A \hookrightarrow X \sqcup X$ are inclusion maps, and $q : X_1 \sqcup X_2 \to X$, $q' : A \sqcup A \to A$ are natural continuous surjections. Recall that the map $\lambda : \mathrm{Ext}(X_1) \oplus \mathrm{Ext}(X_2) \to \mathrm{Ext}(X_1 \sqcup X_2)$ given by

$$\lambda([\tau_1],[\tau_2]) = [\tau_1] \sqcup [\tau_2]$$

satisfies $\beta = q_* \lambda$ (see Definition 3.2.1 and the discussion following (3.2.3) and (3.2.4)). It follows that

$$q_*([\tau_1] \sqcup [\tau_2]) = \beta([\tau_1],[\tau_2]) = 0.$$

However then Proposition 3.5.1 guarantees the existence of $[\tau_{A \sqcup A}]$ in $\text{Ext}(A \sqcup A)$ such that

$$j_*[\tau_{A \sqcup A}] = [\tau_1] \sqcup [\tau_2] \text{ and } q'_*[\tau_{A \sqcup A}] = 0. \tag{3.6.23}$$

By Proposition 3.2.1,

$$\tau_{A \sqcup A} = \tau'_A \sqcup \tau''_A \text{ for some } \tau'_A, \tau''_A \text{ in } \text{Ext}(A). \tag{3.6.24}$$

It follows from (3.6.23) that

$$0 = q'_*(\tau_{A \sqcup A}) = q'_*([\tau'_A] \sqcup [\tau''_A]) = [\tau'_A] + [\tau''_A].$$

Hence, $[\tau'_A]$ and $[\tau''_A]$ are invertible $*$-monomorphisms satisfying $-[\tau'_A] = [\tau''_A]$.

For notational simplicity, we identify X_1, X_2 as subsets of $X_1 \sqcup X_2$ (see the discussion prior to Definition 3.2.1). Let $i_k : A_k \hookrightarrow X \sqcup X$ be the natural inclusion maps with $A_k = A$, $k = 1, 2$. For f in $C(X_1 \sqcup X_2)$, we have

$$
\begin{aligned}
((i_{1*}\tau'_A) \sqcup (i_{2*}\tau''_A))(f) &= \tau'_A(f \circ i_1) \oplus N_{X_1}(f \circ i_1) + \tau''_A(f \circ i_2) \oplus N_{X_2}(f \circ i_2) \\
&= j_*(\tau'_A \sqcup \tau''_A)(f),
\end{aligned}
$$

where $[N_{X_j}]$ denotes the trivial element of $\text{Ext}(X_j)$, $j = 1, 2$. This together with (3.6.24) shows that

$$i_{1*}[\tau'_A] \sqcup i_{2*}[\tau''_A] = j_*([\tau'_A] \sqcup [\tau''_A]) = j_*[\tau_{A \sqcup A}]. \tag{3.6.25}$$

Let $\mu : \text{Ext}(X_1 \sqcup X_2) \to \text{Ext}(X_1) \oplus \text{Ext}(X_2)$ be the isomorphism given by

$$\mu([\tau_1] \sqcup [\tau_2]) = ([\tau_1], [\tau_2]),$$

as ensured by Proposition 3.2.1. Then by (3.6.20),

$$
\begin{aligned}
\alpha[\tau'_A] &= (i_{1*}[\tau'_A], i_{2*}(-[\tau'_A])) = (i_{1*}[\tau'_A], i_{2*}[\tau''_A]) \\
&= \mu(i_{1*}[\tau'_A] \sqcup i_{2*}[\tau''_A]) \overset{(3.6.25)}{=} \mu(j_*[\tau'_{A \sqcup A}]) \\
&\overset{(3.6.23)}{=} \mu([\tau_1] \sqcup [\tau_2]) = ([\tau_1], [\tau_2]).
\end{aligned}
$$

This shows that $\ker\beta \subseteq \text{Im}\,\alpha$, which completes the proof. $\qquad\square$

Here is a key principle in the deduction of the fact that $\text{Ext}(X)$ is a group.

Corollary 3.6.1

Let X be a compact metrizable space. If $\text{Ext}(X)$ is a group, then $\text{Ext}(B)$ is also a group for any closed subset B of X.

Proof Suppose that $\text{Ext}(X)$ is a group. Take $X_1 = B$, $X_2 = X$, and let $j_1 : B \hookrightarrow X$, $j_2 : X \hookrightarrow X$ be inclusion maps. By Theorem 3.6.1, the sequence

$$\text{Ext}_1(B) \overset{\alpha}{\to} \text{Ext}(B) \oplus \text{Ext}(X) \overset{\beta}{\to} \text{Ext}(X)$$

is exact. Let $[\tau]$ be any extension in $\text{Ext}(B)$. As $\text{Ext}(X)$ is a group, $j_{1*}[\tau]$ is invertible in $\text{Ext}(X)$. Moreover, as $j_2 \circ j_1 = j_1$, by (3.6.21),

$$\beta([\tau], -j_{i*}[\tau]) = j_{1*}([\tau]) + j_{2*}(-j_{i*}[\tau]) = 0.$$

In view of $\text{Im}\,\alpha = \ker\beta$, there exists $\tau' \in \text{Ext}_1(B)$ such that $\alpha([\tau']) = ([\tau], -j_{1*}[\tau])$. However, by (3.6.20),

$$\alpha([\tau']) = ((i_{B,B})_*[\tau'], (i_{B,X_2})_*(-[\tau'])) = ([\tau'], (i_{B,X_2})_*(-[\tau'])),$$

and hence, $[\tau] = [\tau']$ is invertible in $\text{Ext}(B)$.   □

Recall that any compact metric space is homeomorphic to a closed subset of $[0,1]^{\mathbb{N}}$ provided $[0,1]^{\mathbb{N}}$ is endowed with the product topology (see Proposition A.1.3). In view of Corollary 3.6.1, the task of showing that Ext is a group now reduces to showing that $\text{Ext}([0,1]^{\mathbb{N}})$ is a group. In the next chapter, we will see that $\text{Ext}([0,1]^{\mathbb{N}})$ is indeed trivial.

3.7   Notes and Exercises

- The treatment in this chapter follows closely the expositions of [25, 26, 27].

Exercises

Exercise 3.7.1   Show that the Toeplitz extension $\tau : C(\mathbb{T}) \to \mathcal{Q}(\mathcal{H})$ as induced by (1.6.27) does not split non-trivially with respect to any closed cover $\{X_1, X_2\}$ of $\mathbb{T}$.
Hint: Consult Exercise 2.8.23.

Exercise 3.7.2   Deduce Proposition 2.3.2 from Corollary 3.3.1.

Exercise 3.7.3   Show by an example that Corollary 3.3.1 no longer holds true if $A \cap B$ is doubleton.
Hint: Take A and B to be upper and lower unit semicircles.

Exercise 3.7.4   Show that in Example 3.4.1, if one replaces $\overline{\mathbb{D}}$ by any compact subset of the complex plane containing 1, we get the same description for X/A (up to homeomorphism) and X_n.

Exercise 3.7.5 ([24])   Let U be as defined in Exercise 2.8.11 and U_+ be the unilateral shift on $\ell^2(\mathbb{N})$. For $n \in \mathbb{Z}$, define $U^{(n)}$ by

$$U^{(n)} = \begin{cases} U_+^n & \text{if } n > 0, \\ U & \text{if } n = 0, \\ (U^*)^{-n} & \text{if } n < 0. \end{cases}$$

Verify the following assertions:

(i) For any sequence $s = \{s_n\}_{n \in \mathbb{N}}$ of integers, $T_s = \oplus_{n \in \mathbb{N}} \frac{1}{n} \left(I + U^{(s_{n+1} - s_n)} \right)$ defines a bounded linear essentially normal operator on $\oplus_{n \in \mathbb{N}} \mathcal{H}_n$, where $\mathcal{H}_0 = \ell^2(\mathbb{Z})$ and $\mathcal{H}_n = \ell^2(\mathbb{N})$ otherwise.

(ii) The essential spectrum of T is the "Hawaiian earring" X formed as the union of all the circles of radius $1/n$ centered at $(1/n, 0)$.

Conclude that $\mathrm{Ext}(X)$ is uncountable.

Hint: Note that $[T_s] \in \mathrm{Ext}(X)$ for every sequence s of integers, and for $s \neq t$, T_s is not essentially equivalent to T_t.

Exercise 3.7.6 Give an example of a non-normal essentially normal operator with $\sigma_e(T) = \overline{\mathbb{D}}$.

Exercise 3.7.7 Show that the norm of a non-zero self-adjoint projection in a C^*-algebra is equal to 1.

Hint: Use the C^*-identity.

Exercise 3.7.8 Let P, Q, R be orthogonal projections in $\mathcal{L}(\mathcal{H})$. Verify the following statements:

(i) If P is of finite codimension k in Q, then $\mathrm{ran}\, P \subseteq \mathrm{ran}\, Q$ and

$$k = \dim(\mathrm{ran}\, Q \ominus \mathrm{ran}\, P).$$

(ii) If P is of finite codimension in Q and Q is of finite codimension in R, then P is of finite codimension in R.

(iii) If P and Q are of finite codimension in R, then the orthogonal projection $\tilde{P}$ of $\mathcal{H}$ onto $P\mathcal{H} \cap Q\mathcal{H}$ is also of finite codimension in R.

Hint: To see the dimension formula in (i), decompose P and Q simultaneously with respect to $\mathcal{H} = \mathrm{ran}\, P \oplus (\mathrm{ran}\, Q \ominus \mathrm{ran}\, P) \oplus (\mathcal{H} \ominus \mathrm{ran}\, Q)$. The parts (ii) and (iii) can be deduced from (i) or on similar lines (see Lemma 1.5.2).

Exercise 3.7.9 Let X be a compact metrizable space and let A be a closed subset of X. Consider the ideal $\mathcal{Z}(A)$ of functions in $C(X)$ vanishing on A and the subalgebra $r^* C(A) = \{g \circ r : g \in C(A)\}$, where $r : X \to A$ is a retraction map. Show that the decomposition $C(X) = \mathcal{Z}(A) \oplus r^* C(A)$ is unique, where $\oplus$ is understood as the linear direct sum.

Hint: Note that $\mathcal{Z}(A) \cap r^* C(A) = \{0\}$.

Exercise 3.7.10 ([94]) Let $N_1, \ldots, N_k \in \mathcal{L}(\mathcal{H})$ be finitely many commuting normal operators. Let $\sigma(N)$ be the maximal ideal space of the unital C^*-algebra generated by $N_1, \ldots, N_k$ and let $\sigma_e(N)$ be the maximal ideal space of the unital C^*-algebra generated by $\pi(N_1), \ldots, \pi(N_k)$. Show that $\mathrm{Ext}(\sigma(N)) \cong \mathrm{Ext}(\sigma_e(N))$.

Hint: Argue as in the proof of Corollary 3.4.1.

4

Determination of Ext(X) as a Group for Planar Sets

The first splitting lemma allowed us to split every extension $[\tau]$ in Ext(X) with respect to the closed cover $\{A, B\}$ of X, where $X = A \cup B$ and $A \cap B = \{x_0\}$. However, we will actually need a stronger form of splitting, one that allows any extension $[\tau]$ in Ext(X) to split with respect to the closed cover $\{A, B\}$ of X such that $A \cap B$ is homeomorphic to a closed interval rather than a point. The precise statement is given at the end of first section (see Corollary 4.1.1).

4.1 Second Splitting Lemma

We begin with generalizations of Lemmas 2.2.1 and 2.2.2. The first one is attributed to von Neumann.

Lemma 4.1.1

Let $H \in \mathcal{L}(\mathcal{H})$, M be a finite dimensional subspace of $\mathcal{H}$ and $\epsilon > 0$. If H is a self-adjoint operator, then there exists a finite dimensional subspace $M' \supseteq M$ and a compact self-adjoint operator $K \in \mathcal{L}(\mathcal{H})$ such that $H + K$ is reduced by M' and $\|K\| < \epsilon$.

Proof   The proof is a slight modification of that of Lemma 2.2.1. Assume that H is a self-adjoint operator. Let $\{\Delta_i\}_{i=1}^{N}$ be a decomposition of $\sigma(H)$ into a finite number of disjoint intervals Δ_i of length less than ϵ and let

$$M' = \sum_{i=1}^{N} \mathbb{E}(\Delta_i)M,$$

where $\mathbb{E}$ is the spectral resolution of H. Clearly, $\mathcal{M}'$ contains $\mathcal{M}$. For $i = 1, \ldots, N$, let E_i denote the orthogonal projection of $\mathcal{H}$ onto $\mathbb{E}(\Delta_i)\mathcal{M}$, $E = \sum_{i=1}^{N} E_i$ and $K = -E^{\perp}HE - EHE^{\perp}$. Note that $H + K$ commutes with E. In fact,

$$E(H + K) = EHE = (H + K)E.$$

Moreover, since E is of finite rank, K is compact. To complete the proof, we must check that $\|E^{\perp}HE\| < \epsilon/2$. To see that, let λ_i be the mid-point of Δ_i. Then,

$$E^{\perp}HE = E^{\perp} \sum_{i=1}^{n} HE_i = E^{\perp} \sum_{i=1}^{n} (H - \lambda_i)E_i,$$

and hence, we obtain

$$\|E^{\perp}HE\| \leq \left\| \sum_{i=1}^{N} (H - \lambda_i)E_i \right\| \leq \max_{i=1}^{N} \|(H - \lambda_i)E_i\|$$

$$\leq \max_{i=1}^{N} \|(H - \lambda_i)\mathbb{E}(\Delta_i)\| < \epsilon/2.$$

This completes the proof of the lemma.   □

Before we extend the last lemma to several compact self-adjoint operators, we formally define n-diagonal operator matrices.

Definition 4.1.1

Let n be a non-negative integer and let $A_{i,j} \in \mathcal{L}(\mathcal{H})$ for every integer $i, j \geq 0$. An operator matrix $(A_{i,j})_{i,j \geq 0}$ is said to be *n-diagonal* if $A_{ij} = 0$ for integers $i, j \geq 0$ such that $|i - j| > n$.

Remark 4.1.2   Note that 0-diagonal operator matrix is diagonal. Let $n \geq 1$ be an integer. Let $A \in \mathcal{L}(H)$ be a self-adjoint operator with operator matrix decomposition $(A_{ij})_{i,j \geq 0}$ with respect to the decomposition $\mathcal{H} = \oplus_{k \geq 0}\mathcal{H}_k$. Then, $(A_{ij})_{i,j \geq 0}$ is n-diagonal if and only if

$$A\mathcal{H}_k \subseteq \mathcal{H}_0 \oplus \cdots \oplus \mathcal{H}_{k+n} \text{ for all integers } k \geq 0$$

(see Exercise 4.6.1).

The following lemma ensures a decomposition of compact self-adjoint operators into an orthogonal sum of n-diagonal operators modulo compacts.

Lemma 4.1.2

For any compact self-adjoint operators $H_0, H_1, \ldots$ in $\mathcal{L}(\mathcal{H})$, there exist compact self adjoint operators $K_0, K_1, \ldots$ in $\mathcal{L}(\mathcal{H})$ and an orthogonal decomposition $\mathcal{H} = \oplus_{k \geq 0}\mathcal{H}_k$ into finite dimensional subspaces $\mathcal{H}_0, \mathcal{H}_1, \ldots$ relative to which the operator matrix for $H_0 + K_0$ is diagonal and that for $H_n + K_n$ is $(n+1)$-diagonal, $n \geq 1$.

Proof   Fix an orthonormal basis $\{\varphi_{ij} \mid 0 < j < i < \infty\}$ for $\mathcal{H}$.

Step I

Apply Lemma 4.1.1 to H_0 with $M = \text{span}\{\varphi_{00}\}$ and $\epsilon = 1$ to get a finite dimensional subspace M_{00} containing φ_{00} and a compact self-adjoint operator K_{00} with $\|K_{00}\| < 1$ such that $H_0 + K_{00}$ is reduced by M_{00}.

Step II

Choose a finite dimensional subspace M_{10} containing $M_{00} + \varphi_{10}$ and compact operator K_{10} with $\|K_{10}\| < 1/4$ such that $(H_0 + K_{00}) + K_{10}$ is reduced by both M_{00} and M_{10} (apply Lemma 4.1.1 to $(H_0 + K_{00})|_{M_{00}^{\perp}}$). Similarly, choose a finite dimensional subspace M_{11} containing $M_{10} + \varphi_{11}$ and compact operator K_{11} with $\|K_{11}\| < 1/4$ such that $H_1 + K_{11}$ is reduced by M_{11}. Iteration of this procedure making n applications of Lemma 4.1.1 at the nth step produces finite dimensional subspaces M_{ij} and compact operators K_{ij}, $0 \leqslant j \leq i < \infty$ such that

(i) $\varphi_{ij} \in M_{ij}$.

(ii) $\|K_{ij}\| < 1/(i+1)^2$.

(iii) $H_n + \sum_{m=n}^{\infty} K_{mn}$ is reduced by M_{mn}, for all integers $m \geqslant n$.

(iv) $M_{ij} \subseteq M_{i,j+1}$ and $M_{ij} \subseteq M_{i+1,0}$

We now complete the proof. Note that the operator $K_n = \sum_{m=n}^{\infty} K_{mn}$ is the limit of compact operators in the operator norm; therefore, it is compact. We put $\mathcal{H}_0 = M_{0,0}$ and $\mathcal{H}_k = M_{k,0} \ominus M_{k-1,0}$ for integers $k \geqslant 1$. By (i) and (iv), $\mathcal{H} = \oplus_{k \geqslant 0} \mathcal{H}_k$, and, by (iii), this decomposition reduces $H_0 + K_0$. Moreover,

$$\mathcal{H}_k \subseteq M_{k,0} \subseteq M_{k+n,n} \subseteq M_{k+n+1,0} = \mathcal{H}_0 \oplus \cdots \oplus \mathcal{H}_{k+n+1},$$

and hence by (iii),

$$(H_n + K_n)\mathcal{H}_k \subseteq \mathcal{H}_0 \oplus \cdots \oplus \mathcal{H}_{k+n+1}$$

for all integers $k \geqslant 0$ and $n \geqslant 1$. As $H_n + K_n$ is self-adjoint, it follows that it is $(n+1)$ diagonal. $\square$

The following characterizes the compactness of the cross-commutator of a diagonal self-adjoint and n-diagonal self-adjoint operator with finite dimensional blocks.

Lemma 4.1.3

Let n be a nonnegative integer and let H, K be two self-adjoint operators in $\mathcal{L}(\mathcal{H})$. If the operator matrix of H is diagonal and that of K is n-diagonal with respect to the decomposition $\mathcal{H} = \oplus_{k \geqslant 0} \mathcal{H}_k$ of $\mathcal{H}$ into mutually orthogonal finite dimensional subspaces $\mathcal{H}_k$, then the operator matrix of $[H, K]$ is n-diagonal with respect to the same decomposition. Further, $[H, K]$ is compact if and only if

$$\lim_{k \to \infty} \|[H, K]|_{\mathcal{H}_k}\| = 0,$$

where $[H, K] = HK - KH$ *and* $\|[H, K]|_{\mathcal{H}_k}\| = \sup_{h \in \mathcal{H}, \|h\|=1} \|[H, K]h\|$ *(the restriction does not mean that* $\mathcal{H}_k$ *is invariant under* $[H, K]$*).*

Proof   For any integer $k \geqslant 0$, note that $[H, K]\mathcal{H}_k \subseteq \mathcal{H}_0 \oplus \cdots \oplus \mathcal{H}_{k+n}$. Since $[H, K]^* = -[H, K]$, it follows that $[H, K]$ is n-diagonal with respect to the decomposition $\mathcal{H} = \oplus_{k \geqslant 0} \mathcal{H}_k$. For $q = 0, \ldots, n$, let $M_q = \bigoplus_{k \geqslant 0} \mathcal{H}_{nk+q}$. Thus, $[H, K] = \oplus_{q=0}^{n-1}[H, K]|_{M_q}$. Hence, $[H, K]$ is compact if and only if $[H, K]|_{M_q}$ is compact for every $q = 0, \ldots, n-1$. However, each of the sum $[H, K]|_{M_q}$ is essentially an orthogonal sum, and hence, $[H, K]$ is compact if and only if $\|[H, K]|_{\mathcal{H}_k}\|$ tends to 0 as $k \to \infty$. $\square$

We now present the main result of this section.

Theorem 4.1.3 (Second Splitting Lemma)

For self-adjoint elements $h_0, h_1, \ldots$ *of* $Q(\mathcal{H})$ *such that* h_0 *commutes with all* h_n, *there exists* $c \in Q(\mathcal{H}), 0 \leqslant c \leqslant 1$ *such that* c *commutes with* h_n *for integers* $n \geqslant 0$ *and*

(a) $cf(h_0) = f(h_0)$ *for all continuous* f *vanishing on* $[1/2, \infty)$,

(b) $cf(h_0) = 0$ *for all continuous* f *vanishing on* $[-\infty, 1/2)$.

Proof   By Lemma 4.1.2, there exist self-adjoint operators $H_n \in \mathcal{L}(\mathcal{H})$ with $\pi(H_n) = h_n$ and decomposition $\mathcal{H} = \oplus_{k \geqslant 0} \mathcal{H}_k$ into finite dimensional subspaces $\mathcal{H}_0, \mathcal{H}_1, \ldots$ relative to which $H_0 = \oplus_{k \geqslant 0} H_{0k}$ is diagonal and H_n is $(n+1)$-diagonal, $n \geqslant 1$. We construct a sequence $\{\varphi_k\}_{k \geqslant 0}$ of continuous functions $\varphi_k : \mathbb{R} \to [0, 1]$ such that

(i) φ_k decreases to the characteristic function of $(-\infty, 1/2]$ and vanishes on $[2, \infty)$.

(ii) $\|\varphi_k - \varphi_{k+1}\|_\infty \to 0$ as $k \to \infty$,

(iii) $\lim_{k \to \infty} \|[\varphi_k(H_0), H_n]|_{\mathcal{H}_k}\| = 0$ for every integer $n \geqslant 1$.

To construct the sequence $\{\varphi_k\}_{k \geqslant 0}$ with desired properties, for every positive integer j, let $f_j : \mathbb{R} \to [0, 1]$ be the continuous map given by

$$f_j(x) = \begin{cases} 1 & \text{for } x \in (-\infty, 1/2], \\ 0 & \text{for } x \in (1/2 + 1/j, \infty), \\ j(\frac{1}{2} - x) + 1 & \text{otherwise.} \end{cases}$$

Then, $\{f_j\}_{j \geqslant 1}$ decreases to the characteristic function of $(-\infty, 1/2]$ and

$$\|f_j - f_{j+1}\|_\infty = \sup_{0 \leqslant x - 1/2 \leqslant 1/j} \|x - 1/2\| \to 0 \text{ as } j \to \infty.$$

As $[H_0, H_n]$ is compact by hypothesis, it follows that $[f(H_0), H_n]$ is compact for all continuous f, and hence, by Lemma 4.1.3, $\|[f(H_0), H_n]|_{\mathcal{H}_k}\| \to 0$ as $k \to \infty$. By induction on j, one may choose positive integers $N_1 < N_2 < \ldots$, such that for every $j \geqslant 1$,

$$\|[f_j(H_0), H_n]|_{\mathcal{H}_k}\| \leqslant \frac{1}{j} \quad \text{for } k \geqslant N_j \text{ and } n \leqslant j.$$

Then, the sequence $\{\varphi_k\}_{k\geqslant 0}$ defined by

$$\varphi_k = \begin{cases} f_1 & \text{if } k < N_1, \\ f_j & \text{if } N_j \leq k < N_{j+1} \text{ for some integer } j \geqslant 1 \end{cases}$$

has properties (i)–(iii).

We now check that $c = \pi(C)$ has properties (a) and (b), where

$$C = \bigoplus_{j=0}^{\infty} \varphi_j(H_{0j}).$$

Obviously, $0 \leqslant C \leqslant I$ and $[C, H_0]$ is compact (indeed C commutes with H_0). To see that $[C, H_n]$ is compact, note that

$$[C, H_n]|_{\mathcal{H}_k} = [C - \varphi_k(H_0), H_n]|_{\mathcal{H}_k} + [\varphi_k(H_0), H_n]|_{\mathcal{H}_k},$$

where the second term tends to zero by (iii); the first term is dominated in norm by

$$2\|H_n\| \sup\{\|\varphi_j - \varphi_k\|_\infty : |j - k| \leqslant n + 1\}$$

(as H_n is $(n+1)$-diagonal) which tends to zero by (ii). Moreover, if f vanishes on $[1/2, \infty)$, then $\varphi_k f = f$ for all $k \geqslant 0$, so that

$$\begin{aligned} Cf(H_0) &= \Big(\bigoplus_{j\geqslant 0}\varphi_j(H_{0j})\Big)\Big(\bigoplus_{j\geqslant 0}f(H_{0j})\Big) \\ &= \bigoplus_{j\geqslant 0}(\varphi_j f)(H_{0j}) \\ &= \bigoplus_{j\geqslant 0}f(H_{0j}) \\ &= f(H_0). \end{aligned}$$

If f vanishes on $(-\infty, \frac{1}{2}]$, then $\|\varphi_k f\|_\infty \to 0$ as $k \to \infty$, and therefore by the calculations above, $Cf(H_0)$ is compact. $\square$

We are now ready to present a useful corollary of the second splitting lemma (cf. Corollary 3.5.1).

Corollary 4.1.1 (Second Splitting Theorem)

Let I denote the unit interval $[0,1]$ in the real line and let $X \subseteq I \times I$ be a closed set containing $\{(1/2, y) : 0 \leqslant y \leqslant 1\}$. Consider the closed subsets A and B of X given by

$$A = X \cap ([0, 1/2] \times I) \text{ and } B = X \cap ([1/2, 1] \times I)$$

(see Figure 4.1). Then any $[\tau]$ in Ext(X) splits with respect to A and B.

$$[0,1] \times [0,1]$$

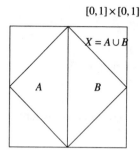

Figure 4.1 *Pictorial representation of the situation of second splitting theorem*

***Remark* 4.1.4**   The foregoing result shows that for any τ in $\mathrm{Ext}(X)$, there exist $\tau_A \in \mathrm{Ext}(A)$ and $\tau_B \in \mathrm{Ext}(X)$ such that

$$[\tau] = (i_{A,X})_*[\tau_A] + (i_{B,X})_*[\tau_B].$$

This representation in general is not unique (see Exercise 4.6.7).

Proof   Let $[\tau] \in \mathrm{Ext}(X)$. As $X \subseteq \mathbb{C}$ (up to a homeomorphic image), there exists an essentially normal operator N such that $\sigma_e(N) = X$ and $\tau(f) = f(\pi(N))$, where $\pi(N) = n = h_0 + ih_1$, n is normal and h_0, h_1 are self-adjoint in $Q(\mathcal{H})$. Note that $\sigma(h_0 + ih_1) \subseteq I \times I$. It follows from the projection property for the spectrum that

$$\sigma(1/2 + ih_1) \subseteq \{1/2\} \times I \subseteq X.$$

It follows that $\eta : C(X) \to Q(\mathcal{H})$ defined by

$$\eta(f) = f(1/2 + ih_1), \quad f \in C(X)$$

is a well defined $*$-homomorphism of $C(X)$. Although η is not necessarily a $*$-monomorphism, it can be perturbed by a trivial element to get a trivial element (see the proof of Corollary 2.7.1). Thus, we may assume that η is a $*$-monomorphism and in this case, $\tau' = \tau + \eta$ is a $*$-monomorphism equivalent to τ.

By Theorem 4.1.3, there exists c in $Q(\mathcal{H})$, $0 \leqslant c \leqslant 1$, commuting with h_0 and h_1 such that

$$cf(h_0) = 0 \quad \text{for all continuous functions } f \text{ vanishing on } (-\infty, 1/2],$$

$$cf(h_0) = f(h_0) \quad \text{for all continuous functions } f \text{ vanishing on } [1/2, \infty).$$

Consider the self-adjoint element e in $M_2(Q(\mathcal{H}))$ (set of 2×2 matrices with entries from $Q(\mathcal{H})$) given by

$$e = \begin{bmatrix} c & (c(1-c))^{1/2} \\ (c(1-c))^{1/2} & 1-c \end{bmatrix}.$$

A straightforward matrix computation shows that e is an idempotent, and hence, an orthogonal projection. Identifying $\tau'(f)$ with $\mathrm{diag}(\tau(f), \eta(f))$ in $M_2(Q(\mathcal{H}))$ for every $f \in C(X)$, we claim that e commutes with $\mathrm{Im}\,\tau'$. This amounts to verifying

$$(c(1-c))^{1/2}\eta(f) = \tau(f)(c(1-c))^{1/2} \text{ for all } f \in C(X). \tag{4.1.1}$$

To see this, define $f_1, f_2 : \mathbb{R} \to \mathbb{R}$ by

$$f_1(x) = \begin{cases} 0 & x \in [1/2, \infty), \\ x - 1/2 & x \in [0, 1/2], \\ -1/2 & x \in (-\infty, 0], \end{cases}$$

$$f_2(x) = \begin{cases} 1/2 & x \in [1, \infty), \\ x - 1/2 & x \in [1/2, 1], \\ 0 & x \in (-\infty, 1/2]. \end{cases}$$

Note that $cf_1(h_0) = f_1(h_0)$ and $cf_2(h_0) = 0$. In particular, $c((f_1 + f_2)(h_0)) = f_1(h_0)$. Note however that $(f_1 + f_2)(x) = x - 1/2$ for $x \in I$. By the functional calculus for h_0, $c(h_0 - 1/2) = f_1(h_0)$, and hence, $c^n(h_0 - 1/2) = f_1(h_0)$ for every positive integer n. It follows that

$$\begin{aligned} (h_0 + ih_1)c^n &= c^n(h_0 - 1/2 + ih_1 + 1/2) \\ &= c^n(h_0 - 1/2) + c^n(1/2 + ih_1) \\ &= f_1(h_0) + c^n(1/2 + ih_1). \end{aligned}$$

Thus, for any polynomial g, we have

$$(h_0 + ih_1)g(c) = g(1)f_1(h_0) + g(c)(1/2 + ih_1).$$

As the function $(x(1-x))^{1/2}$ can be approximated by polynomials vanishing at 1 uniformly on $[0, 1]$, it follows that

$$(h_0 + ih_1)(c(1-c))^{1/2} = (c(1-c))^{1/2}(1/2 + ih_1).$$

However, since $\tau(f) = f(h_0 + ih_1)$ and $\eta(f) = f(1/2 + ih_1)$, by arguing as earlier, we obtain

$$\tau(f)(c(1-c))^{1/2} = (c(1-c))^{1/2}\eta(f).$$

This completes the verification of (4.1.1), and hence, the claim stands verified. Therefore, the projection e commutes with $\mathrm{Im}\,\tau'$. Consider the $*$-monomorphism $\tau_e^d : C(X) \to Q(E\mathcal{H})$ given by

$$\tau_e^d(f) = e\tau'(f), \quad f \in C(X).$$

By Lemma 3.1.2, for a compact subset X_1 of X,

$$\ker \tau_e^d = \{f \in C(X) : f|_{X_1} = 0\}.$$

Let f in $C[0,1]$ be such that f vanishes precisely on $[0,1/2]$. We extend f continuously and boundedly to $\mathbb{R}$, so that f vanishes on $(-\infty, 1/2]$. Define a continuous function g on X by setting $g(x+iy) = f(x)$, $(x,y) \in X$. Note that

$$\tau'(g) = g(h_0 + ih_1) \oplus g(1/2 + ih_1) = f(h_0) \oplus f(1/2) = f(h_0) \oplus 0.$$

Since $cf(h_0) = 0$,

$$\tau_e^d(g) = e\tau'(g) = e(f(h_0) \oplus 0) = 0,$$

which shows that g is in $\ker \tau_e^d$. As the zero set of g equals A, it follows that $X_1 \subseteq A$. Similarly, it can be shown using the $*$-monomorphism τ_{1-e}^d that $X_2 \subseteq B$ (see Lemma 3.1.2 for the definition of τ_{1-e}^d). On the other hand, by Lemma 3.2.1,

$$[\tau] = (i_{X_1,X*})[\tau_1] + (i_{X_2,X*})[\tau_2]$$

for some $\tau_j \in \text{Ext}(X_j)$, $j = 1, 2$. As $i_{A,X} \circ i_{X_1,X} = i_{X_1,X}$ and $i_{B,X} \circ i_{X_2,X} = i_{X_2,X}$,

$$i_{A,X*}[i_{X_1,X*}\tau_1] + i_{B,X*}[i_{X_2,X*}\tau_2] = [\tau].$$

We have thus shown that τ splits with respect to A and B. $\square$

Example 4.1.5 (**Example 3.3.2 continued**)   If X denotes the theta curve (see Figure 3.1), then X can be decomposed as $A \cup B$, where A and B are homeomorphic to the unit circle and $A \cap B$ is homeomorphic to an interval. By the second splitting theorem, $\beta : \text{Ext}(A) \oplus \text{Ext}(B) \to \text{Ext}(X)$ is a surjection. However, $\text{Ext}(A)$ and $\text{Ext}(B)$ can be identified with a subset of $\mathbb{Z}$ (see Remark 2.3.6). Thus, $\text{Ext}(X)$ can be embedded into $\mathbb{Z} \oplus \mathbb{Z}$.   ∎

As a consequence of BDF theorem, one may deduce that $\text{Ext}(X) \cong \mathbb{Z} \oplus \mathbb{Z}$ (see Exercise 4.6.6).

4.2   Projective Limits and Iterated Splitting Lemma

In this section, we study the projective limits of compact Hausdorff topological spaces and that of semigroups. In particular, we relate $\varprojlim(\text{Ext}(X_n), p_{n*})$ and $\text{Ext}(\varprojlim(X_n, p_n))$. This will be used to obtain the iterated splitting lemma.

Definition 4.2.1

Consider a sequence $\{X_n\}_{n \geqslant 0}$ of non-empty topological spaces and also a sequence $\{p_n\}_{n \geqslant 0}$ of continuous maps $p_n : X_{n+1} \to X_n$. The *projective limit* (or *inverse limit*) $\varprojlim(X_n, p_n)$ of (X_n, p_n) is a topological space X together with *projection maps* (or *bonding maps*) $\pi_n : X \to X_n$ such that the following hold:

(1) $p_n \circ \pi_{n+1} = \pi_n$ for all $n \geqslant 0$;

(2) if X' is another topological space together with projection maps $\pi'_n : X' \to X_n$ satisfying $p_n \circ \pi'_{n+1} = \pi'_n$ for all $n \geqslant 0$, then the diagram

$$
\begin{array}{ccc}
 & & X \\
 & \overset{\varphi}{\nearrow} & \big\downarrow \pi_n \\
X' & \underset{\pi'_n}{\longrightarrow} & X_n
\end{array}
$$

is commutative, that is, there exists a unique continuous map $\varphi : X' \to X$ and $\pi'_n = \pi_n \circ \varphi$.

Remark 4.2.2   The projective limit of (X_n, p_n) always exists.   In fact, we may take the projective limit X to be

$$X = \varprojlim(X_n, p_n) = \{x = (x_n) \in \prod_{n \geqslant 0} X_n : p_n(x_{n+1}) = x_n \text{ for all } n \geqslant 0\}$$

(endowed with the product topology) and π_n to be defined by $\pi_n(x) = x_n$ for $x = (x_n)$. Clearly (1) in the aforementioned definition holds whereas (2) is satisfied with $\varphi(x') = (\pi'_n(x'))$. Note that the projection map $\pi_n : X \to X_n$ is always an open continuous mapping. In what follows, we will take the representation of X given here as the working definition for the projective limit.

We now list some elmentary properties of projective limits (refer to [78] for a comprehensive account on projective limits).

Lemma 4.2.1

Let $\{X_n\}_{n \geqslant 0}$ be a sequence of topological spaces and let $\{p_n\}_{n \geqslant 0}$ be a sequence of continuous maps $p_n : X_{n+1} \to X_n$. The following statements are true:

(1) If each X_n is Hausdorff, then the projective limit $\varprojlim(X_n, p_n)$ is closed and Hausdorff.

(2) If each X_n is compact, then $\varprojlim(X_n, p_n)$ is compact.

(3) If each X_n is a finite set, then $\varprojlim(X_n, p_n)$ is totally disconnected.

Proof   Note that the product of Hausdorff spaces (endowed with the product topology) is again Hausdorff. As the subspace of a Hausdorff space is Hausdorff, $\varprojlim(X_n, p_n)$ is Hausdorff. Moreover, as p_n and π_n are all continuous,

$$A_m = \{x \in \prod_{n \geqslant 0} X_n : p_m \circ \pi_{m+1}(x) = \pi_m(x)\}, \quad m \geqslant 0$$

is closed. However, $\varprojlim(X_n, p_n) = \cap_{m \geqslant 0} A_m$, and hence, $\varprojlim(X_n, p_n)$ is closed. Part (2) now follows from Tychonoff's theorem. Finally, part (3) follows from the fact that directed products and subspaces of totally disconnected spaces are totally disconnected.   □

***Remark* 4.2.3**   Under the hypotheses of the preceding lemma, the inverse limit $\varprojlim(X_n, p_n)$ is always non-empty (see [78, Theorems 108 and 111]). We discuss here a special case of this fact in which all topological spaces X_n are identical. To see that $\varprojlim(X_n, p_n)$ is non-empty, define $p_{n,m} : X_m \to X_n$ by

$$p_{n,m} = \begin{cases} p_n \circ p_{n+1} \circ \cdots \circ p_{m-1} & \text{if } n < m, \\ \text{id} & \text{if } n = m, \end{cases}$$

where id denotes the identity mapping. Set

$$G_n = \Big\{ x \in \prod_{n \geqslant 0} X_n : p_i(x_{i+1}) = x_i \text{ for } i \leqslant n \Big\},$$

and note that $G_{n+1} \subseteq G_n$ and $\varprojlim(X_n, p_n) = \cap_n G_n$. Further, if $x \in X_n$ for all $n \geqslant 0$, then

$$(p_{1,n+1}(x), p_{2,n+1}(x), \ldots, p_{n,n+1}(x), x, x, \ldots,) \in G_n.$$

As each G_n is compact, by Cantor's intersection theorem, $\varprojlim(X_n, p_n) \neq \emptyset$.

Note also that in general, the projective limit can be empty (see [76, Example 109] or Exercise 4.6.2).

Definition 4.2.1 extends naturally to groups (semigroups) and homomorphisms. In particular, if $p_n : X_{n+1} \to X_n$ is continuous, then $\varprojlim(\text{Ext}(X_n), p_{n*})$, defined as in Definition 4.2.1, is a semigroup, where $p_{n*} : \text{Ext}(X_{n+1}) \to \text{Ext}(X_n)$ is the homomorphism induced by p_n. Moreover, if each X_n is a compact Hausdorff space, then by Lemma 4.2.1, one can talk about $\text{Ext}(\varprojlim(X_n, p_n))$.

The following lemma connects projective limit of Ext and Ext of projective limit.

Lemma 4.2.2

Let $\{X_n\}_{n \geqslant 0}$ be a sequence of topological spaces and let $\{p_n\}_{n \geqslant 0}$ be a sequence of continuous maps $p_n : X_{n+1} \to X_n$. For $[\tau_X] \in \text{Ext}(\varprojlim(X_n, p_n))$, define $P : \text{Ext}(\varprojlim(X_n, p_n)) \to \varprojlim(\text{Ext}(X_n), p_{n*})$ by

$$P([\tau_X]) = ([\tau_X \circ \pi_n^*]),$$

where $\pi_n^* : C(X) \to C(X_n)$ is the pushback map of $\pi_n : X \to X_n$. Then, P is a well-defined homomorphism.

Proof   To see that $P([\tau_X])$ is in $\varprojlim(\text{Ext}(X_n), p_{n*})$, we need to check that

$$p_{n*}[\tau_X \circ \pi_{n+1}^*] = [\tau_X \circ \pi_n^*], \quad n \geqslant 0.$$

However, as $p_n \circ \pi_{n+1} = \pi_n$, for every $f \in C(X_n)$,

$$p_{n*}(\tau_X \circ \pi_{n+1}^*)(f) = \tau_X(f \circ p_n \circ \pi_{n+1}) = \tau_X(f \circ \pi_n) = \tau_X \circ \pi_n^*(f).$$

Moreover, if τ_X and τ_X' are equivalent, then $\tau_X \circ \pi_n^*$ and $\tau_X' \circ \pi_n^*$ are equivalent, so that P is a well-defined homomorphism. $\square$

In general, the map P, as given in Lemma 4.2.2, may have a non-trivial kernel (see [25, Page 976]). However, for our purposes, it is important to show that the map P is surjective. Although this is always true (see [47, Theorem 8]), we prove it under somewhat restrictive hypothesis.

Theorem 4.2.4

Given a sequence $\{X_n\}_{n \geqslant 0}$ of compact Hausdorff spaces X_n with continuous and surjective maps $p_n : X_{n+1} \to X_n$, the homomorphism $P : \mathrm{Ext}(\varprojlim(X_n, p_n)) \to \varprojlim(\mathrm{Ext}(X_n), p_{n*})$, as defined in Lemma 4.2.2, is surjective.

Proof Let $([\tau_n])_{n \geqslant 0}$ be in $\varprojlim(\mathrm{Ext}(X_n), p_{n*})$, that is,

$$p_{n*}[\tau_{n+1}] = [\tau_n], \quad n \geqslant 0.$$

We claim that some representative in $[\tau_n]$ can be chosen in such a way that the diagram

is commutative, that is,

$$\tau_{n+1} \circ p_n^* = \tau_n \quad \text{for all } n. \tag{4.2.2}$$

We proceed inductively. Let τ_1 be arbitrary. Given $p_{1*}[\tau_2] = [\tau_1]$, the map

$$g \to \tau_2(g \circ p_1), \quad g \in C(X_1)$$

(which is a $*$-monomorphism as p_1 is surjective) is equivalent to τ_1. Thus, there exists a unitary U such that $\tau_1(g) = \pi(U)^* \tau_2(g \circ p_1) \pi(U)$ for every $g \in C(X_1)$. Define

$$\tau_2'(f) = \pi(U)^* \tau_2(f) \pi(U),$$

and note that

$$(\tau_2' p_1^*)(g) = \tau_2'(g \circ p_1) = \pi(U)^* \tau_2(g \circ p_1) \pi(U) = \tau_1(g).$$

This yields the conclusion for $n = 1$. Assuming that $\tau_1', \ldots, \tau_n'$ are obtained with the desired property, one can repeat the argument given above to find τ_{n+1}' such that $\tau_{n+1}' \circ p_n^* = \tau_n'$. Here we need the assumption that p_n is surjective. For simplicity, we denote τ_n' by τ_n only. A particular consequence of this is the inclusion $\mathrm{Im}\,\tau_n \subseteq \mathrm{Im}\,\tau_{n+1}$ for every n.

Let $\mathcal{P}$ denote the algebra generated by $\bigcup_n \pi_n^*(C(X_n))$, where $\pi_n : \varprojlim(X_n, p_n) \to X_n$ is the projection. An application of the Stone–Weierstrass theorem shows that $\mathcal{P}$ is a dense

*-subalgebra of $C(X)$. Thus, the extensions $[\tau_n]$ determine a map $\tau : \mathcal{P} \to Q(\mathcal{H})$ given by

$$\tau(g \circ \pi_n) = \tau_n(g).$$

In view of $(g \circ p_n) \circ \pi_{n+1} = g \circ \pi_n$, one must check however that the definition of τ is independent of the representation. Indeed, as $p_{n*}\tau_{n+1} = \tau_n$,

$$\tau(g \circ p_n \circ \pi_{n+1}) = \tau_{n+1}(g \circ p_n) = \tau_n(g) = \tau(g \circ \pi_n), \quad g \in C(X_n).$$

Moreover, it is not difficult to see that if $f \circ \pi_m = g \circ \pi_n$ for some f, g and m, n, then $\tau_m(f) = \tau_n(g)$. Thus, the map τ is a well defined *-homomorphism. By construction, $P([\tau]) = (\pi_{n*}[\tau])_{n \geqslant 0} = ([\tau_n])_{n \geqslant 0}$.

To complete the proof, we must check that τ is injective. To see that, consider the C^*-algebra C generated by $\cup_n \operatorname{Im} \tau_n$. By (4.2.2), C is commutative. Consider the Gelfand transform $\Gamma : C \to C(Y)$, where Y is the maximal ideal space of C. We check that Y is equal to $X = \lim_{\leftarrow}(X_n, p_n)$ (up to homeomorphism). If $y \in Y$, then $y \circ \tau_n$ belongs to the maximal ideal space of $C(X_n)$, and hence, it is given by evaluation at a point, say, $\beta_n(y) \in X_n$. This allows us to define the map $\varphi : Y \to X$ by $\varphi(y) = (\beta_n(y))$. To see that $\varphi(y)$ belongs to the projective limit X, note that for all $g_n \in C(X_n)$,

$$
\begin{aligned}
g_n(p_n(\beta_{n+1}(y))) &= g_n \circ p_n(\beta_{n+1}(y)) \\
&= y \circ \tau_{n+1}(g_n \circ p_n) \\
&\overset{(4.2.2)}{=} y \circ \tau_n(g_n) \\
&= g_n(\beta_n(y)),
\end{aligned}
$$

and hence, $p_n(\beta_{n+1}(y)) = \beta_n(y)$ for all n. Thus, φ is well-defined. Clearly, φ is continuous and one-to-one. Moreover, if (x_n) belongs to the projective limit X, then $y : C \to \mathbb{C}$ given by

$$y \circ \tau_n(g_n) = g_n(x_n) \text{ with } g_n \in C(X_n)$$

extends to an element in the maximal ideal space of C. Moreover, $\varphi(y) = (x_n)$, so that X and Y are homeomorphic. For simplicity, we identify X with Y. Then, for every $g_n \in C(X_n)$ and every $y \in Y$,

$$
\begin{aligned}
(\Gamma \circ \tau(g_n \circ \pi_n))(y) &= (\Gamma \circ \tau_n(g_n))(y) \\
&= y(\tau_n(g_n)) \\
&= g_n(\beta_n(y)) \\
&= g_n \circ \pi_n(y).
\end{aligned}
$$

Thus, $\Gamma \circ \tau$ agrees with the identity map on a dense set. As Γ is isometric, so is τ on a dense set, and hence, τ extends to a $*$-monomorphism. $\qquad\qquad\qquad\qquad\qquad\qquad\square$

We end the discussion of inverse limits with the iterated splitting lemma. Let X be any compact metrizable space and $[\tau]$ be any extension in Ext(X). Set $X_0 = X$ and $\tau_0 = \tau$. For every positive integer n, let F_n denote the set of multi-indices $\varepsilon : \{1,\dots,n\} \to \{a,b\}$ of size n and let

$$X_n = \bigsqcup_{\varepsilon \in F_n} X_\varepsilon \text{ and } \tau_n = \bigsqcup_{\varepsilon \in F_n} \tau_\varepsilon \text{ for } n \geqslant 1, \qquad (4.2.3)$$

where $X_\varepsilon = X$ (up to homeomorphism) and $\tau_\varepsilon \in \text{Ext}(X_\varepsilon)$ is given for all $\varepsilon \in F_n$ (refer to Definition 3.2.1 and the discussion prior to it). Each X_n contains 2^n disjoint copies of X, and hence, there is a *natural map* $p_n : X_{n+1} \to X_n$, a continuous surjection which maps each X_ε in X_{n+1} onto half of some $X_{\varepsilon'}$ in X_n. If X_a is an interval or a square, X_{aa} and X_{ab} are mapped homeomorphically to two different halves of X_a, and so on (see Figure 4.2). In this case, the maximum of diameters of components in X_n tends to 0 as n tends to ∞.

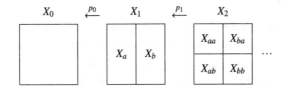

Figure 4.2 *Pictorial representation of the situation of iterated splitting lemma*

Lemma 4.2.3 (Iterated Splitting)

Let $X_0 = X$ be a compact metric space and let $X_n = \sqcup_{\varepsilon \in F_n} X_\varepsilon$, $n \geqslant 1$. Let $\{p_n\}_{n \geqslant 0}$ be a sequence of continuous maps $p_n : X_{n+1} \to X_n$. and assume that the maximum of diameters of components in X_n goes to zero as $n \to \infty$. For an integer $n \geqslant 0$, consider $p_{n*} : \text{Ext}(X_{n+1}) \to \text{Ext}(X_n)$ given by $p_{n*}(\tau_{n+1}) = \tau_n$. If p_{n*} is surjective for every integer $n \geqslant 0$, then Ext(X) is trivial.

Proof Assume that p_{n*} is surjective for every integer $n \geqslant 0$. We contend that $\varprojlim(X_n, p_n)$ is totally disconnected. Let $(x_n), (y_n) \in \varprojlim(X_n, p_n)$ be distinct. Thus, for some positive integer N, $x_N \neq y_N$. As $p_n(x_{n+1}) = x_n$ for any integer $n \geqslant 0$, we must have $x_n \neq y_n$ for every $n \geqslant N$. Further, as maximum of diameters of components in X_n goes to zero as $n \to \infty$, there exists a positive integer $N' \geqslant N$ such that $x_{N'}$ belongs to a component $C_{N'}$ of $X_{N'}$ and $y_{N'} \notin C_{N'}$. Thus, $(x_n) \in Y = \prod_{n \geqslant 0} Y_n$ and $(y_n) \in Z = \prod_{n \geqslant 0} Z_n$ are clopen disjoint subsets of the product space $\prod_{n \geqslant 0} X_n$, where

$$Y_n = \begin{cases} C_{N'} & \text{if } n = N', \\ X_n & \text{otherwise,} \end{cases} \qquad Z_n = \begin{cases} X_{N'} \backslash C_{N'} & \text{if } n = N', \\ X_n & \text{otherwise.} \end{cases}$$

It follows that (x_n) and (y_n) are separated by the disconnection $\{Y \cap \varprojlim(X_n, p_n), Z \cap \varprojlim(X_n, p_n)\}$ of $\varprojlim(X_n, p_n)$, and hence, $\varprojlim(X_n, p_n)$ is totally disconnected.

Assume now that p_{n*} is surjective for every integer $n \geqslant 0$. In view of the preceding paragraph, an application of Corollary 2.7.3 shows that $\mathrm{Ext}(\lim(X_n, p_n))$ is trivial. It can be concluded immediately from Theorem 4.2.4 that the homomorphism P : $\mathrm{Ext}(\lim(X_n, p_n)) \rightarrow \lim(\mathrm{Ext}(X_n), p_{n*})$ is surjective, and consequently, the projective limit $\lim_{\leftarrow}(\mathrm{Ext}(X_n), p_{n*})$ is also trivial. Define a homomorphism $\tilde{P} : \lim(\mathrm{Ext}(X_n), p_{n*}) \rightarrow \mathrm{Ext}(X)$ by $\tilde{P}(([\tau_n])_{n \geqslant 0}) = [\tau_0]$. As each $p_{n*} : \mathrm{Ext}(X_{n+1}) \rightarrow \mathrm{Ext}(X_n)$ is surjective, by induction on integers $n \geqslant 0$, for every $\tau_0 \in \mathrm{Ext}(X)$, there exists $(\tau_n) \in \lim(\mathrm{Ext}(X_n), p_{n*})$ such that $\tilde{P}(([\tau_n])_{n \geqslant 0}) = [\tau_0]$. It follows that $\tilde{P}$ is surjective, and hence, $\mathrm{Ext}(X)$ is trivial. $\qquad \square$

Remark 4.2.5 Note that p_{n*} is surjective if and only if each τ_ε with every $\varepsilon \in F_n$ splits. Thus, the method of Lemma 4.2.3 can be applied to any $[\tau] \in \mathrm{Ext}(X)$ for which τ_ε, $\varepsilon \in F_n$ splits for all n to infer that $[\tau]$ is trivial.

4.3   Ext(X) is a Group

In this section, we will show that $\mathrm{Ext}(X)$ is a group for any compact metric space X. First, we will establish that $\mathrm{Ext}([0,1]^{\mathbb{N}})$ is trivial, where the product space $[0,1]^{\mathbb{N}}$ is endowed with the product topology (see Theorem A.1.3). In particular, it would follow that $\mathrm{Ext}([0,1]^{\mathbb{N}})$ is a group and hence, we would have shown that $\mathrm{Ext}(\Lambda)$ is a group for any closed subset Λ of $[0,1]^{\mathbb{N}}$ (see Corollary 3.6.1). However, any compact metric space X is homeomorphic to a closed subset of $[0,1]^{\mathbb{N}}$ (see Proposition A.1.3), and therefore, $\mathrm{Ext}(X)$ is seen to be a group for any such X.

To use the technique of the proof of the iterated splitting lemma for the Hilbert cube $[0,1]^{\mathbb{N}}$ and to show that $\mathrm{Ext}([0,1]^{\mathbb{N}})$ is a group, we have to discuss the infinite sum of extensions. Although the sum of two extensions was defined as

$$(\tau_1 + \tau_2)(f) = \pi(T_1^f \oplus T_2^f) \quad \text{if} \quad \tau_j(f) = \pi(T_j^f), \ j = 1, 2,$$

there is no obvious way to define the sum of infinitely many extensions, for the simple reason that an infinite sum of compact operators need not be compact. However, sometimes, it is possible to define such a sum. First an observation.

Lemma 4.3.1

*Let X be a compact metrizable space. Let $X_1, X_2, \ldots$ be closed subsets of X such that diameter of X_m converges to zero and let $f \in C(\overline{\cup_{m \geqslant 0} X_m})$. Suppose there exist *-monomorphisms τ_m : $C(X_m) \rightarrow Q(\mathcal{H}_m)$ and points x_m in X_m such that*

$$\pi(T_m^f) = \tau_m(f|_{X_m}), \quad \|T_m^f - f(x_m)\| \rightarrow 0. \tag{4.3.4}$$

Then $\pi(\oplus_{m \geqslant 0} T_m^f)$ is independent of the choice of T_m^f.

Proof   For any other choice of points y_m in X_m and operators S_m^f such that $\pi(S_m^f) = \tau_m(f|_{X_m})$ and $\|S_m^f - f(y_m)\| \rightarrow 0$, we have

$$\|T_m^f - S_m^f\| \leqslant \|T_m^f - f(x_m)\| + \|S_m^f - f(y_m)\| + |f(x_m) - f(y_m)| \rightarrow 0,$$

where we used the assumption that the diameter of X_m converges to 0. Therefore, $\oplus_{m \geqslant 0} T_m^f - \oplus_{m \geqslant 0} S_m^f$ is compact and $\pi(\oplus_{m \geqslant 0} T_m^f)$ is independent of the choice of T_m^f. $\qquad \square$

Remark 4.3.1 Since $\lim_{m \to \infty} \operatorname{diam} X_m = 0$, it follows that for any $x_m \in X_m$,

$$\|\pi(T_m^f - f(x_m))\| = \|\tau_m(f|_{X_m} - f(x_m))\| = \|f|_{X_m} - f(x_m)\|_\infty \to 0.$$

Thus, there exists a subsequence $\{k_n\}_{n \geqslant 1}$ of positive integers such that for every positive integer n, $\|\pi(T_{k_n}^f - f(x_{k_n}))\| < 2/n$, and hence, one may find a compact operator K_n such that $\|T_{k_n}^f - f(x_{k_n}) + K_n\| < 1/n$. Thus, after passing to a subsequence, one may ensure the existence of $\{T_m^f\}_{m \geqslant 1}$ satisfying (4.3.4).

We can now define the sum of infinitely many extensions using Lemma 4.3.1.

Definition 4.3.2

Let X be a compact metrizable space. Let $X_1, X_2, \ldots$ be closed subsets of X such that the diameter of X_m converges to zero and let $f \in C(\overline{\cup_{m \geqslant 1} X_m})$. For every positive integer m, let $\tau_m : C(X_m) \to Q(\mathcal{H}_m)$ be a $*$-monomorphism and let x_m be in X_m such that (4.3.4) holds (see Lemma 4.3.1 and Remark 4.3.1). We define the extension $\tau = \sum_{m=1}^\infty \tau_m : C(\overline{\cup_{m \geqslant 1} X_m}) \to Q(\oplus_{m \geqslant 0} \mathcal{H}_m)$ by

$$\tau(f) = \pi(\oplus_{m \geqslant 1} T_m^f), \quad f \text{ in } C(\overline{\cup_{m \geqslant 1} X_m}).$$

Note that $\sum_{m=1}^\infty \tau_m$ defines a $*$-monomorphism. We now record the following observation required in the proof of Theorem 4.3.3.

Lemma 4.3.2

Let X, Y be compact metrizable spaces. Let $X_1, X_2, \ldots$ be closed subsets of X such that diameter of X_n converges to zero. Let $\{\tau_n\}_{n \geqslant 1}$ be a sequence of $*$-monomorphisms $\tau_n : C(X_n) \to Q(\mathcal{H}_n)$. If $q : X \to Y$ is continuous, $Y_n = q(X_n)$, and $q_n : X \to Y_n$ is given by $q_n = q|_{X_n}$ and $q' = q|_{\overline{\cup_{n \geqslant 1} X_n}}$, then

$$q'_* \left(\sum_{n=1}^\infty (i_{X_n, X})_* \tau_n \right) = \sum_{n=1}^\infty q_{n*}(\tau_n).$$

Proof Note that for any $f \in C(\overline{\cup_{n \geqslant 1} X_n})$,

$$
\begin{aligned}
q'_* \left(\sum_{n=0}^\infty (i_{X_n, X})_* \tau_n \right)(f) &= \sum_{n=1}^\infty ((i_{X_n, X})_* \tau_n)(f \circ q') \\
&= \sum_{n=1}^\infty \tau_n(f \circ q|_{X_n}) \\
&= \sum_{n=1}^\infty q_{n*}(\tau_n)(f)
\end{aligned}
$$

This completes the verification. $\qquad \square$

Now we are ready to prove that the Ext of the Hilbert cube is trivial.

Theorem 4.3.3

$Ext([0,1]^N)$ *is trivial. In particular,* $Ext([0,1]^N)$ *is a group.*

Proof   Let $X_0 = J = [0,1[^N$ and $J_n = \prod_{j>n}[0,1]$ for a positive integer n. Let

$$X_a = [0,1/2] \times J_1, \quad X_b = [1/2,1] \times J_1, \quad X_1 = X_a \sqcup X_b$$

and $p_0 : X_1 \to X_0$ be the natural map (see the discussion following (4.2.3)). For an integer $n \geqslant 0$, let $\pi_n((x_m)) = x_n$ denote the coordinate maps in $C(J)$. Let $[\tau_0]$ in $Ext(X_0)$ be any extension and let $h_n = \tau_0(\pi_n)$, $n \geqslant 0$. Then the map $\eta_0 : C(X_0) \to Q(\mathcal{H})$ defined by

$$\eta_0(f) = f(1/2, h_1, h_2, \ldots)$$

is a $*$-homomorphism, which, after perturbing by a trivial element, can be assumed to be a $*$-monomorphism. Using Corollary 3.5.1 (alternatively, as in the proof of the second splitting lemma), $\tau_0 + \eta_0$ splits, that is, there exists $[\tau_a]$ in $Ext(X_a)$ and $[\tau_b]$ in $Ext(X_b)$ such that

$$p_{0*}[\tau_a \sqcup \tau_b] = [\tau_0 + \eta_0].$$

As both X_a and X_b are homeomorphic to X_0, we may iterate this procedure to obtain closed sets X_n and natural maps $p_n : X_{n+1} \to X_n$ such that

(1) X_n has 2^n components each homeomorphic to X_0 and the maximum of diameters of the 2^n components in X_n goes to zero as $n \to \infty$,

(3) if $\tau_n = \bigsqcup_{\varepsilon \in F_n} \tau_\varepsilon$ and $\eta_n = \bigsqcup_{\varepsilon \in F_n} \eta_\varepsilon$ for every integer $n \geqslant 1$, then $p_{n*}[\tau_{n+1}] = [\tau_n] + [\eta_n]$, where η_n is a trivial element.

By (1), (2) and Remark 4.3.1, we may define the following infinite sum as an element in $Ext(X_n)$:

$$\tau'_n = \tau_n + \eta_n + p_{n*}(\eta_{n+1}) + p_{n*} \circ p_{n+1*}(\eta_{n+2}) + \cdots, \quad n \geqslant 0. \tag{4.3.5}$$

By Lemma 4.3.2 and (2), for any integer $n \geqslant 0$,

$$\begin{aligned} p_{n*}(\tau'_{n+1}) &= p_{n*}(\tau_{n+1}) + p_{n*}(\eta_{n+1}) + p_{n*} \circ p_{n+1*}(\eta_{n+2}) + \cdots \\ &= \tau_n + \eta_n + p_{n*}(\eta_{n+1}) + p_{n*} \circ p_{n+1*}(\eta_{n+2}) + \cdots \\ &= \tau'_n. \end{aligned} \tag{4.3.6}$$

As $[\tau_n] = [\tau'_n]$, this shows that p_{n*} is surjective for every $n \geqslant 0$. The desired conclusion now follows from the iterated splitting (see Lemma 4.2.3).   □

Remark 4.3.4   By (4.3.6), $([\tau'_n])_{n \geqslant 0}$ is an element of $\varprojlim(Ext(X_n), p_{n*})$. As the projective limit of X_n is totally disconnected, by Theorem 4.2.4, τ'_0 is trivial. Hence, by (4.3.5), we obtain an expression for the inverse of τ_0.

The following is immediate from the last theorem and Corollary 3.6.1.

Corollary 4.3.1

For any compact metric space X, Ext(X) is a group.

Ext need not be a group in general. Indeed, there exists a separable C^*-algebra $\mathcal{A} \subseteq Q(\mathcal{H})$ and an extension of $\mathcal{A}$ by $C(\mathcal{H})$ without an inverse (cf. [6]). Further, note that there is a short and elegant proof of Corollary 4.3.1 based on a lifting theorem due to Arveson [9]. In fact, one may recapture this proof by combining a lifting theorem of Choi and Effros (cf. [108]) (refer to Section 6.1 of Epilogue for more details).

We conclude this section by rephrasing a special case of Corollary 4.3.1 (the case of a planar set X) in purely operator–theoretic terms.

Corollary 4.3.2

For any essentially normal $A \in \mathcal{L}(\mathcal{H})$, there exists an essentially normal $B \in \mathcal{L}(\mathcal{K})$ such that $\sigma_e(B) = \sigma_e(A)$ and $A \oplus B$ is unitarily equivalent to $N + K$ for some normal operator $N \in \mathcal{L}(\mathcal{H})$ and a compact operator $K \in \mathcal{L}(\mathcal{H})$.

4.4 γ_X is Injective

Let X be a compact metrizable space and let $[\tau] \in \mathrm{Ext}(X)$. Recall that for any $f \in C(X)$, $\tau(f) = \pi(T^f)$, where the operator T^f is unique up to compact perturbation. If f is nowhere-vanishing, then T^f is Fredholm. In particular, the Fredholm index of $\tau(f)$ defined as the Fredholm index of T^f is independent of the representative T^f. If $\mathcal{N} : C(X) \to Q(\mathcal{H})$ is trivial with trivializing map $\mathcal{N}_0 : C(X) \to \mathcal{L}(\mathcal{H})$, then for nowhere-vanishing f in $C(X)$, $\mathcal{N}_0(f)$ is invertible and hence, $\mathrm{ind}\,\mathcal{N}(f) = 0$. Further, for nowhere-vanishing functions $f, g \in C(X)$ and for $*$-monomophisms $\tau, \tau_1, \tau_2 \in \mathrm{Ext}(X)$,

$$\mathrm{ind}\,\tau(fg) = \mathrm{ind}\,\tau(f)\tau(g) = \mathrm{ind}\,\tau(f) + \mathrm{ind}\,\tau(g), \tag{4.4.7}$$

$$\mathrm{ind}(\tau_1 + \tau_2)(f) = \mathrm{ind}\,(\tau_1(f) \oplus \tau_2(f)) = \mathrm{ind}\,\tau_1(f) + \mathrm{ind}\,\tau_2(f). \tag{4.4.8}$$

With these observations, we are now ready to introduce the map γ_X.

Definition 4.4.1

Let X be a compact metrizable space. Let $\pi^1(X)$ be the first cohomotopy group of X, that is, $\pi^1(X)$ is the group of (pointed) homotopy classes of continuous mappings from X to the unit circle $\mathbb{T}$ and $\mathrm{Hom}(\pi^1(X), \mathbb{Z})$ denotes the group of homomorphisms from $\pi^1(X)$ into $\mathbb{Z}$. Define the map $\gamma_X : \mathrm{Ext}(X) \to \mathrm{Hom}(\pi^1(X), \mathbb{Z})$ by

$$(\gamma_X[\tau])([f]) = \mathrm{ind}\,\tau(f), \quad [\tau] \in \mathrm{Ext}(X),\ [f] \in \pi^1(X),$$

where $\mathrm{ind}\,\tau(f)$ is the Fredholm index of any Fredholm operator T^f satisfying $\tau(f) = \pi(T^f)$.

Remark 4.4.2   The fact that γ_X is well-defined needs justification. Firstly, by the discussion prior to the definition, the map γ_X is independent of the choice of the representative of $[\tau]$.

Secondly, if f is homotopic to g, then $\mathrm{ind}\,\tau(f) = \mathrm{ind}\,\tau(g)$. Indeed, as f and g belong to same connected component of $C(X)$, $\tau(f) = \pi(T^f)$ and $\tau(g) = \pi(T^g)$ belong to same connected component of the Calkin algebra, and one may apply Lemma 1.3.4. Thus, the definition of γ_X is independent of the representative of $[f]$. Further, in view of (4.4.7), $\gamma_X[\tau] : \pi^1(X) \to \mathbb{Z}$ is a homomorphism. Finally, (4.4.8) shows that γ_X is a homomorphism.

***Example* 4.4.3**   Let X be a compact subset of $\mathbb{R}$. Then, as $\mathbb{C} \setminus X$ is connected, by Corollary 1.3.8, $\pi^1(X) = \{0\}$, and hence, γ_X is identically 0. As $\mathrm{Ext}(X)$ is also trivial (see Example 2.3.4), γ_X is clearly injective. If X is the unit circle $\mathbb{T}$, then $\mathrm{Ext}(X) \cong \mathbb{Z}$ and $\pi^1(X) \cong \mathbb{Z}$ (see Theorem 2.3.5 and Remark 1.3.12). In this case, γ_X is the identity map.   ∎

We ask whether γ_X is injective. An affirmative answer will characterize the trivial maps. In 1972, Brown, Douglas, and Fillmore showed that for X, a compact subset of the complex plane $\mathbb{C}$, the map γ_X is, in fact, injective.

We have now all the ingredients to prove that the map γ_X is injective for compact subset X of $\mathbb{C}$. Injectivity of γ_X is established by showing that any extension $[\tau]$ in $\ker \gamma_X$ splits into $[\tau_1]$ and $[\tau_2]$ with respect to some closed cover $\{X_1, X_2\}$ of X such that $[\tau_k]$ is in $\ker \gamma_{X_k}$ for $k = 1, 2$. We iterate this procedure and apply Remark 4.2.5 to see that $[\tau] = 0$. The inductive step in this argument depends on injectivity of $\gamma_{[0,1]/A}$, where A is an arbitrary closed subset of $[0,1]$. The injectivity of the map γ_X, in this special case, in turn depends on the following lemma.

Lemma 4.4.1

Let X be a compact metrizable space and let $\tau : C(X) \to Q(\mathcal{H})$ be a $*$-monomorphism such that the extension $[\tau]$ is in $\ker \gamma_X$. Suppose that τ admits a splitting into $[\tau_1]$ and $[\tau_2]$ with respect to a closed cover $\{X_1, X_2\}$ of X. If $X_1 \cap X_2 = \{x_0\}$ for some $x_0 \in X$, then $[\tau_k]$ is in $\ker \gamma_{X_k}$ for $k = 1, 2$.

Proof   Assume that $X_1 \cap X_2 = \{x_0\}$. Let T^f be any operator such that $\tau(f) = \pi(T^f)$. By assumptions, τ splits and $\gamma_X([\tau]) = 0$. Then there exist operators $T_1^{f_1}$ and $T_2^{f_2}$ inducing $*$-monomorphisms

$$\tau_j(f_j) = \pi(T_j^{f_j}), \quad f_j \in C(X_j), \; j = 1, 2,$$

$$\tau(f) = \pi(T_1^{f|X_1} \oplus T_2^{f|X_2}), \quad f \in C(X).$$

Let $f_1 : X_1 \to \mathbb{C}\setminus\{0\}$ be continuous. Define $f : X \to \mathbb{C}\setminus\{0\}$ by

$$f(x) = \begin{cases} f_1(x) & x \in X_1, \\ f_1(x_0) & x \in X_2, \end{cases}$$

and note that,

$$\tau(f) = \pi\left(T_1^{f_1} \oplus T_2^c\right),$$

where c is the (invertible) constant function $c(x) = f_1(x_0) \neq 0$ for all x in X_2. Moreover, as T_2^c is essentially normal operator with essential spectrum $c(X_2) = \{f_1(x_0)\}$, it is a compact perturbation of a normal operator (see Example 2.3.4). In particular, $\operatorname{ind}\tau_2(c) = 0$. However, $\gamma_X([\tau]) = 0$, so that

$$\operatorname{ind}\tau_1(f_1) + \operatorname{ind}\tau_2(c) = \operatorname{ind}\tau(f) = 0.$$

It follows that

$$(\gamma_{X_1}[\tau_1])(f_1) = \operatorname{ind}\tau_1(f_1) = 0$$

showing that $[\tau_1] \in \ker\gamma_{X_1}$. Similarly, one can show that $[\tau_2] \in \ker\gamma_{X_2}$. This completes the proof. $\qquad\Box$

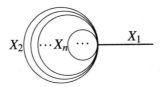

Figure 4.3 *Quotient space* $[0,1]/A$, *where* A *is a closed subset of* $[0,1]$ *such that* $0 \notin A$ *and* $1 \in A$

We already noticed that γ_X is injective if X is a compact subset of real numbers or the unit circle. The following provides another example illustrating this fact.

Proposition 4.4.1

If $X = [0,1]/A$ for some closed subset A of $[0,1]$, then γ_X is injective.

Proof Consider the decomposition $\cup_{n=1}^{\infty} Z_n$ of the complement $[0,1]\backslash A$ of A in $[0,1]$ into the connected components Z_n and let $([0,1]\backslash A) \cup \{\infty\}$ be the one point compactification of $[0,1]\backslash A$. By Corollary A.1.3, the quotient space $X = [0,1]/A$ can be written (up to homeomorphism) as the union of closed sets $X_n = Z_n \cup \{\infty\}$, $n \geqslant 1$ such that

(i) if there are infinitely many components Z_n of $X\backslash A$, then $\lim_{n\to\infty} \operatorname{diam} X_n = 0$ (see Exercise 4.6.3),

(ii) each X_n is homeomorphic to a closed interval or a circle,

(iii) $X_m \cap X_n = \{\infty\}$ for all positive integers $m \neq n$.

Moreover, X can be regarded as a subset of $\mathbb{C}$. For example, in case $0 \notin A$ and $1 \in A$, one may think of X as a stemmed flower with petals of diameter tending to 0, where the stem X_1 is a closed interval whereas the petals X_n, $n \geqslant 2$ are circles (see Figure 4.3).

By (ii) and Example 4.4.3, γ_{X_m} is injective for every integer $m \geqslant 1$. Let $\tau : C(X) \to Q(\mathcal{H})$ be a *-monomorphism, $[\tau] \in \ker\gamma_X$ and let $Y_n = \cup_{m=n+1}^{\infty} X_m$. Note that Y_n is a closed subset of X. As X_1 and Y_1 intersect in a single point, by the first splitting lemma, $[\tau]$ splits into $[\tau_1]$ and $[\tau_{Y_1}]$ with respect to the closed cover $\{X_1, Y_1\}$ of X. If we write Y_1 as $X_2 \cup Y_2$, then once again by the first splitting lemma, the extension $[\tau_{Y_1}]$ will split into $[\tau_2]$ and $[\tau_{Y_2}]$ with respect to the closed cover $\{X_2, Y_2\}$ of Y_2. Continuing this, we obtain

$$\tau(f) = \tau_1(f|_{X_1}) \oplus \cdots \oplus \tau_n(f|_{X_n}) \oplus \tau_{Y_n}(f|_{Y_n}).$$

By Lemma 4.4.1 and Example 4.4.3, each $[\tau_k]$ is in $\ker \gamma_{X_k}$, and therefore $[\tau_k]$ is trivial. In case $X \backslash A$ has finitely many components, then Y_n is empty for large enough n, and hence, γ_X is injective. Assume now that there are infinitely many components of $X \backslash A$.

Consider the collection $\mathscr{D}_n$ of all functions f in $C(X)$, which are constant on Y_n. If $f \in C(X)$ and $\epsilon > 0$, then by (i), one may choose m large enough such that $\sup_{x \in X_n} |f(x) - f(\infty)| < \epsilon$ for all integers $n > m$. Hence, the functions $f_n : X \to \mathbb{C}$ given by

$$f_n = \begin{cases} f & \text{on } \cup_{k=1}^n X_k \\ f(\infty) & \text{on } Y_n \end{cases}, \quad n \geqslant 1,$$

belong to $C(X)$. Moreover, $\|f_n - f\|_{\infty, X} < \epsilon$ for all integers $n > m$. It follows that $\mathscr{D} = \cup_n \mathscr{D}_n$ is dense in $C(X)$. Define τ_0 on $\mathscr{D}$ by

$$\tau_0(f) = \tau_{10}(f|_{X_1}) \oplus \cdots \oplus \tau_{n0}(f|_{X_n}) \oplus f(\infty), \quad f \in \mathscr{D}_n,$$

where τ_{k0} is the trivializing map for τ_k. As τ_0 is defined on the dense subset $\mathscr{D}$ of $C(X)$, it has a continuous extension to $C(X)$. However, $\pi \circ \tau_0 = \tau$ on $\mathscr{D}$, and hence, $\pi \circ \tau_0 = \tau$ on all of $C(X)$. Therefore, τ is trivial and the proof is complete. $\qquad\square$

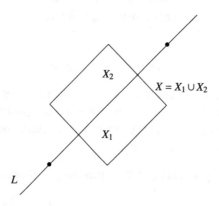

Figure 4.4 *Pictorial representation of the situation of Lemma 4.4.2*

Just as we needed Lemma 4.4.1 to prove injectivity of γ_X in this special case, we would need the following lemma to prove injectivity of γ_X in general.

Lemma 4.4.2

Let $X \subseteq \mathbb{C}$ be compact and X_1, X_2 denote the intersections of X with the closed half planes determined by a straight line L (so that $X = X_1 \cup X_2$ and $X_1 \cap X_2 = L \cap X$; see Figure 4.4). If $\beta : Ext(X_1) \oplus Ext(X_2) \to Ext(X)$ is as defined in (3.2.3), then we have the inclusion

$$\ker \gamma_X \beta \subseteq \ker \gamma_{X_1} \oplus \ker \gamma_{X_2}.$$

Proof   Let $([\tau_1], [\tau_2])$ belong to $\ker \gamma_X \beta$. Then,

$$\mathrm{ind}(\tau_1(f|_{X_1}) \oplus \tau_2(f|_{X_2})) = 0, \quad f \in C(X). \tag{4.4.9}$$

For $g : X_1 \to \mathbb{C} \backslash \{0\}$, define $g' : X_1 \cup L \to \mathbb{C} \backslash \{0\}$ by extending g linearly on the components of $L \backslash (L \cap X)$ while taking care to avoid the origin, if necessary. Let $p : X \to X_1 \cup L$ denote the continuous map

$$p(x) = \begin{cases} \text{projection onto } L & \text{if } x \in X_2, \\ x & \text{if } x \in X_1, \end{cases}$$

and let $f = g' \circ p$.

Assume that $L \cap X_2$ is connected. As $f(X_2) = g'(L \cap X_2)$ and $g'|_{L \cap X_2} = id$ is null-homotopic, $f|_{X_2}$ is null homotopic, and hence, $\mathrm{ind}(\tau_2(f|_{X_2})) = 0$. It follows from (4.4.9) that $\mathrm{ind}(\tau_1(g)) = \mathrm{ind}(\tau_1(f|_{X_1})) = 0$. As g was arbitrary, $[\tau_1]$ is in $\ker \gamma_{X_1}$. In case $L \cap X_2$ is not connected, $f|_{X_2}$ is sum of finitely many null-homotopic maps, and hence, $\mathrm{ind}(\tau_2(f|_{X_2})) = 0$ in this case also. Similarly, we can show $[\tau_2]$ is in $\ker \gamma_{X_2}$ and the proof is complete. $\square$

To derive the injectivity of the map γ_X, we need some of its elementary properties.

Lemma 4.4.3

Let X and Y be compact Hausdorff spaces and let $i : X \to Y$ be a continuous map. The following statements are true:

(1) We have the inclusion $i_* \ker \gamma_X \subseteq \ker \gamma_Y$.

(2) If i is an isomorphism, then $\ker \gamma_X$ and $\ker \gamma_Y$ are isomorphic.

Proof   Let $[\tau] \in \ker \gamma_X$. Then, for any $f \in C(Y)$,

$$\gamma_Y(i_*[\tau])(f) = \mathrm{ind}\, \tau(f \circ i) = \gamma_X[\tau](f \circ i) = 0.$$

This yields (1). To see (2), note that if i is an isomorphism, then so is i_* (see Corollary 2.7.1), and apply (1) to i and its inverse. $\square$

The elements in $\ker \gamma$ admit splitting with components belonging to respective kernels. Here is the precise statement.

Theorem 4.4.4

Let $X \subseteq \mathbb{C}$ be compact and let X_1 and X_2 be the intersections of X with the closed half planes determined by a straight line L. If $[\tau]$ is any extension in $\ker \gamma_X$, then τ splits into $[\tau_1]$ and $[\tau_2]$ with respect to the closed cover $\{X_1, X_2\}$ of X. Furthermore, $[\tau_1]$ is in $\ker \gamma_{X_1}$ and $[\tau_2]$ is in $\ker \gamma_{X_2}$.

Proof   We will need the following diagram and inclusion maps. Let J be any compact interval on L that contains $X \cap L$ (see Figure 4.4, where J is the closed interval on L enclosed by •). Consider

$$X_1 \cup J \xrightarrow{i'_1} X \cup J \xleftarrow{i'_2} X_2 \cup J$$

$$\uparrow j_1 \qquad\qquad \uparrow j \qquad\qquad \uparrow j_2$$

$$X_1 \xrightarrow{i_1} X \xleftarrow{i_2} X_2$$

where all the connecting maps are inclusions. Let $[\tau]$ be in $\ker \gamma_X$. Then, by Lemma 4.4.3, $j_*[\tau]$ is in $\ker \gamma_{X \cup J}$. Further, by the second splitting lemma, there exists $\tau_{X_1 \cup J}$ and $\tau_{X_2 \cup J}$ such that

$$j_*[\tau] = i'_{1*}[\tau_{X_1 \cup J}] + i'_{2*}[\tau_{X_2 \cup J}]. \tag{4.4.10}$$

However, by Lemma 4.4.2,

$$\tau_{X_1 \cup J} \in \ker \gamma_{X_1 \cup J} \text{ and } \tau_{X_2 \cup J} \in \ker \gamma_{X_2 \cup J}. \tag{4.4.11}$$

Fix $k = 1, 2$ and let $q_k : X_k \cup J \to (X_k \cup J)/X_k$ be the quotient maps. By (4.4.11), $q_{k*}[\tau_{X_k \cup J}]$ belongs to the $\ker \gamma_{(X_k \cup J)/X_k}$. However, $(X_k \cup J)/X_k$ is homeomorphic to $J/(X_1 \cap X_2)$ (see Figure 4.4), and hence, by Propositions 4.4.1 and 2.3.1, $\gamma_{(X_k \cup J)/X_k}$ is injective. Therefore, $q_{k*}[\tau_{X_k \cup J}]$ is trivial. Further, by Theorem 3.5.1,

$$\text{Ext}(X_k) \xrightarrow{j_{k*}} \text{Ext}(X_k \cup J) \xrightarrow{q_{k*}} \text{Ext}((X_k \cup J)/X_k)$$

is exact, so that there exists $[\tau_{X_k}]$ such that $j_{k*}[\tau_{X_k}] = [\tau_{X_k \cup J}]$. It is now immediate from (4.4.10) that

$$j_* \beta([\tau_{X_1}], [\tau_{X_2}]) = j_*[\tau]. \tag{4.4.12}$$

As $\gamma_X([\tau]) = 0$, by Lemma 4.4.2, it suffices to show that $\beta([\tau_{X_1}], [\tau_{X_2}]) = [\tau]$. In turn, in view of (4.4.12), it is enough to show j_* is injective. However, by Theorem 3.6.1, the Mayer–Vietoris sequence

$$\text{Ext}(X \cap J) \xrightarrow{\alpha} \text{Ext}(X) \oplus \text{Ext}(J) \xrightarrow{\beta} \text{Ext}(X \cup J)$$

is exact, that is, $\text{Im}\,\alpha = \ker \beta$ (see (3.6.20) and (3.6.21)). As $\text{Ext}(X \cap J) = \{0\}$, β and hence, j_* is injective. This completes the proof. $\square$

As in proof of Theorem 4.3.3, Theorem 4.4.4 allows us to apply the iterated splitting argument, establishing the injectivity of γ_X.

Corollary 4.4.1

For any compact subset X of the complex plane $\mathbb{C}$, the map $\gamma_X : Ext(X) \to Hom(\pi^1(X), \mathbb{Z})$ is injective.

It is actually possible to show that γ_X is a surjective mapping [26] (see Theorem 5.1.1 for a special case). Therefore, the equivalence classes of essentially normal operators with essential spectrum X is obtained by prescribing arbitrary integers for the bounded components of $\mathbb{C}\backslash X$. In particular, for a compact subset X of $\mathbb{C}$,

$$\text{Ext}(X) \simeq \text{Hom}(\pi^1(X), \mathbb{Z}).$$

The conclusion of Corollary 4.4.1 is no longer true if X is non-planar.

***Example* 4.4.5**   Let $\mathbb{S}$ be the unit sphere in $\mathbb{C}^2$ and let $L^2(\mathbb{S})$ denote Hilbert space of functions $f : \mathbb{S} \to \mathbb{C}$ square-integrable with respect to the normalized surface area measure σ on $\mathbb{S}$. Let $H^2(\mathbb{S})$ denote the closure of complex polynomials in complex variables z and w in $L^2(\mathbb{S})$. Recall the formula

$$\int_{\mathbb{S}} z^m w^n \bar{z}^l \bar{w}^k d\sigma(z) = \begin{cases} \frac{m!\,n!}{(m+n+1)!} & \text{if } m = l, \ n = k \in \mathbb{N}, \\ 0 & \text{otherwise.} \end{cases} \qquad (4.4.13)$$

Note that $H^2(\mathbb{S})$ is invariant under the multiplication by z and w. The linear operators defined by

$$(M_z f)(z,w) = zf(z,w), \quad (M_w f)(z,w) = wf(z,w), \ f \in H^2(\mathbb{S}),$$

are easily seen to be bounded. Indeed, one can check using (4.4.13) that $\|zf\|^2 + \|wf\|^2 = \|f\|^2$ for any polynomial f in z, w, and hence, the operators M_z and M_w can be extended as bounded linear operators to $H^2(\mathbb{S})$ of norm at most 1. In the following discussion, we need some elementary facts pertaining to the multiplication operators M_z and M_w :

(1) $M_z^* M_z + M_w^* M_w = I$ and $M_z M_w = M_w M_z$.

(2) $[M_z^*, M_z]$, $[M_w^*, M_w]$ and $[M_z^*, M_w]$ are compact.

The assertions (1) and (2) can be deduced easily from the fact that $\{z^m w^n\}_{m,n\in\mathbb{N}}$ forms an orthogonal basis for $H^2(\mathbb{S})$ (see Exercise 4.6.9).

For $f \in C(\mathbb{S})$, define the *Toeplitz operator* $T_f : H^2(\mathbb{S}) \to H^2(\mathbb{S})$ by

$$T_f(g) = P(fg), \quad g \in H^2(\mathbb{S}),$$

where P denotes the orthogonal projection of $L^2(\mathbb{S})$ onto $H^2(\mathbb{S})$. Note that T_f is a bounded linear operator with $\|T_f\| \leqslant \|f\|_\infty$. Define $\tau : C(\mathbb{S}) \to Q(H^2(\mathbb{S}))$ by $\tau(f) = \pi(T_f)$, $f \in C(\mathbb{S})$. As $T_f^* = T_{\bar{f}}$ and $T_{fg} - T_f T_g$ is compact, τ is a $*$-homomorphism. Moreover, since T_f is compact if and only if $f = 0$, τ is a $*$-monomorphism (see Exercise 4.6.11 for elementary properties of Toeplitz operators on $H^2(\mathbb{S})$).

We claim that τ is a non-trivial element in $\text{Ext}(\mathbb{S})$. To see that, define a bounded linear operator T on $H^2(\mathbb{S}) \oplus H^2(\mathbb{S})$ by

$$T = \begin{bmatrix} M_z & M_w \\ -M_w^* & M_z^* \end{bmatrix}.$$

A routine calculation using (1) shows that

$$T^*T = \begin{bmatrix} M_z^* M_z + M_w M_w^* & M_z^* M_w - M_w M_z^* \\ M_w^* M_z - M_z M_w^* & M_w^* M_w + M_z M_z^* \end{bmatrix},$$

$$TT^* = \begin{bmatrix} M_z M_z^* + M_w M_w^* & 0 \\ 0 & I \end{bmatrix}.$$

It now follows from (2) that $\pi(T)$ is essentially unitary. Assume that τ is trivial. Then $\tau(M_z) = 0$, that is, $M_z = N_1 + K_1$ for some normal operator N_1 and compact operator K_1. Similarly, one can see that $M_w = N_2 + K_2$ for some normal operator N_2 and compact operator K_2. Moreover, as M_z and M_w commute, N_1 and N_2 commute modulo compact operators. It now follows that T must be a compact perturbation of a normal operator on $H^2(\mathbb{S}) \oplus H^2(\mathbb{S})$. Thus, to conclude that τ is non-trivial, it suffices to check that the Fredholm index of T is nonzero. Note that for $f, g \in H^2(\mathbb{S})$,

$$\begin{bmatrix} f \\ g \end{bmatrix} \in \ker T \text{ if and only if } M_z f + M_w g = 0, \ M_z^* g - M_w^* f = 0.$$

Let $f(z, w) = \sum_{k,l=0}^{\infty} a_{kl} z^k w^l$ and $g(z, w) = \sum_{k,l=0}^{\infty} b_{kl} z^k w^l$ be in $H^2(\mathbb{S})$. Then, $M_z f + M_w g = 0$ yields

$$\sum_{k=1}^{\infty} \sum_{l=0}^{\infty} a_{k-1 l} z^k w^l + \sum_{k=0}^{\infty} \sum_{l=1}^{\infty} b_{k l-1} z^k w^l = 0.$$

As $\{z^k w^l\}_{k,l \geqslant 0}$ forms an orthogonal sequence in $H^2(\mathbb{S})$, we must have $a_{k0} = 0 = b_{0l}$ for all integers $k, l \geqslant 0$, and $a_{k-1 l} = -b_{k l-1}$ for all integers $k, l \geqslant 1$, and hence, $f(z, w) = w h(z, w)$ and $g(z, w) = -z h(z, w)$ for some $h \in H^2(\mathbb{S})$ in this case. Combining this with $M_z^* g - M_w^* f = 0$ yields that $h = 0$, that is, $\ker T = \{0\}$. As

$$\begin{bmatrix} 1 \\ 0 \end{bmatrix} \in \ker T^*,$$

T is not essentially equivalent to a normal operator. Hence, $\text{Ext}(\mathbb{S})$ is non-trivial. As $\pi^1(\mathbb{S})$ is trivial, $\gamma_{\mathbb{S}} : \text{Ext}(\mathbb{S}) \to \text{Hom}(\pi^1(\mathbb{S}), \mathbb{Z})$ is not injective. ∎

The reader is referred to [67] for a complete classification of essentially commutative C^*-algebras with essential spectrum homeomorphic to $\mathbb{S}$. It is worth noting that the so-called $\mathscr{C}_{n-1}$-smooth extensions of $C(\mathbb{S}^{2n-1})$ are necessarily trivial if $n \geqslant 2$ (see [57, Proposition 3]), where $\mathbb{S}^{2n-1}$ denotes the unit sphere in $\mathbb{C}^n$.

4.5   BDF Theorem and Its Consequences

BDF theorem determines the essential unitary equivalence classes of essentially normal operators with essential spectrum $X \subseteq \mathbb{C}$. In its proof, we need a lemma pertaining to the first cohomotopy group. Recall that the first cohomotopy group $\pi^1(X)$ of X is the group of pointed homotopy classes of continuous mappings from X to the unit circle $\mathbb{T}$. Equivalently, it is the quotient G/G_0, where G is the multiplicative group of continuous, nowhere-vanishing functions on X and G_0 is the component of G containing 1. Thus, $\pi^1(X)$ is the abstract index group of G (see Proposition 1.3.1). We however need an alternate description of $\pi^1(X)$ in the proof of the BDF theorem.

Lemma 4.5.1

Let X be a compact subset of $\mathbb{C}$ and let

$$\mathbb{C}\backslash X = O_\infty \cup O_1 \cup O_2 \cup \cdots,$$

where $O_1, O_2, \ldots,$ are bounded components and O_∞ is the unbounded component of $\mathbb{C}\backslash X$. If I is a set consisting of exactly one point from each O_i, $i \geqslant 1$, then the abstract index group $\Lambda(X)$ is freely generated as an abelian group by $\{[\varphi_\lambda] : \lambda \in I\}$, where $\varphi_\lambda(z) = z - \lambda$, $z \in X$ (see Figure 4.5). In particular, the first cohomotopy group $\pi^1(X)$ of X is freely generated as an abelian group with one generator $[O_i]$ for each bounded component.

Proof In view of Proposition 1.3.1, it suffices to check that the abstract index group $\Lambda(X)$ is freely generated as an abelian group by $\{[\varphi_\lambda] : \lambda \in I\}$. Let $f : X \to \mathbb{C}\backslash\{0\}$ be a continuous function. We divide the proof into two steps:

Step I *X is a finite union of disjoint compact sets, each bounded by finitely many disjoint smooth Jordan curves:*

In this case, f is of the form e^g if and only if the variation of $\arg(f)$ around each of these curves is zero (see Example 1.3.11). If this is the case, then f is a trivial element in $\Lambda(X)$. In case f is not a trivial element in $\Lambda(X)$, one can find φ_λ with the same variation of argument $\arg(f)$ around each curve. It follows that f is a product of finitely many φ_λ, $\lambda \in I$. Further, any non-trivial product of finitely many φ_λ's has nonzero variation $\arg(f)$ around some curve, and hence, it yields a non-trivial element of the abstract index group $\Lambda(X)$.

Step II *$X = \cap_{n \geqslant 1} X_n$, where $\{X_n\}_{n \geqslant 1}$ is a decreasing sequence of compact sets X_n of the form given in Step I.*

We extend $f : X \to \mathbb{C}\backslash\{0\}$ to $g : \mathbb{C} \to \mathbb{C}\backslash\{0\}$ by setting $g = 0$ outside X. Then f is nowhere vanishing on X_n for some $n \geqslant 1$. Otherwise, there exists a sequence $\{x_n\}_{n \geqslant 0}$ such that $x_n \in X_n$ and $f(x_n) = 0$ for every $n \geqslant 1$. However, for every $N \geqslant 1$, $\{x_n\}_{n \geqslant N} \subseteq X_N$, and X_N being compact, $\{x_n\}_{n \geqslant N}$ has a convergent subsequence converging to $x \in X_N$. It follows that $x \in X$ with $f(x) = 0$, which contradicts the fact that f is nowhere vanishing on X. By Step I, $f = e^g h$ on X_n, where $g \in C(X_n)$ and h is a product of finitely many φ_μ's for $\mu \in \mathbb{C}\backslash X_n$. However for each such μ, μ belongs to the same component of $\mathbb{C}\backslash X$ as some $\lambda \in I$. Clearly, $\varphi_\lambda/\varphi_\mu$ admits a logarithm in $C(X)$, that is, $\varphi_\lambda/\varphi_\mu = e^{h_{\lambda,\mu}}$ for some $h_{\lambda,\mu} \in C(X)$. It follows that f, as an element in $\Lambda(X)$, is a product of finitely many φ_λ, $\lambda \in I$.

To complete the proof, it suffices to show that $X = \cap_{n \geq 1} X_n$, where $\{X_n\}_{n \geqslant 1}$ is decreasing and each X_n is bounded by finitely many disjoint smooth Jordan curves. For a positive integer m, let C_m be a finite cover of X consisting of closed discs of radius $1/m$, each meeting X. Let Y_n be the union of the members from C_n and let $X_n = \cap_{k=1}^n Y_k$. Clearly, $\{X_n\}_{n \geqslant 1}$ is a decreasing sequence and $\cap_{n \geqslant 1} X_n = \cap_{n \geqslant 1} \cap_{k=1}^n Y_k = \cap_{n \geqslant 1} Y_n = X$ (as X is contained in each Y_k). $\quad\square$

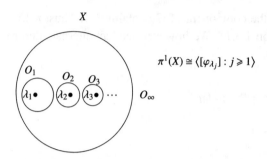

Figure 4.5 *Cohomotopy group $\pi^1(X)$ of a compact set X*

We now present the celebrated theorem of Brown–Douglas–Fillmore [26, Theorem 11.1].

Theorem 4.5.1 (BDF Theorem)

Two essentially normal operators T_1 and T_2 in $\mathcal{L}(\mathcal{H})$ are essentially equivalent if and only if

$$\sigma_e(T_1) = X = \sigma_e(T_2),$$
$$ind(T_1 - \lambda) = ind(T_2 - \lambda), \ \lambda \in \mathbb{C}\backslash X. \tag{4.5.14}$$

Proof The necessity part follows from Theorems 1.2.8 and 1.3.1. To see the sufficiency part, let τ_1 and τ_2 be $*$-monomorphism associated with T_1 and T_2, respectively. Assume that (4.5.14) holds. It suffices to check that τ_1 and τ_2 are equivalent, that is, $[\tau_1] = [\tau_1]$. Let $\mathbb{C}\backslash X = O_\infty \cup O_1 \cup \cdots$, where O_∞ is the unbounded component and $O_1, O_2 \ldots$ are bounded components. Then, by Lemma 4.5.1, $\pi^1(X)$ is the free abelian group with one generator $[O_i]$ for each bounded component. If $[\tau]$ is the extension corresponding to an essentially normal operator T, then the map $\gamma_X([\tau])$ satisfies

$$\gamma_X([\tau])([O_i]) = ind(T^{[O_i]}) = ind(T - \lambda_i), \quad \lambda_i \in O_i, \ i \geq 1.$$

If we now assume that T_1, T_2 have the same essential spectra and same index data, then $\gamma_X([\tau_1])$ and $\gamma_X([\tau_2])$ agree on generators of $\pi^1(X)$, and hence, by the injectivity of γ_X (Corollary 4.4.1), we obtain $[\tau_1] = [\tau_1]$.   □

Let us discuss some immediate consequences of the BDF theorem, the first of which had been conjectured in [110].

Corollary 4.5.1

An essentially normal operator $T \in \mathcal{L}(\mathcal{H})$ is a compact perturbation of a diagonal operator if and only if its index function is identically zero.

Proof Let T be an essentially normal operator with essential spectrum X and let D be the diagonal operator with essential spectrum equal to X as constructed in Example 1.3.8. By Theorem 4.5.1, T is essentially equivalent to D if and only if its index function is identically zero.   □

It turns out that Corollary 4.5.1 is equivalent to the BDF theorem [22] (see Exercise 4.6.4). The following shows that up to essential equivalence there is only one essentially normal operator for which the complement of the essential spectrum is connected.

Corollary 4.5.2

If an operator $T \in \mathcal{L}(\mathcal{H})$ is essentially normal such that the complement of $\sigma_e(T)$ in $\mathbb{C}$ is connected, then T is a compact perturbation of a diagonal operator. In particular, for any compact set X with a connected complement in $\mathbb{C}$, Ext(X) is trivial.

Proof If X is a compact subset of $\mathbb{C}$ with a connected complement, then $\pi^1(X)$ is trivial (see Lemma 4.5.1), and hence, so is Ext(X).   □

Remark 4.5.2 The preceding corollary applies to any essentially normal operator with essential spectrum equal to the closed unit disc. For instance, if S is the unilateral shift and N_z is the multiplication operator on $L^2(\mathbb{D})$, then $S \oplus N_z$ is a compact perturbation of a diagonal operator. This fact was first proved by Deddens and Stampli by an entirely different method [43].

An application of the BDF theorem to Toeplitz operators is given as follows.

Corollary 4.5.3

If $\varphi, \psi \in C(\mathbb{T})$, then T_φ is essentially equivalent to T_ψ if and only if $\varphi(\mathbb{T}) = X = \psi(\mathbb{T})$ and $wind_\lambda(\varphi) = wind_\lambda(\psi)$ for every $\lambda \in \mathbb{C}\backslash X$. In particular, if $wind_\lambda(\varphi) = 0$ for every $\lambda \in \mathbb{C}\backslash\varphi(\mathbb{T})$, then T_φ is a compact perturbation of a normal operator.

Proof By Corollary 1.6.2, T_φ and T_ψ are essentially normal. The desired equivalence now follows from the BDF theorem, Corollary 1.6.3 and Theorem 1.6.7. The remaining part is a particular case of Corollary 4.5.2.   □

We conclude this section with an unexpected consequence of the BDF theorem.

Corollary 4.5.4

The collection

$$\mathcal{N} + C = \{N + K \in \mathcal{L}(\mathcal{H}) : N \text{ is normal and } K \text{ is compact}\}$$

is norm-closed in $\mathcal{L}(\mathcal{H})$.

Proof By Corollary 4.5.1, an essentially normal operator is a compact perturbation of a normal operator if and only its index is zero at every point outside the essential spectrum. Thus, if there exists a sequence $\{N_m + K_m\}_{m\in\mathbb{N}}$ in $\mathcal{N} + C$ converging to S then by the continuity of Fredholm index, ind $S = 0$. Moreover, as $m \to \infty$,

$$[N_m^* + K_m^*, N_m + K_m] \to [S^*, S],$$

and hence, S is essentially normal. Therefore, $S \in \mathcal{N} + C$.   □

Suppose that X, Y are any two compact subsets of $\mathbb{C}$. Let $[\tau] \in$ Ext(X). If $p, q : X \to Y$ are homotopic, then to show that $p_*([\tau]) = q_*([\tau])$, by the injectivity of γ_Y, it suffices to check that

$\gamma_Y(p_*[\tau]) = \gamma_Y(q_*[\tau])$. However, for any $g \in C(Y)$, $g \circ p$ and $g \circ q$ are homotopic, and hence, by Remark 4.4.2,

$$\gamma_Y(p_*[\tau])(g) = \mathrm{ind}(\tau(g \circ p)) = \mathrm{ind}(\tau(g \circ q)) = \gamma_Y(q_*[\tau])(g).$$

As $[\tau]$ was arbitrary, we get $p_* = q_*$. We have therefore proved the homotopy invariance of $\mathrm{Ext}(X)$ for planar compact sets X.

Proposition 4.5.1

Suppose X, Y are planar sets and $p, q : X \to Y$ are homotopic continuous maps. Then, $p_* = q_*$.

One immediate consequence of Proposition 4.5.1 is that $\mathrm{Ext}(X)$ is trivial if X is *contractible* (recall that X is contractible if the identity map on X is homotopic to a constant map), see Theorem E.1.2 and Corollary E.1.1 for generalizations of this fact to arbitrary compact metric spaces.

4.6 Notes and Exercises

- Elmentary properties of projective limits including Lemma 4.2.1 are borrowed from [78] whereas, Example 4.4.5 and Lemma 4.5.1 are taken from [42].

- The treatment of the remaining part of this chapter follows closely the expositions [25, 26, 27].

Exercises

Exercise 4.6.1 Let $n \geqslant 1$ be an integer. Let $A \in \mathcal{L}(H)$ be a self-adjoint operator with operator matrix decomposition $(A_{ij})_{i,j \geqslant 0}$ with respect to the decomposition $\mathcal{H} = \oplus_{k \geqslant 0} \mathcal{H}_k$. Show that $(A_{ij})_{i,j \geqslant 0}$ is n-diagonal if and only if

$$A\mathcal{H}_k \subseteq \mathcal{H}_0 \oplus \cdots \oplus \mathcal{H}_{k+n} \quad \text{for all integers } k \geqslant 0.$$

Exercise 4.6.2 ([78]) For every positive integer n, let $X_n = (0,1)$, and $p_n = f$, where $f : (0,1) \to (0,1)$ is given by $f(x) = x/2$, $x \in (0,1)$. Show that

$$\varprojlim(X_n, p_n) = \emptyset.$$

Hint: If n is a positive integer, then no point of $\varprojlim(X_n, p_n)$ can have a point with first coordinate greater than $1/2$.

Exercise 4.6.3 Let X be a compact metric space. Suppose that there exists $x_0 \in X$ such that $X \setminus \{x_0\} = \cup_{n=1}^{\infty} X_n$ for some clopen subsets X_n of X. Show that the diameter of X_n tends to 0 as $n \to \infty$.

Exercise 4.6.4 ([22]) Let T_1 and T_2 be two essentially normal operators such that (4.5.14) holds. Let T_1' denote the real transpose of T_1. Verify the following statements:

(i) $T_1 \oplus T_1'$ and $T_1' \oplus T_2$ are essentially equivalent to a normal operator.

(ii) T_1 and T_2 are essentially equivalent.

Conclude that Corollary 4.5.1 is equivalent to the BDF theorem.

Hint: For the first part, apply Corollary 4.5.1, whereas the second part may be concluded from the absorption lemma.

Exercise 4.6.5 Deduce that for any compact subset σ of $\mathbb{T}$, $\text{Ext}(\sigma)$ is given by

$$\text{Ext}(\sigma) = \begin{cases} \mathbb{Z} & \text{if } \sigma = \mathbb{T}, \\ \{0\} & \text{otherwise.} \end{cases}$$

Exercise 4.6.6 Let X denote the theta curve (see Figure 3.1). Show that $\text{Ext}(X)$ is isomorphic to $\mathbb{Z} \oplus \mathbb{Z}$.

Hint: Combine Example 4.1.5 with the BDF theorem.

Exercise 4.6.7 Show that the representation $[\tau] = (i_{A,X})_*[\tau_A] + (i_{B,X})_*[\tau_B]$ of Corollary 4.1.1 in general is not unique.

Hint: Consider $X = \{(1/2, y) : 0 \leqslant y \leqslant 1\}$.

Exercise 4.6.8 ([42]) Show that there exists essentially normal operators T_1 and T_2 satisfying the following:

(i) $T_1 T_2 - T_2 T_1$ is compact,

(ii) there exists commuting normal operators N_1, N_2 and compact operators K_1, K_2 such that $T_j \oplus T_j = N_j + K_j$, $j = 1, 2$,

(iii) T_1 and T_2 are not simultaneously normal modulo compacts.

Hint: If $\mathbb{P}$ denotes the real projective plane (considered as a subset of $\mathbb{C}^2$), then $\text{Ext}(\mathbb{P})$ contains an element of order 2.

Exercise 4.6.9 ([142]) Show that $\{z^m w^n\}_{m,n\in\mathbb{N}}$ forms an orthogonal basis for $H^2(\mathbb{S})$, where $\mathbb{S}$ is the unit sphere in $\mathbb{C}^2$.

Hint: Recall that by definition, $H^2(\mathbb{S})$ is the closure of polynomials in $L^2(\mathbb{S})$. Now verify (4.4.13).

Exercise 4.6.10 Let X be a compact subset of $\mathbb{C}$. Show, by elementary means, that the image of γ_X is a subgroup of $\text{Hom}(\pi^1(X), \mathbb{Z})$. Derive Theorem 4.3.1 from Corollary 4.4.1.

Hint: Let T be an essentially normal operator with essential spectrum X. Let T' be the transpose (with respect to fixed orthonormal basis of $\mathcal{H}$) of T, and note that T' is essentially normal with $\sigma_e(T') = X$. Note that $\gamma_X([\tau_T] + [\tau_{T'}])([id]) = \gamma_X([\tau_{T\oplus T'}])([id]) = \text{ind}\, T + \text{ind}\, T' = 0$. Conclude that $\gamma_X([\tau_T])$ and $\gamma_X[\tau_{T'}]$ are inverses of each other.

Exercise 4.6.11 Let $\mathbb{S}$ be a unit sphere in $\mathbb{C}^2$. For $f \in C(\mathbb{S})$, let $T_f : H^2(\mathbb{S}) \to H^2(\mathbb{S})$ denote the Toeplitz operator. Verify the following statements:

(i) $T_f^* = T_{\bar{f}}$.

(ii) $T_{fg} - T_f T_g$ is a compact operator.

(iii) T_f is a compact operator if and only if $f = 0$.

Exercise 4.6.12 Let X be a compact subset of $\mathbb{R}^2$ and let $\mathcal{H}$ be a complex Hilbert space. Verify the following:

(i) Let $S = (S_1, S_2)$ be a commuting pair of self-adjoint operators in $Q(\mathcal{H})$ such that X is the joint spectrum of (S_1, S_2). Define $\tau_S : C(X) \to Q(\mathcal{H})$ by the continuous functional calculus of (S_2, S_2), that is,

$$\tau_S(f) = f(S_1, S_2), \quad f \in C(X).$$

Then, τ_S defines a $*$-monomorphism.

(ii) Let $\tau : C(X) \to Q(\mathcal{H})$ be a $*$-monomorphism. Then, $(\tau(x_1), \tau(x_2))$ is a commuting pair of self-adjoint operators in $Q(\mathcal{H})$ with joint spectrum X.

Conclude that for any two pairs $S = (S_1, S_2)$ and $T = (T_1, T_2)$ of commuting self-adjoint operators in $Q(\mathcal{H})$, there exists a unitary $U \in \mathcal{L}(\mathcal{H})$ such that

$$\pi(U) S_j = T_j \pi(U), \quad j = 1, 2$$

if and only if τ_S and τ_T are essentially equivalent.

Hint: For the first part, note that the C^*-algebra generated by $(\tau(x_1), \tau(x_2))$ is isometrically isomorphic to $C(X)$. The second part follows from Proposition 2.5.1.

5

Applications to Operator Theory

In this chapter, we discuss several applications of the BDF theorem. These include a model theorem for a family of essentially normal operators, classification of essentially homogeneous operators, description of essential spectra of essentially normal circular operators. We also discuss applications to the theories of hyponormal and m-isometric operators. We conclude this chapter with a fairly long discussion on essentially reductive quotient sub-modules.

5.1 Bergman Operators and Surjectivity of γ_X

Let G be a bounded open subset of the complex plane $\mathbb{C}$ and let dA denote the normalized area measure on G. Let $L^2(G, dA)$ (or $L_a^2(G)$) stand for the Hilbert space of dA-square integrable Lebesgue measurable functions on G (with two functions being identified if they are dA-almost everywhere equal to each other). Let

$$L_a^2(G, dA) = \left\{ f \in L^2(G, dA) : f \text{ is analytic on } G \right\}.$$

The normed linear space $L_a^2(G, dA)$, with the norm inherited from $L^2(G, dA)$, is a reproducing kernel Hilbert space as shown here.

Lemma 5.1.1

For any compact subset K of G, there exists $C_K > 0$ such that

$$\sup_{z \in K} |f(z)| \leqslant C_K \|f\|, \quad f \in L_a^2(G, dA). \tag{5.1.1}$$

In particular, for every $\lambda \in G$, the point evaluation $f \mapsto f(\lambda)$ is bounded on $L_a^2(G, dA)$ and $L_a^2(G, dA)$ is a Hilbert space.

Proof Let $z_0 \in G$ and let $r < \text{dist}(z_0, \partial G)$. By the mean value property for analytic functions,

$$f(z_0) = \frac{1}{\pi r^2} \iint_{\mathbb{D}_r(z_0)} f(z) dA.$$

147

It follows now from the Cauchy–Schwarz inequality that

$$|f(z_0)| \leqslant \frac{1}{r\sqrt{\pi}} \|f\|.$$

As r can be taken to be arbitrarily close to $\mathrm{dist}(z_0, \partial G)$, we get

$$|f(z_0)| \leqslant \frac{1}{\mathrm{dist}(z_0, \partial G)\sqrt{\pi}} \|f\|.$$

Let $z \in \mathbb{D}_r(z_0)$ and note that by applying the estimate above, we obtain

$$\sup_{z \in \mathbb{D}_r(z_0)} |f(z)| \leqslant \frac{1}{\mathrm{dist}(\mathbb{D}_r(z_0), \partial G)\sqrt{\pi}} \|f\|.$$

A routine compactness argument now yields (5.1.1). To see that $L_a^2(G, dA)$ is complete, it suffices to check that $L_a^2(G, dA)$ is closed in $L^2(G, dA)$. By (5.1.1), the convergence in $L_a^2(G, dA)$ yields the uniform convergence on all compact subsets of G. As uniform limit of analytic functions is analytic, the desired conclusion may be derived from the fact that every convergent sequence in $L^2(G, dA)$ has a subsequence converging pointwise almost everywhere. □

By Lemma 5.1.1 and the Riesz representation theorem (see Lemma B.1.8), for any $w \in G$, there exists $k_w \in L_a^2(G, dA)$ such that

$$\langle f, k_w \rangle = f(w) \quad \text{for all } f \in L_a^2(G, dA). \tag{5.1.2}$$

The two-variable function $B(z, w) := k_w(z)$, $z, w \in \Omega$, is uniquely determined by the reproducing property (5.1.2).

We now define the *Bergman operator* B_G as the operator of multiplication by z on $L_a^2(G, dA)$. As G is a bounded subset of $\mathbb{C}$, B_G defines a bounded linear operator on $L_a^2(G, dA)$. Moreover, for every $\lambda \in G$, $\bar{\lambda}$ is an eigenvalue of B_G^*. Indeed, for any $f \in L_a^2(G, dA)$,

$$\langle B_G^* k_\lambda, f \rangle = \langle k_\lambda, zf \rangle = \overline{\bar{\lambda} f(\lambda)} = \bar{\lambda} \langle k_\lambda, f \rangle,$$

that is, $B_G^* k_\lambda = \bar{\lambda} k_\lambda$. This also shows that $\overline{G} \subseteq \sigma(B_G)$, and more importantly, the spectrum of B_G has nonempty interior.

Let us compute the essential spectra and index data of Bergman operators.

Lemma 5.1.2

The essential spectrum of B_G is contained in ∂G. Moreover,

$$\mathrm{ind}(B_G - \lambda) = -1 \quad \text{for every } \lambda \in G.$$

Proof Let $\lambda \in G$. It is easy to see from (5.1.2) that $B_G^* \kappa_\lambda = \bar{\lambda} \kappa_\lambda$. We claim that the range of $B_G - \lambda$ is given by

$$\mathrm{ran}(B_G - \lambda) = L_a^2(G, dA) \ominus [\kappa_\lambda] = \{ f \in L_a^2(G, dA) : f(\lambda) = 0 \},$$

where $[\kappa_\lambda]$ denotes the space spanned by κ_λ. Clearly, $\mathrm{ran}(B_G - \lambda) \subseteq L_a^2(G, dA) \ominus [\kappa_\lambda]$. To see the reverse inclusion, let $f \in L_a^2(G, dA)$ be such that $f(\lambda) = 0$. In particular, the function $f(z)/(z - \lambda)$

has removable singularity at λ, and hence, there exists a holomorphic function $g : G \to \mathbb{C}$ such that $f(z) = (z - \lambda)g(z)$ for all $z \in G$. We must check that $g \in L_a^2(G, dA)$. Let $\mathbb{D}_r(\lambda)$ be a disc around λ with closure contained in G. By (5.1.1), there exists $C_\lambda > 0$ such that $\sup_{z \in \mathbb{D}_r(\lambda)} |f(z)| \leqslant C_\lambda \|f\|$. Clearly,

$$|g(z)| \leqslant C_\lambda \frac{\|f\|}{r} \text{ on } |z - \lambda| = r,$$

and hence, by the maximum modulus principle, $|g(z)| \leqslant C_\lambda \frac{\|f\|}{r}$ on $\mathbb{D}_r(\lambda)$. On the other hand,

$$\int_{G \backslash \mathbb{D}_r(\lambda)} |g(z)|^2 dA(z) \leqslant \frac{\|f\|^2}{r^2},$$

and hence, $g \in L_a^2(G, dA)$. The remaining part follows from the fact that $B_G - \lambda$ is injective for every $\lambda \in \mathbb{C}$. $\qquad\square$

We say that a bounded, open set G is an *admissible domain* if polynomials are dense in $L_a^2(G, dA)$. By a classical result of Farrell and Markusevic [34, Chapter II, Theorem 8.15], every bounded *Carathéodory region* (that is, an open connected set with same boundary and outer boundary) is admissible.

The proof of the first application relies on a special case of the Berger–Shaw theorem for Bergman operators [33], [34]. An operator $T \in \mathcal{L}(\mathcal{H})$ is *hyponormal* if the self-commutator $T^*T - TT^*$ of T is positive.

> **Lemma 5.1.3**

If G is an admissible domain in $\mathbb{C}$, then the Bergman operator B_G is of trace-class. Moreover, $\mathrm{trace}\,[B_G^*, B_G] \leqslant 1$.

Proof   Assume that G is an admissible domain in $\mathbb{C}$ and let N_z denote the operator of multiplication by z on $L^2(G, dA)$. Then, $B_G = N_z|_{L_a^2(G,dA)}$ and $B_G^* = PN_z^*|_{L_a^2(G,dA)}$, where P is the orthogonal projection of $L^2(G, dA)$ onto $L_a^2(G, dA)$. As P is of norm 1, for any $f \in L_a^2(G, dA)$, we have

$$
\begin{aligned}
\langle (B_G^* B_G - B_G B_G^*)f, f \rangle &= \|B_G f\|^2 - \|B_G^* f\|^2 \\
&= \|N_z f\|^2 - \|PN_z^* f\|^2 \\
&\geqslant \|N_z f\|^2 - \|N_z^* f\|^2,
\end{aligned}
$$

which is 0 as N_z is normal. Thus, B_G is hyponormal.

After replacing B_G by a scalar multiple of B_G, we may assume that B_G has norm less than 1. We claim that there exists a Hilbert–Schmidt injection $W : L_a^2(G, dA) \to L_a^2(G, dA)$ with dense range such that $WU_+ = B_G W$, where $U_+(e_k) = e_{k+1}$ is the shift operator for an orthonormal basis $\{e_k\}_{k \geqslant 0}$ of $L_a^2(G, dA)$. Define a linear operator W by

$$W\left(\sum_{k=0}^{n} a_k e_k\right) = \sum_{k=0}^{n} a_k z^k, \quad a_0, \ldots, a_k \in \mathbb{C}.$$

By the Cauchy–Schwarz inequality, $\|W(a)\| \leqslant \dfrac{1}{\sqrt{1-\|B_G\|^2}}\|a\|_2$ holds for any $a = \sum_{k=0}^{n} a_k e_k$ and $n \in \mathbb{N}$. Thus, W extends to a bounded linear operator from $L_a^2(G, dA)$ into $L_a^2(G, dA)$. A similar calculation shows that

$$\sum_{k=0}^{\infty} \|We_n\|^2 \leqslant \sum_{k=0}^{\infty} \|z^n\|^2 = \sum_{k=0}^{\infty} \|B_G^n 1\|^2 \leqslant \frac{\|1\|^2}{1-\|B_G\|^2},$$

and hence, W is a Hilbert–Schmidt operator. Clearly, W has dense range and it satisfies $WU_+ = B_G W$. Moreover, $W(a) = 0$ if and only if $\sum_{k=0}^{\infty} a_k B_G^k = 0$. By spectral mapping theorem, this is possible only if $a = 0$, where we used the fact that $\sigma(B_G)$ has a nonempty interior. Thus, W is one-to-one and the claim stands verified.

For $t \in \mathbb{R}$, let $A_t = (U_+ \oplus B_G)J_t$, where J_t is the orthogonal projection

$$J_t = \begin{bmatrix} W_t & W_t V_t^* \\ V_t W_t & V_t W_t V_t^* \end{bmatrix}, \quad V_t = tW, \quad W_t = (1+t^2 W^* W)^{-1}.$$

We check that $\operatorname{trace}[A_t^*, A_t] = 1$. Write $J_t = L_t + Q_t$ with self-adjoint operator Q_t given by

$$Q_t = \begin{bmatrix} -W_t V_t V_t^* & 0 \\ 0 & V_t W_t V_t^* \end{bmatrix}.$$

As the product of the Hilbert–Schmidt operators is of trace-class (see Corollary 1.4.1), Q_t is of trace-class. Repeated applications of Corollary 1.4.1(3) show that

$$[A_t^*, A_t] = [L_t(U_+ \oplus B_G)^*, (U_+ \oplus B_G)L_t] + R_t,$$

where R_t is a trace-class operator. However, as $\operatorname{trace} AB = \operatorname{trace} BA$ for any trace class operator A and bounded linear operator B (Corollary 1.4.1(3)), it follows that $\operatorname{trace} R_t = 0$. A routine calculation shows that

$$[L_t(U_+ \oplus B_G)^*, (U_+ \oplus B_G)L_t]$$

$$= \begin{bmatrix} P_0 + W_t V_t^* B_G^* B_G V_t W_t - U_+ W_t V_t^* V_t W_t U_+^* & P_0 W_t V_t^* + S_t \\ V_t W_t P_0 + S_t^* & V_t W_t^2 V_t^* - B_G V_t W_t^2 V_t^* B_G^* \end{bmatrix},$$

where P_0 is the orthogonal projection onto the space spanned by e_0 and S_t is a trace-class operator. It is now immediate that $[A_t^*, A_t]$ is of trace-class, and

$$\begin{aligned} \operatorname{trace}[A_t^*, A_t] &= \operatorname{trace} P_0 + \operatorname{trace} W_t V_t^* B_G^* B_G V_t W_t - \operatorname{trace} U_+ W_t V_t^* V_t W_t U_+^* \\ &+ \operatorname{trace} V_t W_t^2 V_t^* - \operatorname{trace} B_G V_t W_t^2 V_t^* B_G^*. \end{aligned}$$

By Corollary 1.4.1(3), we have $\operatorname{trace}[A_t^*, A_t] = 1$.

To complete the proof, we check that $\operatorname{trace}[B_G^*, B_G] \leqslant 1$. Fix an orthonormal basis $\{\phi_n\}_{n \geqslant 0}$ of $L_a^2(G, dA)$, and note that

$$\operatorname{trace}[A_t^*, A_t] \geqslant \sum_{n=0}^{\infty} (\|(U_+ \oplus B_G)J_t(0 \oplus \phi_n)\|^2 - \|J_t(U_+ \oplus B_G)^* J_t(0 \oplus \phi_n)\|^2). \tag{5.1.3}$$

As $\|J_t\| \leqslant 1$ and $U_+ \oplus B_G$ is hyponormal,

$$\|(U_+ \oplus B_G)J_t(0 \oplus \phi_n)\|^2 - \|J_t(U_+ \oplus B_G)^* J_t(0 \oplus \phi_n)\|^2 \geqslant 0.$$

Moreover, by the functional calculus of W^*W, J_t converges in strong operator topology to $0 \oplus I$ as $t \to \infty$. As trace $[A_t^*, A_t] = 1$, it follows from (5.1.3) and Fatou's lemma that trace $[B_G^*, B_G] \leqslant 1$. $\qquad\square$

Here is the main result of this section. In the remaining part, for a positive integer n, the symbol $T^{\oplus n}$ denotes the n-fold direct sum of the operator T.

Theorem 5.1.1

Let $T \in \mathcal{L}(\mathcal{H})$ be such that $\mathbb{C}\backslash\sigma_e(T)$ has finitely many bounded, admissible components. If T is essentially normal, then there exist negative integers $p_1,\ldots,p_k$ and positive integers $q_1,\ldots,q_m$ such that T is essentially equivalent to

$$N \oplus A \oplus B,$$

where N is a normal operator with essential spectrum equal to $\sigma_e(T)$, and

$$A = \oplus_{r=1}^k (B_{U_r}^{\oplus|p_r|}), \quad B = \oplus_{t=1}^m (B_{W_t}^{*\oplus q_t}) \tag{5.1.4}$$

for some bounded, admissible open subsets $U_1,\ldots,U_k$ and $W_1,\ldots,W_m$ of $\mathbb{C}$.

Proof Assume that T is essentially normal. We label the bounded, admissible components of $\mathbb{C}\backslash\sigma_e(T)$ by $U_1,\ldots,U_k$, $V_1,\ldots,V_l$ and $W_1,\ldots,W_m$ so that the index data for T is given by

$$\mathrm{ind}|_{U_r} = p_r < \mathrm{ind}|_{V_s} = 0 < \mathrm{ind}|_{W_t} = q_t,$$
$$r = 1,\ldots,k,\ s = 1,\ldots,l,\ t = 1,\ldots,m.$$

We choose a normal operator N with essential spectrum equal to $\sigma_e(T)$ (see Example 1.3.8). We now check that the essential spectra for T and $N \oplus A \oplus B$ coincide (see (5.1.4)). It is easy to see that

$$\sigma_e(N \oplus A \oplus B) = \sigma_e(N) \cup \sigma_e(A) \cup \sigma_e(B)$$

(see Exercise 1.7.38). As $\sigma_e(N) = \sigma_e(T)$, it suffices to check that $\sigma_e(A) \subseteq \sigma_e(T)$ and $\sigma_e(B) \subseteq \sigma_e(T)$. Let $\lambda \in \mathbb{C}\backslash\sigma_e(T)$. If $\lambda \in V_j$ for some $j = 1,\ldots,l$, then

$$\inf\{\mathrm{dist}(\lambda, U_i), \mathrm{dist}(\lambda, W_j) \mid 1 \leq i \leq k,\ 1 \leq j \leq m\} > 0.$$

In this case, $\lambda \notin \sigma(A) \cup \sigma(B)$ (see Exercise 1.7.37). Almost the same argument shows that if $\lambda \in U_i$, then $\oplus_{r \neq i}(B_{U_r}^{\oplus|p_r|} - \lambda)$ is invertible and $B_{U_i}^{\oplus|p_i|} - \lambda$ is Fredholm (see Lemma 5.1.2). Similarly, one can treat the case in which $\lambda \in V_j$. Note that the index data agree for T and $N \oplus A \oplus B$. The desired conclusion now follows from the BDF theorem provided we check that A and B are essentially normal. However, A (respectively B) being finite direct sum of essentially normal operators is essentially normal (see Lemma 5.1.3). This completes the proof. □

The foregoing model theorem is true without the assumption that the complement of the essential spectrum has finitely many bounded, admissible components. The general version however relies on the Berger–Shaw theorem [34], and can be obtained along similar lines. This general version can be used to prove that any essentially normal operator admits a proper Følner sequence in the sense of [7] (see [7, Theorem 3.5]). What follows is a special case of Theorem 5.1.1.

Corollary 5.1.1

Let $T \in \mathcal{L}(\mathcal{H})$ be an essentially normal operator such that $\mathbb{C}\backslash\sigma_e(T)$ has bounded, admissible components. If for every $\lambda \in \mathbb{C}\backslash\sigma_e(T)$, $\mathrm{ind}(T - \lambda) \leqslant 0$, then there exist positive integers $p_1,\dots,p_m$ such that T is essentially equivalent to

$$N \oplus B_{U_1}^{\oplus p_1} \oplus \cdots \oplus B_{U_m}^{\oplus p_m},$$

where N is a normal operator with essential spectrum equal to $\sigma_e(T)$, and $U_1,\dots,U_m$ are bounded, admissible open subsets of $\mathbb{C}$.

If T is a hyponormal operator, then so is $T - \lambda$ for any $\lambda \in \mathbb{C}$. In particular,

$$\ker(T - \lambda) \subseteq \ker(T^* - \overline{\lambda}),$$

and hence, $\mathrm{ind}(T - \lambda) \leqslant 0$ for every $\lambda \in \mathbb{C}\backslash\sigma_e(T)$ in this case. Combining this with the preceding corollary yields the following:

Corollary 5.1.2

Let $T \in \mathcal{L}(\mathcal{H})$ be an essentially normal operator such that the complement of $\sigma_e(T)$ has finitely many bounded, admissible components. If T is hyponormal, then there exist positive integers $p_1,\dots,p_m$ such that T is essentially equivalent to

$$N \oplus B_{U_1}^{\oplus p_1} \oplus \cdots \oplus B_{U_m}^{\oplus p_m},$$

where N is a normal operator with essential spectrum equal to $\sigma_e(T)$, and $U_1,\dots,U_m$ are bounded, admissible open subsets of $\mathbb{C}$.

5.2 Hyponormal Operators and m-Isometries

In this section, we see some applications of the BDF theorem to the theories of hyponormal operators and m-isometries. We begin with a lemma (refer to Appendix B for the definition of the spectral radius).

Lemma 5.2.1

For any hyponormal operator $T \in \mathcal{L}(\mathcal{H})$, norm $\|T\|$ of T is equal to the spectral radius $r(T)$ of T.

Proof Note that for any positive integer n and any $h \in \mathcal{H}$,

$$\|T^n h\|^2 = \langle T^* T^n h, T^{n-1} h \rangle \le \|T^* T^n h\| \|T^{n-1} h\| \le \|T^{n+1} h\| \|T^{n-1} h\|.$$

Thus, we have $\|T^n\|^2 \le \|T^{n+1}\| \|T^{n-1}\|$ for any positive integer n. We now check by induction on integers $n \ge 1$ that $\|T^n\| \ge \|T\|^n$. Assuming the inductive hypothesis for $k = 1, \ldots, n$, we obtain

$$\|T^{n+1}\| \|T\|^{n-1} = \|T^{n+1}\| \|T^{n-1}\| \ge \|T^n\|^2 = \|T\|^{2n},$$

which yields $\|T^{n+1}\| \ge \|T\|^{n+1}$ completing the proof of induction. By the spectral radius formula $r(T) = \lim_{n \to \infty} \|T^n\|^{1/n}$ (see Lemma B.1.9), we get $r(T) \ge \|T\|$. As $r(T) \le \|T\|$ holds true for any operator T in $\mathcal{L}(\mathcal{H})$ (see (B.1.3)), we obtain the equality $r(T) = \|T\|$. □

Remark 5.2.1 In view of Remark 1.2.6, we can define an element in a C^*-algebra to be hyponormal exactly in the same way as we do in the C^*-algebra of bounded linear operators on a Hilbert space. In particular, the spectral radius formula (see Lemma B.1.9(4)) as well as Lemma 5.2.1 hold (with the same proof) for hyponormal elements in the Calkin algebra..

We say that $T \in \mathcal{L}(\mathcal{H})$ is *essentially hyponormal* if the image $\pi(T)$ of T under the Calkin map π is hyponormal in the Calkin algebra. Similarly, one may say that T is *essentially isometry* if $\pi(T)$ is an isometry.

In the following theorem, we describe all essentially hyponormal operators with essential spectrum contained in the unit circle.

Theorem 5.2.2

Let $T \in \mathcal{L}(\mathcal{H})$ be an essentially hyponormal operator. If $\sigma_e(T)$ is contained in the unit circle $\mathbb{T}$, then T is a compact perturbation of a unitary operator, a shift of multiplicity n or the adjoint of the shift of multiplicity n.

Proof Assume that $\sigma_e(T) \subseteq \mathbb{T}$. Thus, $S = \pi(T)$ is a hyponormal operator in the Calkin algebra with spectrum contained in the unit circle. By Lemma 5.2.1 and Remark 5.2.1, $\|S\| = 1$. As T is Fredholm, S is invertible. Moreover, S^{-1} is a hyponormal operator. For any $\lambda \in \mathbb{C} \setminus \{0\}$, $S^{-1} - \lambda = -\lambda S^{-1}(S - \lambda^{-1})$. It is now easy to see that the spectrum of S^{-1} is also contained in the unit circle. By arguing as earlier, we obtain $\|S^{-1}\| = 1$. Thus, we have $S^*S \le I$ and $(S^*)^{-1}S^{-1} \le I$. However, then S must be an unitary, and hence, T is essentially unitary. The desired conclusion now follows from Theorem 2.3.5. □

We can now identify, up to essential equivalence, all hyponormal operators with essential spectrum contained in the unit circle.

Corollary 5.2.1

Let $T \in \mathcal{L}(\mathcal{H})$ be a hyponormal operator. If $\sigma_e(T) \subseteq \mathbb{T}$, then T is a compact perturbation of a unitary or a shift of multiplicity n.

Proof   Note that $S = \pi(T)$ is a hyponormal operator in the Calkin algebra. As $\ker A \subseteq \ker A^*$ holds for any hyponormal operator $A \in \mathcal{L}(\mathcal{H})$, the index of T is always non-positive. We may now apply Theorem 5.2.2.   □

The preceding two results are applicable to (essentially) isometric Fredholm operators. Here is a general fact, first obtained in [62, Theorem 6.2], applicable to possibly non-Fredholm essentially isometric operators.

Theorem 5.2.3

An operator $T \in \mathcal{L}(\mathcal{H})$ is essentially an isometry if and only if $T = U + K$, where $K \in \mathcal{L}(\mathcal{H})$ is compact and $U \in \mathcal{L}(\mathcal{H})$ is either an isometry or co-isometry (that is, T^* is an isometry) with finite dimensional kernel.

Proof   The sufficiency part is immediate from the definition of an essentially isometric operator.   To see the necessity part, note first that any essentially isometric operator is left-Fredholm. Hence, by Remark 1.2.9, T has a closed range and finite dimensional kernel. The desired conclusion now can be obtained by following the proof of Theorem 2.3.5 (except the application of the absorption lemma). This completes the proof.   □

As the index of a co-isometry is positive unless it is unitary, Theorem 5.2.3 yields the following (see [62, pp. 191]):

Corollary 5.2.2

An operator $T \in \mathcal{L}(\mathcal{H})$ is a compact perturbation of an isometry if and only if T is essentially isometric with $\mathrm{ind}(T) \leqslant 0$ (possibly infinite).

An isometry is said to be *pure* if it has no unitary summand (or equivalently, it is analytic). A result of Douglas says that any non-invertible isometry is a compact perturbation of a pure isometry (refer to [45]).

In the remaining part of this section, we discuss applications of BDF theory to the well-studied class of *m*-isometries. The notion of an *m*-isometry is due to Agler (see [2, pp. 11]).

Definition 5.2.4

Let m be a positive integer. A bounded linear operator $T \in \mathcal{L}(\mathcal{H})$ is said to be an *m-isometry* if $B_m(T) = 0$, where $B_m(T)$ is the defect operator given by

$$B_m(T) = \sum_{k=0}^{m} (-1)^k \binom{m}{k} T^{*k} T^k. \tag{5.2.5}$$

An 1-isometry is just an isometry. We say that T is *norm increasing* if $T^*T \geqslant I$.

Remark 5.2.5   Note that $T \in \mathcal{L}(\mathcal{H})$ is an *m*-isometry if and only if for every $h \in \mathcal{H}$, $f_h(k) = \|T^k h\|^2$ is a polynomial of degree at most $m-1$ (see Exercise 5.6.8). Further, it can be seen that T is an *m*-isometry if and only if $k \mapsto T^{*k} T^k$ is an $\mathcal{L}(\mathcal{H})$-valued polynomial of degree at most $m-1$. It follows from Lemma B.1.9(4) that the spectral radius of an *m*-isometry is necessarily 1.

Lemma 5.2.2

Let $T \in \mathcal{L}(\mathcal{H})$ be an m-isometry. Then the following are true:

(i) *The approximate point spectrum of T is contained in the unit circle.*

(ii) *If T is norm increasing and $\lambda \in \mathbb{T}$ is an eigenvalue of T, then $\ker(T - \lambda)$ is a reducing subspace for T.*

Proof   Let $\lambda \in \sigma_{ap}(T)$. Then there exists a sequence $\{x_n\}_{n \geqslant 1}$ of unit vectors in $\mathcal{H}$ such that $\|Tx_n - \lambda x_n\| \to 0$ as $n \to \infty$. Since

$$T^k - \lambda^k = (T - \lambda) \sum_{j=0}^{k-1} T^j \lambda^{k-1-j}, \quad k \geqslant 1,$$

$\|(T^k - \lambda^k)x_n\| \to 0$ as $n \to \infty$. This yields that

$$0 = \langle B_m(T)x_n, x_n \rangle \to \sum_{k=0}^{m} (-1)^k \binom{m}{k} |\lambda|^{2k} = (1 - |\lambda|^2)^m,$$

and hence $|\lambda| = 1$ completing the verification of (i).

To see (ii), assume that T is norm increasing and let $\lambda \in \mathbb{T}$ be such that $Tx = \lambda x$ for some nonzero $x \in \mathcal{H}$. Since $\bar{\lambda}T$ is an m-isometry, after replacing T by $\bar{\lambda}T$, we may assume that $\lambda = 1$, that is, $Tx = x$. Note that $\langle (T^*T - I)x, x \rangle = 0$, and since T is norm increasing, $T^*Tx = x$. Moreover, since T is left-invertible, T^*T is invertible (see Exercise 1.7.3), and hence $T' = T(T^*T)^{-1}$ defines an operator in $\mathcal{L}(\mathcal{H})$. Note that T' is a contraction such that $T'x = x$, and hence

$$\|T'^*x - x\|^2 = \|T'^*x\|^2 - 2\Re\langle T'^*x, x\rangle + \|x\|^2 = \|T'^*x\|^2 - \|x\|^2 \leqslant 0.$$

This implies that $T'^*x = x$ or $T^*x = x$ completing the proof.   □

We need a preliminary result in the classification, up to essential equivalence, of all finitely multicyclic m-isometries, which are norm increasing (see Exercise 1.7.36 for the definition of multicyclic operators).

Lemma 5.2.3

Let $T \in \mathcal{L}(\mathcal{H})$ be finitely multicyclic. If T is a norm increasing m-isometry, then either T is unitary or $\sigma_e(T) = \mathbb{T}$.

Proof   Assume that T is a norm increasing m-isometry. We consider here two cases.

Case I *T is invertible:*

As an invertible isometry is unitary, we may assume that $m \geqslant 2$. Let $B_m(\cdot)$ be as given in (5.2.5). Note that $S = T^{-1}$ is also an m-isometry. Further, as $S^*S \leqslant I$, $\{S^{*n}S^n\}_{n \geqslant 0}$ is a monotonically decreasing sequence bounded below by 0. Hence, there exists a positive operator A in $\mathcal{B}(\mathcal{H})$ such that

$$S^{*n}S^n \to A \text{ (wot) as } n \to \infty, \tag{5.2.6}$$

where wot stands for the convergence in the weak operator topology (see Exercise 5.6.9). Moreover, as $B_m(S) = B_{m-1}(S) - S^* B_{m-1}(S)S$, one has

$$B_{m-1}(S) = S^* B_{m-1}(S)S. \tag{5.2.7}$$

An inductive argument on k shows that $B_{m-1}(S) = S^{*k} B_{m-1}(S)S^k$, $k \geqslant 1$. We claim that $S^{*k} B_{k-1}(S)S^k \to 0$ (wot) as $k \to \infty$. Note that for any integer $k \geqslant 1$,

$$S^{*k} B_{m-1}(S)S^k = \sum_{0 \leqslant p \leqslant m-1} (-1)^p \binom{m-1}{p} S^{*p+k} S^{p+k}.$$

Letting $k \to \infty$, we obtain

$$S^{*k} B_{m-1}(S)S^k \to \sum_{0 \leqslant p \leqslant m-1} (-1)^p \binom{m-1}{p} A = 0 \ (\text{wot}).$$

Thus, the claim stands verified and we conclude from (5.2.7) that S is an $(m-1)$-isometry. By a finite inductive argument, S must be an isometry, and hence, T is unitary.

Case II $0 \in \sigma(T)$:

In view of Lemma 5.2.2, we may assume that T has no eigenvalues. As $T^* T \geqslant I$, for any $\lambda \in \mathbb{D}$,

$$\|Tx - \lambda x\| \geqslant (1 - |\lambda|)\|x\|, \quad x \in \mathcal{H}.$$

Moreover, since T is finitely cyclic, $\dim \ker(T^* - \lambda)$ is finite for every $\lambda \in \mathbb{C}$. This together with Lemma 5.2.2 shows that

$$\sigma_e(T) = \sigma_{ap}(T) \subseteq \mathbb{T} \tag{5.2.8}$$

(cf. Exercise 1.7.36). Additionally, as the boundary of the spectrum is contained in $\sigma_{ap}(T)$ (see Remark B.1.16) and $0 \in \sigma(T)$, thererefore $\sigma_{ap}(T)$ must be $\mathbb{T}$. It now follows from (5.2.8) that $\sigma_e(T) = \mathbb{T}$. □

Theorem 5.2.6

Let $T \in \mathcal{L}(\mathcal{H})$ be finitely multicyclic. If T is a norm increasing m-isometry, then T is a compact perturbation of a unitary operator or a shift of multiplicity n according as $\operatorname{ind}(T) = 0$ or $\operatorname{ind}(T) = -n < 0$.

Proof Assume that T is a norm increasing m-isometry. In view of the preceding lemma, we may assume that $0 \in \sigma(T)$ and $\sigma_e(T) = \mathbb{T}$. Note first that $\pi(T)$ is an invertible, m-isometry. Hence, by Case I of the proof of the preceding lemma, $\pi(T)$ is unitary. In particular, T is essentially unitary with $\sigma_e(T) = \mathbb{T}$. As the index of T is negative, the desired conclusion is now immediate from Theorem 2.3.5. □

The following is the special case $m = 2$, in which the assumption $T^* T \geqslant I$ is redundant (see [115, Lemma 1(a)]).

Corollary 5.2.3

Let $T \in \mathcal{L}(\mathcal{H})$ be finitely multicyclic. If T is a 2-isometry, then T is a compact perturbation of a unitary operator or a shift of multiplicity n according as $\mathrm{ind}(T) = 0$ or $\mathrm{ind}(T) = -n < 0$.

Proof Assume that T is a 2-isometry. In view of the last theorem, it suffices to check that T is norm increasing. Suppose the contrary. Thus, for some unit vector $h \in \mathcal{H}$, $\|Th\| < 1$. One may verify by induction on $n \in \mathbb{N}$ that

$$\|T^n h\|^2 = 1 + n(\|Th\|^2 - 1), \quad n \in \mathbb{N}.$$

However, $1 + n(\|Th\|^2 - 1)$ is negative for sufficiently large values of n, which is not possible unless $T^*T \geqslant I$. □

The conclusion of the preceding corollary holds for any finitely multicyclic $T \in \mathcal{L}(\mathcal{H})$ for which $I - 2T^*T + T^{*2}T^2 \leqslant 0$. Indeed, for any such T, the commutator $T^*T - TT^*$ is of trace-class (refer to [29]). However, we do not know whether the commutator of a finitely multicyclic norm-increasing m-isometry belongs to the trace-class.

5.3 Essentially Normal Circular Operators

Definition 5.3.1

A bounded linear operator $T \in \mathcal{L}(\mathcal{H})$ is *circular* if for every $\theta \in \mathbb{R}$, $e^{i\theta}T$ is unitarily equivalent to T.

Remark 5.3.2 Note that the spectrum of a circular operator T has the circular symmetry: $z \in \sigma(T)$ if and only if $e^{i\theta}z \in \sigma(T)$ for every $\theta \in \mathbb{R}$. The same remark applies to the essential spectrum of a circular operator.

It is clear that any orthogonal direct sum of circular operators is circular whenever it defines a bounded linear operator. We now exhibit an important class of circular operators (for an excellent reference for the basic properties of weighted shifts, refer to [127]).

Example 5.3.3 (Unilateral weighted shift operators) Let $\mathcal{H}$ be a separable Hilbert space and let $\{e_n\}_{n \geqslant 0}$ be an orthonormal basis for $\mathcal{H}$. An *unilateral weighted shift operator* W in $\mathcal{L}(\mathcal{H})$ with the bounded weight sequence $\{\alpha_n\}_{n \geqslant 0}$ is defined through the relations

$$We_n = \alpha_n e_{n+1}, \quad n \geqslant 0.$$

We contend that W is circular. Fix $\theta \in \mathbb{R}$ and define $U_\theta : \mathcal{H} \to \mathcal{H}$ by setting

$$U_\theta(e_n) = e^{in\theta}e_n, \quad n \geqslant 0,$$

and extending U_θ linearly and continuously to $\mathcal{H}$ as a unitary operator. It is now easy to see that

$$U_\theta W(e_n) = \alpha_n e^{i(n+1)\theta}e_{n+1} = e^{i\theta}W U_\theta(e_n), \quad n \geqslant 0.$$

This shows that W is circular. ∎

Note that not all circular operators are weighted shifts or their direct sums (see Exercise 5.6.4). The essential spectrum of an essentially normal circular operator T can be expressed in terms of the essential spectrum of the self-adjoint operator T^*T.

Theorem 5.3.4

Let T be an essentially normal operator in $\mathcal{L}(\mathcal{H})$. If T is circular, then the essential spectrum of T is given by

$$\sigma_e(T) = \left\{ z \in \mathbb{C} : |z|^2 \in \sigma_e(T^*T) \right\}.$$

Proof Assume that T is circular. As T is essentially normal, $\pi(T)$ is a normal element in the Calkin algebra. Let $\mathcal{M}$ be the maximal ideal space of the commutative C^*-algebra $C^*(\pi(T))$ generated by $\pi(T)$. Since the spectrum of an element f in $C(X)$ is precisely its range $f(X)$, by the spectral theorem, the essential spectrum of T can be easily seen to be

$$\sigma_e(T) = \{\phi(\pi(T)) : \phi \in \mathcal{M}\}.$$

If $\lambda \in \sigma_e(T)$, then for some $\phi \in \mathcal{M}$,

$$\pi(T^*T - |\lambda|^2 I) = \pi(T^*)\pi(T) - |\phi(\pi(T))|^2.$$

Clearly, ϕ annihilates $\pi(T^*T - |\lambda|^2 I) \in C^*(\pi(T))$. Thus, $\pi(T^*T - |\lambda|^2 I)$ is not invertible, and hence, $|\lambda|^2 \in \sigma_e(T^*T)$.

Conversely, suppose $|\lambda|^2 \in \sigma_e(T^*T)$ for some $\lambda \in \mathbb{C}$. Thus, $\pi(T^*T - |\lambda|^2 I)$ is not invertible in the Calkin algebra, and hence, it is not invertible in $C^*(\pi(T))$ by spectral permanence. Thus, there exists some $\phi_\lambda \in \mathcal{M}$ annihilating $\pi(T^*T - |\lambda|^2 I)$. This gives $|\lambda|^2 = |\phi_\lambda(\pi(T))|^2$. On the other hand, $\phi_\lambda(\pi(T)) \in \sigma_e(T)$. By the circular symmetry of the essential spectrum (Remark 5.3.2), we must have $\lambda \in \sigma_e(T)$. $\qquad\square$

The following is immediate from Example 5.3.3 and Theorem 5.3.4.

Corollary 5.3.1

Let $W \in \mathcal{L}(\mathcal{H})$ be a weighted shift with bounded weight sequence $\{\alpha_n\}_{n \geqslant 0}$. If $\lim_{n \to \infty}(|\alpha_n| - |\alpha_{n-1}|) = 0$, then W is essentially normal with essential spectrum $\sigma_e(W)$ given by

$$\sigma_e(W) = \{z \in \mathbb{C} : |z|^2 \in \sigma_e(diag(|\alpha_n|^2))\}.$$

As a remarkable consequence of the BDF theorem, we see that essentially normal unilateral weighted shifts in $\mathcal{N} + \mathcal{C}$ exist in abundance.

Example 5.3.5 We produce a class of essentially normal weighted shifts, which are compact perturbations of a normal operator. To see this, consider a weighted shift W in $\ell^2(\mathbb{N})$ with weight sequence $\{\alpha_n\}_{n \geqslant 0}$ satisfying

(1) $\lim_{n \to \infty}(|\alpha_n| - |\alpha_{n-1}|) = 0$,

(2) $\{\alpha_n\}_{n \geqslant 0}$ is dense in $[0, 1]$ and

(3) each term in $\{\alpha_n\}_{n \geqslant 0}$ occurs infinitely many times.

For example, $\{\alpha_n\}_{n \geqslant 0}$ given by

$$1, \frac{1}{2}, 1, \frac{3}{4}, \frac{1}{2}, \frac{1}{4}, \frac{1}{2}, \frac{3}{4}, 1, \frac{7}{8}, \ldots,$$

satisfies (1)–(3). By Corollary 5.3.1, W is essentially normal. Further, by Corollary 5.3.1 and Theorem 1.3.16, $\sigma_e(W) = \overline{\mathbb{D}}$. By Corollary 4.5.2, W is a compact perturbation of a normal operator.   ∎

5.4 Essentially Homogeneous Operators

Let Möb($\mathbb{D}$) be the group of bi-holomorphic automorphisms of $\mathbb{D}$. If ϕ is in Möb($\mathbb{D}$), then it is of the form

$$\phi_{\alpha,a}(z) = \alpha \frac{z-a}{1-\bar{a}z}, \quad \alpha \in \mathbb{T}, a \in \mathbb{D}. \tag{5.4.9}$$

It follows that Möb($\mathbb{D}$) is *transitive*, that is, for any $a, b \in \mathbb{D}$, there exists $\phi \in$ Möb($\mathbb{D}$) such that $\phi(a) = b$. Indeed, one can take $\phi = \phi_{1,-b} \circ \phi_{1,a}$.

Definition 5.4.1

A bounded linear operator $T \in \mathcal{L}(\mathcal{H})$ with $\sigma(T) \subseteq \overline{\mathbb{D}}$ is said to be *homogeneous* if $\phi(T)$ is unitarily equivalent to T for every $\phi \in$ Möb($\mathbb{D}$). We say that T is *essentially homogeneous* if the image of T under the Calkin map is homogeneous as an element of the Calkin algebra $Q(\mathcal{H})$ (considered as a C^*-subalgebra of $\mathcal{L}(\mathcal{K})$ for some Hilbert space $\mathcal{K}$).

Remark 5.4.2   An operator $T \in \mathcal{L}(\mathcal{H})$ is essentially homogeneous if there exists a unitary representation U of Möb($\mathbb{D}$) on $\mathcal{H}$ such that

$$U_\phi^* T U_\phi - \phi \cdot T \in C(\mathcal{H}), \quad \phi \in \text{Möb}(\mathbb{D}),$$

where $\phi \cdot T = \phi(T)$. Note that if T is homogeneous, then so is T^*. This may be deduced from $\sigma(T^*) = \{\lambda \in \mathbb{C} : \bar{\lambda} \in \sigma(T)\}$ and $\phi_{\alpha,a}(T^*) = \phi_{\bar{\alpha},\bar{a}}(T)^*$, where $\phi_{\alpha,a}$ is as given in (5.4.9).

For the classification of essentially homogeneous operators, we need a couple of preparatory lemmas.

Lemma 5.4.1

For ϕ in Möb($\mathbb{D}$) and $T \in \mathcal{L}(\mathcal{H})$ with $\sigma(T) \subseteq \overline{\mathbb{D}}$, we have

$$\phi(\pi(T)) = \pi(\phi(T)), \tag{5.4.10}$$

where $\pi : \mathcal{L}(\mathcal{H}) \to Q(\mathcal{H})$ is the Calkin map.

Proof   As π is an algebra homomorphism, the conclusion in (5.4.10) holds for a polynomial ϕ. Moreover, since a Möbius transformation is a power series convergent compactly on some open set containing $\overline{\mathbb{D}}$, and $\sigma(T) \subseteq \overline{\mathbb{D}}$, we obtain the desired conclusion from the continuity of π.   □

It is easy to describe the spectra and essential spectra of homogeneous operators.

> **Lemma 5.4.2**
>
> *The spectrum (resp. essential spectrum) of a homogeneous operator $T \in \mathcal{L}(\mathcal{H})$ is either the unit circle $\mathbb{T}$ or the closed unit disc $\overline{\mathbb{D}}$.*

Proof Let T be a homogeneous operator. By definition, $\sigma(T)$ is contained in $\overline{\mathbb{D}}$. By the spectral mapping property, the spectrum is invariant under the automorphism group Möb($\mathbb{D}$). However, $\mathbb{T}$ and $\overline{\mathbb{D}}$ are the only subsets of $\overline{\mathbb{D}}$ invariant under Möb($\mathbb{D}$), and hence, $\sigma(T)$ is either the unit circle $\mathbb{T}$ or the closed unit disc $\overline{\mathbb{D}}$. As $\pi(T)$ is also homogeneous by the preceding lemma, the same assertion holds for the essential spectrum of T. $\qquad\qquad\square$

Homogeneous normal operators

Let $(X, \mathcal{A})$ and $(Y, \mathcal{B})$ be two measure spaces and $\varphi : X \to Y$ be a measurable function between them. Given σ-finite measure μ on X, define the *push-forward* $\varphi_* \mu$ of the measure μ by the requirement

$$(\varphi_* \mu)(B) := \mu(\varphi^{-1}(B)), \ B \in \mathcal{B}.$$

One of the immediate consequences of this definition is the following change of variable formula.

> **Lemma 5.4.3 (Change of variable)**
>
> *A measurable function f on Y is integrable with respect to the push-forward measure $\varphi_* \mu$ if and only if the composition $f \circ \varphi$ is integrable with respect to the measure μ. For integrable functions f and $f \circ \varphi$, we have the change of variable formula:*
>
> $$\int_Y f\, d(\varphi_* \mu) = \int_X f \circ \varphi\, d\mu. \tag{5.4.11}$$

Recall that a measure ν on a measurable space $(X, \mathcal{A})$ is said to be *absolutely continuous* with respect to a measure μ on the same space if $\mu(A) = 0$ implies $\nu(A) = 0$ for any A in the sigma algebra $\mathcal{A}$. Clearly, if $\nu(A) := \int \chi_A \alpha\, d\mu$ for some measurable non-negative function $\alpha : X \to \mathbb{R}$, then ν is absolutely continuous with respect to μ. The *Radon–Nikódym* theorem says that the converse is true, namely, if μ and ν are σ-finite measures and if ν is absolutely continuous with respect to μ, then there exists a non-negative measurable function $\alpha : X \to \mathbb{R}$ such that $\nu = \alpha\mu$, that is, $\int f d\nu = \int f\alpha d\mu$. The function α is called the *Radon–Nikodym derivative* and is denoted by $\frac{d\nu}{d\mu}$. Thus, if $\varphi_* \mu$ is absolutely continuous with respect to μ, then we have

$$\int f d(\varphi_* \mu) = \int f \circ \varphi d\mu = \int f \frac{d(\varphi_* \mu)}{d\mu} d\mu. \tag{5.4.12}$$

> **Definition 5.4.3**

Suppose that $(X, \mathcal{A}, \mu)$ is a measure space and $\varphi : X \to X$ is a measurable bijective self map such that φ^{-1} is also measurable. If $\varphi_* \mu$ and μ are mutually absolutely continuous, then we say that

φ is an automorphism of the measure space $(X,\mathcal{A},\mu)$. The set of all such automorphisms is a group, which is denoted by $\mathrm{Aut}(X,\mathcal{A},\mu)$.

Let $L^2_{\mathcal{E}}(X,\mu)$ be the space of square integrable functions on X taking values in some Hilbert space $\mathcal{E}$. Now, because of the change of variable formula (5.4.11), for any measurable function $\varphi : X \to X$, the linear operator $U_\varphi : L^2_{\mathcal{E}}(X,\mu) \to L^2_{\mathcal{E}}(X,\mu)$ given by

$$U_\varphi f = \left(\tfrac{d(\varphi_*\mu)}{d\mu}\right)^{1/2} f \circ \varphi^{-1}, \quad f \in L^2_{\mathcal{E}}(X,\mu) \tag{5.4.13}$$

is unitary. Indeed, for any $f \in L^2_{\mathcal{E}}(X,\mu)$, we have

$$\|U_\varphi f\|^2_{L^2_{\mathcal{E}}(X,\mu)} = \int \|f \circ \varphi^{-1}\|^2 d(\varphi_*\mu) = \|f\|^2_{L^2_{\mathcal{E}}(X,\mu)}.$$

Furthermore, if we assume that φ is an automorphism, then using the chain rule for the Radon–Nikódym derivative, it is evident that the map $\varphi \to U_\varphi$ is a homomorphism from $\mathrm{Aut}(\Lambda,\mathcal{A},\mu)$ to the group of unitary operators on $\mathcal{H}$. Moreover, the unitary U_φ intertwines M_φ and M_z, where M_φ and M_z are multiplication operators defined by the function φ and the coordinate function z, repsectively.

Let $\int_\Lambda \oplus \mathcal{H}_\lambda d\mu$ be a direct integral space and $\varphi : \Lambda \to \Lambda$ be an automorphism. The operator M_φ on $\int_\Lambda \oplus \mathcal{H}_\lambda d\mu$ is a normal operator and therefore it must be unitarily equivalent to the operator M_z on some other direct integral space, say $\int_{\Lambda'} \oplus \mathcal{H}'_\lambda d\mu'$. Our goal is to find this representation.

An isomorphism intertwining M_φ and M_z is equivalent to asking for an operator $\widehat{U}_\varphi$ in $\mathcal{L}(\int_\Lambda \oplus \mathcal{H}_\lambda d\mu, \int_{\Lambda'} \oplus \mathcal{H}'_\lambda d\mu')$ such that $\widehat{U}_\varphi(\lambda) : \mathcal{H}_\lambda \to \mathcal{H}'_\lambda$ is invertible for each $\lambda \in \Lambda$. The operator $\widehat{U}_\varphi$ intertwines M_z and M_φ if, for any s in $\int_\Lambda \oplus \mathcal{H}_\lambda d\mu$,

$$(\widehat{U}_\varphi s)(\lambda) = (s \circ \varphi^{-1})(\lambda), \quad \lambda \in \Lambda.$$

Now, if we take the measure μ' to be the push-forward $\varphi_*\mu$ of the measure μ, then by definition, $\widehat{U}_\varphi$ is unitary. Moreover, if we assume that $\varphi_*\mu$ is mutually absolutely continuous with respect to μ, then $\varphi_*\mu = \frac{d(\varphi_*\mu)}{d\mu}$ and $\widehat{U}_\varphi$ can be taken to be of the form given in (5.4.13) without changing the measure.

Theorem 5.4.4

A normal operator $N \in \mathcal{L}(\mathcal{H})$ is homogeneous if and only if there exists $m,m' \in \{0\} \cup \mathbb{N} \cup \{\infty\}$ such that $N = A_m \oplus B_{m'}$, where A_m is the m-fold direct sum of M_z on $L^2(\mathbb{D}, dA)$ and $B_{m'}$ is m'-fold direct sum of M_z on $L^2(\mathbb{T}, d\theta)$.

Proof Assume that the operator N is normal. It is easy to verify that the operators described in the theorem are homogeneous by constructing an explicit unitary operator intertwining M_z and M_φ, $\varphi \in \mathrm{M\ddot{o}b}(\mathbb{D})$. The area measure dA and the arc length measure $d\theta$ are quasi-invariant under the action of the group $\mathrm{M\ddot{o}b}(\mathbb{D})$. Indeed, the Radon–Nikódym derivatives are $\frac{d(\varphi_*dA)}{dA} = |\varphi'|^2$ and

$\frac{d(\varphi_*d\theta)}{d\theta} = |\varphi'|$, respectively. It follows that the direct sum of an appropriate number of copies of the unitary operator defined in (5.4.13) is the intertwining unitary.

Conversely, suppose that N is homogeneous. By the spectral mapping theorem, the spectrum of $\varphi(N)$ is the set $\varphi(\sigma(N)) := \{\varphi(z) : z \in \sigma(T)\}$. As N is homogeneous, it follows that $\sigma(N) = \varphi(\sigma(N))$ for all $\varphi \in \mathrm{M\ddot{o}b}(\mathbb{D})$. The only compact subsets of the plane with this property are $\overline{\mathbb{D}}$ and $\mathbb{T}$.

Now that the spectrum is determined, we consider the two cases, namely. $\sigma(N) = \overline{\mathbb{D}}$ and $\mathbb{T}$. In either case, a scalar-valued spectral measure μ_N supported on $\sigma(N)$ together with the multiplicity function $m_N : \mathbb{C} \to \mathbb{N} \cup \{\infty\}$ completely determines the unitary equivalence class of the operator N.

First, let us fix the spectrum of the homogeneous operator N to be $\overline{\mathbb{D}}$. Take N to be the multiplication operator M_z on the direct integral space $\int_{\overline{\mathbb{D}}} \oplus \mathcal{H}_z d\mu_N(z)$ without loss of generality. Now, for any φ in $\mathrm{M\ddot{o}b}(\mathbb{D})$, note that $\varphi(M_z) = M_\varphi$ by the usual functional calculus. By hypothesis, M_z is unitarily equivalent to M_φ. As we have discussed earlier, we can represent M_φ, up to unitary equivalence, as M_z on the direct integral space $\int_{\overline{\mathbb{D}}} \oplus (\varphi_* \mathcal{H}_z) \varphi_* d\mu_N$, where $\varphi_* \mathcal{H}_z := \{f \circ \varphi^{-1} : f \in \mathcal{H}_z\}$. Thus by the spectral theorem, we have that (a) $\dim \varphi_* \mathcal{H}_z = \dim \mathcal{H}_z$ for all φ in $\mathrm{M\ddot{o}b}(\mathbb{D})$, that is, the multiplicity function is constant. and (b) the spectral measure μ_N must be quasi-invariant.

Now, if the multiplicity is assumed to be 1, then the only possibility is that μ_n is the normalized Lebesgue measure and the Hilbert spaces $\mathcal{H}_z$ are isomorphic to the one-dimensional Hilbert space $\mathbb{C}$. If the multiplicity is > 1, then the measure μ_N must be the sum of several copies of the quasi-invariant measures each having support equal to either $\overline{\mathbb{D}}$ or $\mathbb{T}$, that is, $\mu_N = m_1 dA + m_2 d\theta$. In this case, the direct integral space can be identified with the direct sum of m_1 copies of $L^2(\mathbb{D}, dA)$ and m_2 copies of $L^2(\mathbb{T}, d\theta)$.

It is known from general principles (see [137, Theorem 5.19]) that the class of a quasi-invariant measure on $\overline{\mathbb{D}}$ is uniquely determined. Therefore, we have shown that if N is homogeneous and $\sigma(N) = \overline{\mathbb{D}}$, then it must be of the form stipulated.

The second case, where the spectrum of N equals $\mathbb{T}$ is even simpler as the measure μ_N in this case cannot split once we know that its support must be $\mathbb{T}$. This completes the proof. □

List of essentially normal homogeneous operators

The following examples of non-normal, essentially normal homogeneous operators are implicit in the proof of Theorem 5.4.4. Indeed, the first of the two examples is obtained by restricting the multiplication operator M_z on $L^2(\mathbb{D}, dA)$ to the subspace of holomorphic functions. A similar remark applies to the second example as well.

Example 5.4.5 Let B be the Bergman shift defined on the Bergman space $L_a^2(\mathbb{D})$ as discussed in Example 2.1.3. We claim that the operator B is homogeneous. To see this, define $U_\phi : L_a^2(\mathbb{D}) \to L_a^2(\mathbb{D})$ by $U_\phi f = \phi' \cdot (f \circ \phi)$, $\phi \in \mathrm{M\ddot{o}b}(\mathbb{D})$, The operator U_ϕ is unitary:

$$\int_{\mathbb{D}} |\phi'(z)|^2|(f \circ \phi)(z)| dA(z) = \int_{\mathbb{D}} |f(z)|^2 dA(z), \quad f \in L_a^2(\mathbb{D}).$$

Furthermore, we have

$$U_\phi B f = \phi'((Bf) \circ \phi) = \phi \phi'(f \circ \phi) = \phi(B)U_\phi f, \quad f \in L_a^2(\mathbb{D}),$$

which shows that $U_\phi B = \phi(B)U_\phi$. Similarly, the map $U_\phi f = \sqrt{\phi'} \cdot (f \circ \phi)$ is unitary on $H^2(\mathbb{T})$ (see Exercise 5.6.12). This unitary intertwines the multiplication by the coordinate function and the multiplication by the function ϕ on $H^2(\mathbb{T})$ showing that the multiplication operator M_z on $H^2(\mathbb{T})$ is homogeneous. ∎

We now describe all essentially homogeneous operators within the class of essentially normal operators.

Theorem 5.4.6

If $T \in \mathcal{L}(\mathcal{H})$ is an essentially normal operator, then the following statements are equivalent:

(i) T is essentially homogeneous.

(ii) T is essentially equivalent to a homogeneous operator.

(iii) $\sigma_e(T) = \mathbb{T}$ or $\sigma_e(T) = \overline{\mathbb{D}}$.

Proof Assume that T is an essentially normal operator. Clearly, any compact perturbation of a homogeneous operator is essentially homogeneous. This yields (ii) $\Rightarrow$ (i). Suppose that T is essentially homogeneous. Then, $\pi(T)$ is homogeneous, and hence, by Lemma 5.4.2, $\sigma_e(T) = \mathbb{T}$ or $\sigma_e(T) = \overline{\mathbb{D}}$. This yields the implication (i) $\Rightarrow$ (iii). It now suffices to check that (iii) $\Rightarrow$ (ii). We consider the following cases:

Case I $\sigma_e(T) = \mathbb{T}$:

In this case, by Theorem 2.3.5, either T is a compact perturbation of a unitary U, a shift of multiplicity n, or the adjoint of the shift of multiplicity n. Note that a shift of multiplicity n is essentially equivalent to n copies of the Bergman shift B, and by Example 5.4.5, B is homogeneous. As adjoint of a homogeneous operator is homogeneous, it now suffices to check that any unitary U with $\sigma_e(U) = \mathbb{T}$ is essentially equivalent to a homogeneous operator. However, by the Weyl–von Neumann–Berg theorem, U is essentially equivalent to the multiplication operator N_z on $L^2(\mathbb{T})$, and by Proposition 5.4.4, N_z is homogeneous.

Case II $\sigma_e(T) = \overline{\mathbb{D}}$:

In this case, $\text{Ext}(\sigma_e(T))$ is trivial. Hence, by Corollary 4.5.2, T is essentially equivalent to the multiplication operator M_z on $L^2(\mathbb{D}, dA)$, which is homogeneous by Proposition 5.4.4. □

Corollary 5.4.1

Any essentially normal essentially homogeneous operator in $\mathcal{L}(\mathcal{H})$ with essential spectrum $\mathbb{T}$ is essentially equivalent to orthogonal direct sum of either $\oplus_{n \in \mathbb{G}} N_z$, or $\oplus_{n \in \mathbb{G}} M_z$, or $\oplus_{n \in \mathbb{F}'} M_z^*$ for some finite subsets $\mathbb{F}, \mathbb{F}', \mathbb{G}$ of $\mathbb{N}$, where N_z, M_z are the operators of multiplication by z on $L^2(\mathbb{T})$, $H^2(\mathbb{T})$ respectively. If the essential spectrum is $\overline{\mathbb{D}}$, then such an operator is essentially equivalent to the multiplication operator N_z on $L^2(\mathbb{D}, dA)$.

5.5 Essentially Reductive Quotient and Submodules

The purpose of this section is to discuss the Arveson–Douglas conjecture. The discussion is not exhaustive. Our limited goal here is to develop enough theory so that we can state some interesting open problems at the end of Section 6.3.2

The central role of the BDF theorem is evident from the numerous applications involving the notion of essential normality presented in this section. Now, we discuss some recent developments using the language of Hilbert modules introduced in [55]. These questions involve the notion of *essentially reductive* quotient and submodules, which we describe here. In particular, we discuss the so-called $\mathcal{G}$-homogeneous submodules in the context of the Arveson–Douglas conjecture. We must emphasize that our intention is not to survey the vast body of results surrounding the Arveson–Douglas conjecture. Although we briefly indicate some of the main results, the reader may consult [59, 125, 126] for an excellent and detailed survey of this very interesting topic.

The book [55] describes a very fruitful approach of using algebraic techniques from the structure theory of modules to study multi-variate operator theory.

Fix a Hilbert space $\mathcal{M}$ and a commuting n-tuple $T = (T_1, \ldots, T_n)$ of bounded linear operators on it. Evidently, the map $m : \mathcal{M} \times \mathbb{C}[z] \to \mathcal{M}$,

$$ m_p(h) = p(T)h, \quad p \in \mathbb{C}[z], \, h \in \mathcal{M} $$

defines the module multiplication on $\mathcal{M}$ as the rule $p \mapsto p(T)$ is a homomorphism. It has become clear over the past couple of decades that studying the module $\mathcal{M}$ often provides new insights into the behaviour of the commuting tuple T of operators. If the map m is assumed to be separately continuous in both the variables, where the polynomial ring is equipped with the sup norm on some bounded, polynomially convex domain Ω, then in the terminology of [55], the module $\mathcal{M}$ is called a *Hilbert module*. In this case, the module map extends to the closure with respect to the supremum norm on Ω of the polynomial ring, which is a Banach algebra to be denoted by $\mathcal{A}(\Omega)$.

Familiar examples of Hilbert modules $\mathcal{M}$ are the Hardy and the Bergman spaces defined on some domain $\Omega \subseteq \mathbb{C}^n$ along with the commuting n-tuple $(M_1, \ldots, M_n)$ consisting of multiplication by the coordinate functions $z_1, \ldots, z_n$. Thus, the module multiplication is given by $m_f(h) = fh$, $f \in \mathcal{A}(\Omega)$, $h \in \mathcal{M}$, where $(fh)(z) = f(z)h(z)$, $z \in \Omega$.

Two Hilbert modules, $\mathcal{M}_1$ and $\mathcal{M}_2$ are said to be *isomorphic* if there exists a *module map*, that is, there is a unitary operator $\Gamma : \mathcal{M}_1 \to \mathcal{M}_2$, which intertwines the module multiplications: $\Gamma m_p^1 = m_p^2 \Gamma$. A *submodule* of a Hilbert module $\mathcal{M}$ is a closed linear subspace $\mathcal{S}$ that is invariant under the module multiplication: $m_p(\mathcal{S}) \subseteq \mathcal{S}$. Finally, if $\mathcal{S}$ is a submodule, then we let $Q := \mathcal{M} \ominus \mathcal{S}$ be the quotient module equipped with the multiplication induced by compressing the multiplication on $\mathcal{M}$ to Q: $m_p^Q = Q(m_p|_Q)$, where Q is the projection to the subspace Q.

The notion of an essentially reductive Hilbert module, reproduced in the following, was introduced in [55] and they were termed essentially normal in [13]. These are the modules where the module multiplication is required to be essentially normal. Please note that for most natural examples of Hilbert modules which appear as function spaces, the multiplication by a polynomial is defined as pointwise multiplication. However, if $\mathcal{H}$ is any Hilbert space, then commuting n-tuple $(T_1, \ldots, T_n)$ of operators on $\mathcal{H}$ gives rise to a module multiplication as in the following Definition.

Definition 5.5.1

Let $\mathbb{C}[z]$ denote the ring of complex polynomials in $z_1, \ldots, z_m$. A Hilbert module $\mathcal{M}$ over the polynomial ring $\mathbb{C}[z]$ is said to be *essentially reductive* if the module multiplication m_p, $p \in \mathbb{C}[z]$ is essentially normal, that is,

$$[p(T)^*, p(T)] \in C(\mathcal{M}), \quad p \in \mathbb{C}[z].$$

Let $\mathcal{A}(\Omega)$ be the algebra obtained by taking the closure of the ring $\mathbb{C}[z]$ in the sup norm on Ω. Suppose the module multiplication defined on $\mathcal{M}$ over $\mathbb{C}[z]$ is continuous in both the variables. Then this multiplication extends to the algebra $\mathcal{A}(\Omega)$ and we let $\varphi(T) := m_\varphi$ for $\varphi \in \mathcal{A}(\Omega)$. A Hilbert module over $\mathcal{A}(\Omega)$ is said to be *essentially reductive* if

$$[\varphi(T)^*, \varphi(T)] \in C(\mathcal{M}), \quad \varphi \in \mathcal{A}(\Omega).$$

As we have already seen, the Hardy and the Bergman modules $H^2(\mathbb{D})$ and $L_a^2(\mathbb{D})$ over the disc algebra $\mathcal{A}(\mathbb{D})$ are essentially reductive (see Example 2.1.3). However, neither the Hardy module nor the Bergman module (both these are subspaces of the space of complex-valued holomorphic functions defined on the polydisc $\mathbb{D}^m$) over the algebra $\mathcal{A}(\mathbb{D}^m)$, $m > 1$, is essentially reductive. On the other hand, the Drury–Arveson module H_m^2 is a module over the polynomial ring $\mathbb{C}[z]$. This multiplication does not extend to the ball algebra. Nevertheless, the Drury–Arveson module H_m^2 is essentially reductive over $\mathbb{C}[z]$.

We discuss this phenomenon in some detail before stating the conjecture of Arveson and Douglas. Although, initially, this conjecture was made only for the Drury–Arveson modules of the Euclidean ball $\mathbb{B}_m$, we will discuss in the following the membership of the cross commutators for a much larger class of modules, namely, the ones corresponding to spherical modules, or equivalently, $U(m)$-invariant submodules. Here $U(m)$ is the group of $m \times m$ unitary matrices. These have been studied in detail in the paper [31] and what follows is taken from it. The reader may also consult [68], where $U(m)$-invariant reproducing kernel Hilbert spaces are studied. Although stated in slightly different language, the notion of $U(m)$-invariant modules plays a central role in the study of weighted Bergman spaces defined on a bounded symmetric domain, see [8, 64], which we discuss briefly.

Let $\mathbb{B}_m$ be the open unit ball in $\mathbb{C}^m$ and let $\mathbb{K}$ be the maximal compact subgroup of the bi-holomorphic automorphism group G of $\mathbb{B}_m$. The group $\mathbb{K}$ can be identified with the unitary group $U(m)$; it acts on $\mathbb{B}_m$ by the rule:

$$(U, z) \mapsto U(z), \quad z \in \mathbb{B}_m, \ U \in U(m).$$

Let $T = (T_1, \ldots, T_m)$ be a commuting m-tuple of operators, that is, an m-tuple of commuting bounded linear operators $T_1, \ldots, T_m$ acting on a Hilbert space $\mathcal{H}$. The polynomial functional calculus gives

$$U \cdot T = \Big(\sum_{j=1}^{m} U_{1j} T_j, \dots, \sum_{j=1}^{m} U_{m,j} T_j \Big), \quad U \in K. \tag{5.5.14}$$

The commuting m-tuple of operators T is said to be *spherical*, or equivalently, $U(m)$-*homogeneous* if for each U in $U(m)$, $\Gamma_U^* T \Gamma_U = U \cdot T$ for some unitary Γ_U on $\mathcal{H}$. In general, Γ need not be a unitary representation. However, we will assume that a choice of Γ_U exists such that the map $U \mapsto \Gamma_U$ is a unitary homomorphism. Several basic properties of spherical operators are listed in [31, Remark 1.2].

Let $\mathbf{w} = \big\{ w_n^{(j)} : 1 \le j \le m, n \in \mathbb{N}^m \big\}$ be a multi-sequence of positive numbers. An m-*variable weighted shift* $T = (T_1, \dots, T_m)$ with respect to an orthonormal basis $\{e_n\}_{n \in \mathbb{N}^m}$ of a Hilbert space $\mathcal{H}$ is defined by

$$T_j e_n := w_n^{(j)} e_{n+\epsilon_j}, \quad 1 \le j \le m, \tag{5.5.15}$$

where ϵ_j is the m-tuple with 1 in the jth place and zeros elsewhere. The m-tuple T extends to an m-tuple of bounded linear operators on $\mathcal{H}$ if and only if

$$\sup \big\{ \big| w_n^{(j)} \big| : 1 \le j \le m, n \in \mathbb{N}^m \big\} < \infty.$$

If this happens, then T_j commutes with T_k if and only if $w_n^{(j)} w_{n+\epsilon_j}^{(k)} = w_n^{(k)} w_{n+\epsilon_k}^{(j)}$ for all $n \in \mathbb{N}^m$. We always assume that T is bounded and commuting. Note that T is *cyclic* with *cyclic vector* e_0, that is, $\mathcal{H} = \bigvee \{ T^\alpha e_0 : \alpha \in \mathbb{N}^m \}$.

Let T be an m-variable weighted shift with weight sequence $\mathbf{w}$. Set $\beta_n = \| T^n e_0 \|$, $n \in \mathbb{N}^m$, and consider the Hilbert space $H^2(\beta)$ of formal power series

$$f(z) = \sum_{n \in \mathbb{N}^m} a_n z^n \text{ such that } \| f \|_\beta^2 = \sum_{n \in \mathbb{N}^m} |a_n|^2 \beta_n^2 < \infty.$$

It follows that any m-variable weighted shift T is unitarily equivalent to the m-tuple $M_z = (M_{z_1}, \dots, M_{z_m})$ of operators of multiplication by the co-ordinate functions $z_1, \dots, z_m$ on the corresponding space $H^2(\beta)$. Notice that the linear set of polynomials in $z_1, \dots, z_m$ (that is, formal power series with finitely many non-zero coefficients) is dense in $H^2(\beta)$. Equivalently, M_z is cyclic with cyclic vector 1, that is, the formal series $\sum_{n \in \mathbb{N}^m} a_n z^n$, for which $a_n = 0$ for all non-zero $n \in \mathbb{N}^m$ and $a_0 = 1$. The relation between weights $w_n^{(j)}$ and the sequence β_n is given by

$$w_n^{(j)} = \beta_{n+\epsilon_j} / \beta_n, \quad 1 \le j \le m, \ n \in \mathbb{N}^m. \tag{5.5.16}$$

Here are some important examples of spherical weighted multishifts.

Example 5.5.2   For any real number $p > 0$, let $\mathscr{H}^{(\nu)}$ be the Hilbert module consisting of holomorphic functions on the unit ball $\mathbb{B}$ determined by the positive definite kernel

$$\kappa_\nu(z, w) = \frac{1}{(1 - \langle z, w \rangle)^\nu}, \quad z, w \in \mathbb{B}.$$

If $M_{z,v}$ denotes the multiplication tuple on $\mathcal{H}^{(v)}$, then it is unitarily equivalent to the weighted shift m-tuple with weight sequence

$$w_{n,v}^{(i)} = \sqrt{\frac{n_i + 1}{|n| + v}}, \quad n \in \mathbb{N}^m, i = 1, \cdots, m, \tag{5.5.17}$$

where $|n| = n_1 + \cdots + n_m$. The Hilbert modules $\mathcal{H}^{(m)}, \mathcal{H}^{(m+1)}, \mathcal{H}^{(1)}$ are, respectively, the *Hardy module* $H^2(\partial \mathbb{B})$, the *Bergman module* $A^2(\mathbb{B})$, and the *Drury–Arveson module* H_m^2. The multiplication tuples $M_{z,m}, M_{z,m+1}, M_{z,1}$ are commonly known as the *Szegö m-shift*, the *Bergman m-shift*, and the *Drury–Arveson m-shift* respectively.

Definition 5.5.3

Let T be a spherical m-variable weighted shift with weight multi-sequence $\mathbf{w} = \left\{ w_n^{(j)} : 1 \le j \le m, n \in \mathbb{N}^m \right\}$ and let $\{w_k\}_{k \in \mathbb{N}}$ be a scalar weight sequence. Then the *shift associated with T* is the one-variable weighted shift T_δ with weight sequence $\{\delta_k : k \in \mathbb{N}\}$ if

$$w_n^{(i)} = \delta_k \sqrt{\frac{n_i + 1}{k + m}}, \quad k = |n|, \ n \in \mathbb{N}^m, \ 1 \le i \le m. \tag{5.5.18}$$

The following theorem from [31] is a characterization of spherical m-variable weighted shift operators.

Theorem 5.5.4 (Theorem 2.5, [31])

Given any multi-sequence $\mathbf{w}$, let β be the multi-sequence determined by the requirement $w_n^{(j)} = \frac{\beta_{n+\epsilon_j}}{\beta_n}, n \in \mathbb{N}^m$. Let M_z be the commuting m-tuple of multiplication operators on $H^2(\beta)$. The operator m-tuple M_z is spherical if and only if there is a sequence $\{\tilde{\beta}_k\}_{k=0}^\infty$ such that

$$\beta_n = \tilde{\beta}_{|n|} \sqrt{\frac{(m-1)! \, n!}{(m - 1 + |n|)!}}, \quad n \in \mathbb{N}^m.$$

It follows that an m-variable weighted shift T with weight multi-sequence $\mathbf{w}$ is spherical if and only if the weight sequence can be written in the form (5.5.18).

Recall that an m-tuple T of commuting bounded linear operators $T_1, \ldots, T_m$ is *p-essentially normal* if the cross-commutators $[T_j^*, T_l]$ belong to the Schatten p-class for all $j, l = 1, \ldots, m$. As before, we put $\delta_k = \tilde{\beta}_{k+1}/\tilde{\beta}_k$, $k \in \mathbb{N}$. In what follows, we assume that $m \ge 2$.

An m-variable weighted shift T with weights $\left\{ w_n^{(j)} : 1 \le j \le m, n \in \mathbb{N}^m \right\}$ is compact if and only if $\lim_{|n| \to \infty} w_n^{(j)} = 0$ for all indices $j = 1, \ldots, m$, see [81, Proposition 6]. It follows that a spherical m-variable weighted shift is compact if and only if $\lim_{k \to \infty} \delta_k = 0$. The following theorem provides very detailed information on the essential normality of a spherically symmetric m-tuple of multiplication operators on the Hilbert space $H^2(\beta)$.

Theorem 5.5.5 (Theorem 4.2, [31])

Let M_z be a bounded spherical multiplication m-tuple in $H^2(\beta)$ determined by a sequence $\bar\beta_0, \bar\beta_1, \bar\beta_2, \ldots$, of positive numbers. If $1 \leqslant p \leqslant \infty$, then the following statements are equivalent:

(i) The self-commutators $[M_{z_j}^*, M_{z_j}]$ belong to the Schatten class S_p for all j, $1 \leqslant j \leqslant m$;

(ii) The cross-commutators $[M_{z_j}^*, M_{z_l}]$ belong to the Schatten class S_p for all indices j, l;

(iii)

$$\sum_{k=1}^{\infty} \delta_k^{2p} k^{m-p-1} + \sum_{k=1}^{\infty} \left| \delta_k^2 - \delta_{k-1}^2 \right|^p k^{m-1} < \infty.$$

A reformulation of [14, Proposition A.3] is the claim: If $p > m$ and $\delta_k > \epsilon > 0$, the cross-commutators $[M_{z_j}^*, M_{z_\ell}]$ are all in the Schatten class S_p if and only if $\sum_{k=1}^{\infty} \left| \delta_k^2 - \delta_{k-1}^2 \right|^p k^{m-1}$ converges. This proposition is a direct consequence of the preceding theorem. Moreover, note that Arveson's ρ_k are related to the δ_k by $\rho_k = \sqrt{\frac{k+1}{k+m}} \delta_k$. One may view this proposition as a precursor to the Arveson–Douglas conjecture stated as follows.

Corollary 5.5.1

Suppose that the sequence $\{\delta_k\}$ does not tend to zero and $|\delta_{k+1} - \delta_k| \leqslant C/k$ for some constant C. Then the commutators $[M_{z_j}^*, M_{z_k}]$ belong to S_p if and only if $p > m$.

It is easy to verify the criterion of Corollary 5.51 for the weighted Bergman modules $\mathscr{H}^{(\nu)}$, $\nu > 0$, of the Euclidean ball $\mathbb{B}_m$ as $\delta_\nu(k)$, in this case, is $\sqrt{\frac{k+m}{k+\nu}}$. It follows that all the modules $\mathscr{H}^{(\nu)}$, $\nu > 0$, are *essentially p-reductive, p > m*.

What we have said about the Euclidean ball applies equally well to the case of other bounded symmetric domains $\Omega \subset \mathbb{C}^m$, say of rank r and genus p. For the basic definitions, we refer to the excellent survey article [8]. Let $B : \Omega \times \Omega \to \mathbb{C}$ be the Bergman kernel, namely, the unique function $B(\cdot, w)$ in the Bergman module $L_a^2(\Omega)$ of square integrable holomorphic functions (with respect to the volume measure on Ω) with the reproducing property:

$$f(w) = \int_\Omega f(z) \overline{B_w(z)} \, dV(z), \quad w \in \Omega,$$

where $B_w(z) = B(z, w)$ for $z, w \in \Omega$. Thus, B is a positive definite kernel, that is, $\big((B(z_i, z_j)) \big)_{i,j=1}^{n}$ is positive definite for any subset $\{z_1, \ldots, z_n\}$ of Ω. For any positive real number ν, let $B^{(\nu)} := B^{\frac{\nu}{p}}$ be defined by polarizing the real analytic function $B(z, z)^{\frac{\nu}{p}}$ to obtain the function $B(z, w)^{\frac{\nu}{p}}$ holomorphic in z and anti-holomorphic in w. The set of positive real numbers ν for which $B^{(\nu)}$ remains a positive definite kernel is known (cf. [60]) and designated as the *Wallach set of* Ω. For a fixed but arbitrary ν in the Wallach set, let $\mathscr{H}^{(\nu)}$ denote the Hilbert space determined by the positive definite kernel $B^{(\nu)}$. The bi-holomorphic automorphisms of the domain Ω form a group, say G. Let $\mathbb{K}$ be the maximal compact subgroup of G. One may realize the domain Ω

as the quotient $\Omega = G/\mathbb{K}$ (see [5, pp. 10]). Now, we speak freely of $\mathbb{K}$-homogeneous operators. To describe this more general situation, we recall some basic notions from the representation theory of the group $\mathbb{K}$.

For a positive integer r, let $s \in \mathbb{N}^r$ be a partition ($s = (s_1 \geq \dots \geq s_r \geq 0)$) of length r. Let $\mathcal{P}_s$ denote the space of irreducible $\mathbb{K}$-invariant homogeneous polynomials of isotypic type s, having total degree $|s|$. These are mutually inequivalent as $\mathbb{K}$-modules and $\mathcal{P} = \sum_{s \in \mathbb{N}^r_+} \mathcal{P}_s$ is the Peter–Weyl decomposition of the polynomials $\mathcal{P}$ under the action of the group $\mathbb{K}$, see [8, Equation (3.1), pp. 21]. Now, equip the submodules $\mathcal{P}_s$ with the Fischer–Fock inner product

$$(p|q)_s = (q^*(\partial)(p))(0), \quad p, q \in \mathcal{P}_s,$$

where $q^*(z) = \overline{q(\bar{z})}$ and $\partial = (\partial_1, \dots, \partial_m)$. Let E^s be the reproducing kernel of the finite dimensional space $\mathcal{P}_s$. Then the Faraut–Korányi formula ([60]) for the reproducing kernel $K^{(v)}$ of the Hilbert space $\mathcal{H}^{(v)}$ is

$$B^{(v)} = \sum_{s \in \mathbb{N}^r_+} (v)_s E^s, \tag{5.5.19}$$

where $(v)_s := \prod_{j=1}^r (v - \frac{a}{2}(j-1))_{s_j}$ are the generalized *Pochhammer symbols* and $a \in \mathbb{R}^+$ is a certain invariant associated with the domain Ω. The commuting tuple of multiplication operators M on the Hilbert space $\mathcal{H}^{(v)}$ is $\mathbb{K}$-homogeneous. Indeed, the natural action of $\mathbb{K}$ on $\mathcal{H}^{(v)}$ given by $f \mapsto f(k^{-1} \cdot z)$, $k \in \mathbb{K}$, $f \in \mathcal{H}^{(v)}$ is a unitary homomorphism intertwining $k \cdot M$ and M. What are the other $\mathbb{K}$-homogeneous operators? As $\mathcal{P}_s$ is an irreducible module under the action of the group $\mathbb{K}$, it follows that the Hilbert space $\mathcal{H}^{(a)}$, obtained by setting $K^{(a)} = \sum_{s \in \mathbb{N}^r_+} a_s E^s$ for an arbitrary choice of positive numbers a_s is a weighted direct sum of the K modules $\mathcal{P}_s$. Hence, the commuting tuple of multiplication operators M on $\mathcal{H}^{(a)}$ is $\mathbb{K}$-homogeneous. It is shown in [64], under some additional hypothesis, that these are the only $\mathbb{K}$-homogeneous operators. This is analogous to Theorem 5.5.4.

Suppose that v in the Wallach set is chosen such that that the commuting tuple of operators M_{z_i} of multiplication by the coordinate functions z_i ($1 \leq i \leq m$) on the weighted Bergman spaces $\mathcal{H}^{(v)}$ are bounded. Then the commutators $[M_{z_i}^*, M_{z_i}]$ ($1 \leq i \leq m$) of these multiplication operators are compact if and only if the rank of Ω is 1, see [64]. Hence, there is a dichotomy, namely, if the rank is 1, then the weighted Bergman modules are always essentially reductive, whereas in all other cases, they are not. Apart from asking if a Hilbert module is essentially reductive, one may also ask which submodules and quotient modules are essentially reductive. Determining which of the quotient or the submodules of the weighted Bergman modules $\mathcal{H}^{(v)}$ over a bounded symmetric domain are essentially reductive appears to be a difficult question. For instance, the unit ball Ω (with respect to the operator norm) in the $n \times n$, $n \in \mathbb{N}$, matrices is a bounded symmetric domain. The Euclidean ball $\mathbb{B}_n$ embeds into it. Let $\mathbf{z} \in \mathbb{B}_n$ map to the $n \times n$ matrix whose first column is $\mathbf{z}$ and each of the remaining $(n-1)$ columns is the zero vector of size n. If we consider the submodule S of functions vanishing on $\mathbb{B}_n$ in the module $\mathcal{H}^{(v)}(\Omega)$, then the corresponding quotient module Q can be identified with the restriction of the functions in $\mathcal{H}^{(v)}(\Omega)$ to the zero set, namely, $\mathbb{B}_n$. In consequence, the quotient module is isomorphic to the weighted Bergman module $\mathscr{H}^{(\nu)}(\mathbb{B}_n)$ and these are essentially reductive, as we have pointed out earlier. This idea might even work, via the jet construction [54, 63], even if we start with a submodule of functions vanishing on a variety to order $k > 1$.

Many results obtained starting with the work of Arveson [12, 13] show that the question of essential reductivity for a submodule has an interesting answer, if the module itself is essentially reductive. We have seen that weighted Bergman modules of the polydisc $\mathbb{D}^m$ are not essentially reductive whereas the weighted Bergman modules of the Euclidean ball $\mathbb{B}_m$ are essentially reductive. In his paper [12], Arveson had established that the module H_m^2 is essentially reductive and then in the paper [13], he established that the submodules $[I]$ obtained by taking the closure of a polynomial ideal I of H_m^2 are essentially reductive if I is generated by monomials. A generalization to finitely generated homogeneous submodules of $H_m^2 \otimes \mathbb{C}^n$, $n \in \mathbb{N}$, is also obtained in [13]. He then made the following conjecture, which has been the subject of intense research.

Arveson's Conjecture Let $\mathscr{I}$ be a homogeneous ideal (that is, ideal generated by homogeneous polynomials) of $\mathbb{C}[z]$, and let $[\mathscr{I}]$ denote the closure in H_m^2 (this is the Hilbert module $\mathscr{H}^{(1)}$ over the polynomial ring $\mathbb{C}[z]$ in m-variables). Then, $[\mathscr{I}]$ is k-essentially reductive for every $k > \dim \mathscr{I}$.

What is known as the Arveson–Douglas conjecture extends the original conjecture of Arveson to other domains in $\mathbb{C}^m$ as well as to other spaces like the weighted Bergman space. Indeed, a more quantitative version of the Arveson conjecture involving the dimension of the zero set

$$V(I) := \{z \in \mathbb{C}^m : p(z) = 0, p \in I\}.$$

of the ideal $I \subset \mathbb{C}[z]$ is known as the geometric Arevson–Douglas conjecture:

Geometric Arveson–Douglas Conjecture Let $\mathbb{A}(\mathbb{B}_m)$ be the Bergman module. If the ideal I is homogeneous, that is, generated by homogeneous polynomials, then the quotient module $[I]^\perp$ is k-essentially reductive for every $k > \dim V(I)$.

An affirmative solution to the Geometric Arveson–Douglas Conjecture also provides a model for noncommutative polynomials. Let $\mathscr{T}([I])$ be the C^*-algebra generated by $[I]$ and the compact operators $C([I])$. Then a positive solution to the Arveson–Douglas conjecture implies that

$$0 \to C([I]) \to \mathscr{T}([I]) \to C(V(I) \cap \partial\mathbb{B}_m) \to 0$$

is a short exact sequence. One of the main questions is to identify the K-homology class $[\mathscr{T}([I])]$ defined by this extension. In the case that $I = \{0\}$ and $m = 1$, the Toeplitz index theorem for $\mathbb{T}$ gives the answer to this question (cf. Corollary 1.6.1).

Guo and Wang [71] verified both the conjectures for a) I is a principal homogeneous submodule, b) I is homogeneous and dim $V(I) \leq 1$ and c) homogeneous submodules for $m < 3$.

The Berger–Shaw theorem asserts that if T is hyponormal and finitely cyclic, then $[T^*, T]$ is in trace class. This prompts the question: If S is a finitely generated submodule of the Drury–Arveson module H_m^2, then does it follow that it is k-essentially reductive for some $k \in \mathbb{N}$?

It has been shown in [66] that there exists submodules $\mathcal{M}$ of H_m^2 such that the essential spectrum of $M_z|_\mathcal{M}$ intersects with the open unit ball, where M_z denotes the multiplication d-tuple on H_m^2. This fact combined with the following lemma shows that there exist submodules of H_m^2, which are not essentially normal [126]. The proof of the following lemma is not entirely self-contained.

Lemma 5.5.1

Let M_z denote the Drury–Arveson d-shift on the Drury–Arveson module H_m^2. Let M be a nonzero z_j-invariant submodule of H_m^2 for $j = 1, \ldots, m$. If $S = M_z|_M$ is essentially normal, then $\sigma_e(S) \subseteq \sigma_{ap}(S) \subseteq \partial \mathbb{B}$.

Proof We may assume that $m > 1$ (see Example 1.5.6). Let $S_i = M_{z_i}|_M$ for $i = 1, \ldots, d$ and let $\mathscr{C}_S$ denote commutator ideal of $C^*(S)$. Let $C(\mathcal{H})$ denote the ideal of compact operators in $\mathcal{L}(\mathcal{H})$, where $\mathcal{H} = H_m^2$. We claim that

$$C(\mathcal{H}) \subseteq \mathscr{C}_S \subseteq C^*(S). \tag{5.5.20}$$

As $I - \sum_{i=1}^d M_{z_i}^* M_{z_i}$ is compact (see (2) of Example 4.4.5), so is $I - \sum_{i=1}^d S_i^* S_i$. Moreover, $C^*(S)$ is irreducible [68, Theorem 5.1]. As $m > 1$, $I - \sum_{i=1}^d S^* S_i$ is nonzero, and hence, it follows from the general theory that $C(\mathcal{H}) \subseteq C^*(S)$ (see [34, pp. 91]). Moreover, the inclusion $C(\mathcal{H}) \subseteq \mathscr{C}_S$ follows from the fact that $C(\mathcal{H})$ admits no nonzero characters (that is, multiplicative linear functionals sending I to 1) in view of

$$\mathscr{C}_S = \cap \{ \rho^{-1}(0) : \rho \text{ is a character on } C^*(S) \}$$

(refer to the discussion following [28, Corollary 4]). This completes the verification of the claim.

It is easy to see that each M_{z_i} is hyponormal, so is each $S_i = M_{z_i}|_M$. Thus, S consists of commuting hyponormal operators $S_1, \ldots, S_m$. By [28, Corollary 4], $C^*(S)/\mathscr{C}_S$ is isomorphic, via say ψ_1, to the C^*-algebra $C(\sigma_{ap}(S))$ of continuous functions on $\sigma_{ap}(S)$, where $\sigma_{ap}(S)$ denotes the approximate-point spectrum of S. Suppose that $S = M_z|_M$ is essentially normal. Notice that $C^*(S)/C(\mathcal{H})$ is generated by normal elements $S_1 + C(\mathcal{H}), \ldots, S_m + C(\mathcal{H})$, and hence, it is abelian. In particular, $C(\mathcal{H})$ contains $\mathscr{C}_S$. By (5.5.20), we obtain $C(\mathcal{H}) = \mathscr{C}_S$. Moreover, by the spectral permanence [37, pp. 23], $C^*(S)/C(\mathcal{H})$ is isomorphic, via say ψ_2, to $C(\sigma_e(S))$. Both the aforementioned isomorphisms ψ_1, ψ_2 send polynomials $p(S, S^*)$ in S and S^* to $p(z, \bar{z})$.

We claim that $\sigma_{ap}(S)$ contains $\sigma_e(S)$. Suppose $\lambda \notin \sigma_{ap}(S)$. As $\sigma_{ap}(S)$ is a closed subset of $\mathbb{C}^d$, there exists $\alpha > 0$ such that $\sum_{i=1}^d |z_i - \lambda_i|^2 \geq \alpha$ for every $z \in \sigma_{ap}(S)$. It follows that $\sum_{j=1}^m (S_j^* - \overline{\lambda_j} I)(S_j - \lambda_j I) + \mathscr{C}_S$ is invertible in $C^*(S)/\mathscr{C}_S$, and hence, $\sum_{j=1}^m (S_j^* - \overline{\lambda_j} I)(S_j - \lambda_j I) + C(\mathcal{H})$ is invertible in $C^*(S)/C(\mathcal{H})$. This implies that $S - \lambda I$ is Fredholm (see [36, Corollary 3.9]), so that $\lambda \notin \sigma_e(S)$. This proves that $\sigma_e(S) \subseteq \sigma_{ap}(S)$. Finally, as $\sum_{j=1}^d S_j^* S_j - I$ is compact, the function

$\sum_{j=1}^{d} |z_j|^2 - 1$ vanishes on $\sigma_{ap}(S)$, and hence, $\sigma_{ap}(S)$ is contained in the unit sphere. Therefore the essential spectrum of S is contained in the unit sphere.   □

In [52], it is shown that some quotient modules of the Hardy module over the algebra $\mathcal{A}(\mathbb{D}^2)$ are essentially reductive and some are not. No nonzero submodule of the Hardy module can be essentially reductive as the coordinate functions define a pair of commuting isometries, both of infinite multiplicity.

Remark 5.5.6   Let $\{e_\alpha\}_{\alpha \in \mathbb{N}^d}$ be an orthonormal basis for $\mathcal{H}$ and define a commuting d-tuple U by

$$U_j e_\alpha = e_{\alpha + \epsilon_j}, \quad \alpha \in \mathbb{N}^d, \; j = 1, \ldots, d,$$

where $\mathbb{N}^d$ denotes the set of d-tuples of non-negative integers and ϵ_j is the d-tuple with 1 in the jth place and zeros elsewhere. If $d \geqslant 2$, then

$$\sum_{j=1}^{d} U_j U_j^* e_{k\epsilon_1} = e_{k\epsilon_1}, \quad k \geqslant 1,$$

and hence, the range of $\sum_{j=1}^{d} (U_j^* U_j - U_j U_j^*)$ contains an infinite dimensional closed subspace. One may apply Corollary 1.2.2 to conclude that $\sum_{j=1}^{d} (U_j^* U_j - U_j U_j^*)$ is not compact.

Before we continue, let us point out that when the module $\mathcal{M}$ is reductive, unlike the case of the Hardy module over the polydisc, the distinction between the sub and the quotient module disappears. This was first shown in [13]. The following proof is taken from [49].

Lemma 5.5.2

Let $\mathcal{M}$ be an essentially reductive Hilbert module over $\mathbb{C}[z]$, $\mathcal{N}$ be a submodule of $\mathcal{M}$ and $\mathcal{Q} = \mathcal{M} \ominus \mathcal{N}$ be the corresponding quotient module. The submodule $\mathcal{N}$ is essentially reductive if and only if so is $\mathcal{Q}$.

Proof   For $p \in \mathbb{C}[z]$, let $\begin{pmatrix} A & B \\ C & D \end{pmatrix}$ be the matrix representation of m_p relative to the decomposition $\mathcal{N} \oplus \mathcal{Q}$. As $\mathcal{N}$ is invariant for m_p, we have $C = 0$. Moreover, the action of p on $\mathcal{N}$ defines the operator A, whereas the action of p on $\mathcal{Q}$ defines an operator unitarily equivalent to D. It is then not hard to find the matrix representation for the operator

$$[m_p^*, m_p] = \begin{pmatrix} [A^*,A] - BB^* & A^*B - BD^* \\ B^*A - DB^* & [D^*,D] + B^*B \end{pmatrix}.$$

As $\mathcal{M}$ is essentially reductive, $[m_p^*, m_p]$ is compact and hence, so are the operators $[A^*, A] - BB^*$ and $[D^*, D] + B^*B$. If $\mathcal{N}$ is essentially reductive, then $[A^*, A]$ is compact and hence, BB^* is compact. This implies that B^*B is compact (see Exercise 1.7.13 and Theorem 1.1.3) and that $[D^*, D]$ is compact. As this is true for every $p \in \mathbb{C}[z]$, we see that $\mathcal{Q}$ is essentially reductive. A similar argument shows that $\mathcal{Q}$ being essentially reductive implies that $\mathcal{N}$ is also essentially reductive.   □

What about the Hardy and the weighted Bergman modules over the algebra $\mathcal{A}(\mathbb{D}^n)$? The calculations in [52] appear to have anticipated the Arveson's conjecture for these modules.

Rather than survey the vast literature in this area (cf. [70, 139, 140]), we reproduce the original computations from the paper [52].

As before, let $[I]$ be the closure of an ideal $\mathscr{I} \subseteq \mathbb{C}[\mathbf{z}]$ in the Hardy module $H^2(\mathbb{D}^n)$ and $V(I)$ be the common zero set of the polynomials in the ideal I. If the zero set $V(I)$ is discrete and finite, then the quotient module $[I]^\perp$ is essentially reductive as the quotient module is finite-dimensional (cf. [8]). In [140], it is proved that if the homogeneous quotient module $H^2(\mathbb{D}^n) \ominus [I]$ is essentially reductive, then $\dim V(I) = 1$.

Here we want to consider the case when $\mathscr{I}$ is the principal ideal in $\mathbb{C}[w, z]$ generated by a polynomial $p(w, z)$. If Q_p denotes the quotient module $H^2(\mathbb{D}^2) \ominus [I]$, then we are interested in describing the properties of Q_p in terms of those of p. Here we examine some very simple examples.

Let $p(w, z) = w - z$ and set, $Q_0 := Q_{w-z}$. For $k \in \mathbb{N}$, let $\mathcal{P}_k$ denote the closed subspace of homogeneous polynomials in $H^2(\mathbb{D}^2)$ spanned by $\{w^{k-j}z^j\}_{j=0}^k$. Then, we have $H^2(\mathbb{D}^2) = \bigoplus_{k=0}^{\infty} \mathcal{P}_k$

and $Q_0 = \bigoplus_{k=0}^{\infty} (\mathcal{P}_k \cap Q_0)$. Moreover, each $\mathcal{P}_k \cap Q_0$ is one-dimensional and is spanned by the

polynomial $e_k = \sum_{j=0}^{k} w^{k-j}z^j$. The set $\{e_k\}_{k=0}^{\infty}$ forms an orthogonal basis for Q_0 and $\|e_k\| = \sqrt{k+1}$.

Therefore, we have the orthonormal basis $\left\{\frac{1}{\sqrt{k+1}} e_k\right\}_{k=0}^{\infty}$ for Q_0.

What about the module multiplication on Q_0? It is enough to calculate the action of w as it is identical to that of z on Q_0. However

$$w \cdot \frac{1}{\sqrt{k+1}} e_k = \sqrt{\frac{k+1}{k+2}} \frac{1}{\sqrt{k+2}} e_{k+1}$$

in Q_0 and hence, Q_0 is isomorphic to the Bergman module on which the $\mathcal{A}(\mathbb{D}^2)$ action is pulled back via the map $\mathbb{D} \longrightarrow \mathbb{D}^2$ defined by $z \longrightarrow (z, z)$. In particular, the operator action is essentially normal and hence, the module Q_0 is what was called (cf. [52]) an essentially reductive *Silov module* for $\mathcal{A}(\mathbb{D}^2)$.

Not all quotient modules for $H^2(\mathbb{D}^2)$ are essentially reductive. The principal ideal in $\mathbb{C}[w, z]$ generated by $p(w, z) = w^2$ does not yield an essentially reductive module as multiplication by w is not essentially normal. Let us consider the case of $(w - z)^2$.

Let Q_1 be the quotient module for $H^2(\mathbb{D}^2) \ominus I$, where I is the principal ideal generated by the polynomial $(z - w)^2$ and S_2 is the closure of I. Observe that we have $Q_1 = \oplus_{k=0}^{\infty} (\mathcal{P}_k \cap Q_1)$ and, in fact, we have $\mathcal{P}_k \cap Q_0 \subset \mathcal{P}_k \cap Q_1$. Moreover, if we ignore the anomalous cases $\mathcal{P}_0$ and $\mathcal{P}_1$, which are contained in Q_1, we see that

$$\dim(\mathcal{P}_k \cap \mathcal{M}_1) \ominus (\mathcal{P}_k \cap Q_0) = 1 \quad \text{for} \quad k \geqslant 2.$$

Thus, we have the orthogonal basis for $\mathcal{P}_k \cap \mathcal{Q}_1$ consisting of e_k and f_k, where

$$
f_k = \begin{cases}
mw^k + (m-1)w^{k-1}z + \cdots + w^{m+1}z^{m-1} \\
\quad -(w^{m-1}z^{m+1} + 2w^{m-2}z^{m+2} + \cdots + mz^k) & \text{if } k = 2m \geqslant 2, \\
kw^k + (k-2)w^{k-1} + \cdots + w^{m+1}z^m \\
\quad -(w^m z^{m+1} + 3w^{m-1}z^{m+2} + \cdots + kz^k) & \text{if } k = 2m+1 \geqslant 3.
\end{cases}
$$

It can be checked that e_k and f_k are orthogonal and in turn are orthogonal to S_2. Finally, we have

$$
\|f_k\|^2 = \begin{cases}
\frac{2}{3}m(m+1)(2m+1) & \text{if } k = 2m \geqslant 2, \\
\frac{4}{3}(m+1)(2m+1)(2m+3) & \text{if } k = 2m+1 \geqslant 3.
\end{cases}
$$

Now the action of z again agrees with that of w and as $w\mathcal{P}_k \subset \mathcal{P}_{k+1}$, we have

$$
w \cdot \frac{e_k}{\sqrt{k+1}} = \alpha_k \frac{e_{k+1}}{\sqrt{k+2}} + \beta_k \frac{f_{k+1}}{\|f_{k+1}\|},
$$

$$
w \cdot \frac{f_k}{\|f_k\|} = \gamma_k \frac{e_{k+1}}{\sqrt{k+2}} + \Delta_k \frac{f_{k+1}}{\|f_{k+1}\|}.
$$

Calculating this, we have the matrix,

$$
\begin{pmatrix} \alpha_k & \gamma_k \\ \beta_k & \Delta_k \end{pmatrix} = \begin{pmatrix} \sqrt{\dfrac{k+1}{k+2}} & 0 \\[2mm] \sqrt{\dfrac{3}{2}\dfrac{1}{(k+2)(k+3)}} & \sqrt{\dfrac{k}{k+3}} \end{pmatrix} \quad \text{for } k \geqslant 2.
$$

Therefore, the module multiplication on $\mathcal{Q}_1$ is essentially normal and hence, $\mathcal{Q}_1$ is an essentially reductive module as modulo compacts; multiplication by w is the unilateral shift of multiplicity two. Moreover, the semi-invariant module $S_2 \ominus S_1$, where the module multiplication induced by the shift $\left\{ \sqrt{\frac{k}{k+3}} \right\}_{k=1}^{\infty}$, is essentially reductive as well.

The conclusion of essential reductivity of the quotient module $\mathcal{Q}_1$ of the Hardy module over the bi-disc algebra remains true in somewhat greater generality by replacing the $H^2(\mathbb{D}) \otimes H^2(\mathbb{D})$ with the weighted Bergman module $H^{(\lambda)}(\mathbb{D}) \otimes H^{(\mu)}(\mathbb{D})$, $\lambda, \mu > 0$, where $H^{(\lambda)}(\mathbb{D})$ is the weighted Bergman space determined by the kernel function $K^{(\lambda)}(z, w) = (1 - z\bar{w})^{-2\lambda}$, $z, w \in \mathbb{D}$. The details of the quotient module corresponding to the submodule of functions vanishing to order 2 on the diagonal subset $\{(z, z) : z \in \mathbb{D}\} \subset \mathbb{D}^2$ in the weighted Bergman spaces $H^{(\lambda)}(\mathbb{D}) \otimes H^{(\mu)}(\mathbb{D})$ have been carried out in [53, Section 6.4]. Here the weighted Bergman spaces are identified as the subspace of the space of holomorphic functions on $\mathbb{D}^2$ via the map $f \otimes g \mapsto f \cdot g$, where $f \cdot g(z, w) = f(z)g(w)$, $z, w \in \mathbb{D}$. Again, the semi-sub-modules are essentially reductive.

After analyzing the quotient module obtained from a homogeneous ideal carefully in the paper [140], a complete characterization of all essentially reductive quotient modules of the Hardy module $H^2(\mathbb{D}^n)$ over the algebra $\mathcal{A}(\mathbb{D}^n)$ is given. In this paper, the case of weighted Bergman module is also discussed.

5.6 Notes and Exercises

- The first section is based partly on [33] and [34]. An elegant proof of a special case of the Berger–Shaw theorem for Bergman operators is borrowed from [33]. The applications of the BDF theory are taken from [26].

- Sections 2 and 3 are essentially based on [3, 4, 5, 29, 34, 62, 127].

- The treatment of Section 4 follows, in part, the survey article [19].

- Section 5 is in three parts, the first part on spherical operator tuples is entirely based on the paper [31]. We recall the Arveson–Douglas conjecture in the second part following [49, 50] and [125]. In the third part, we discuss quotient modules of Hardy module on the bidisc and reproduce some computations from [52].

Exercises

Exercise 5.6.1 Assuming the fact that the trace of the self-commutator of the Bergman operator B_G is equal to $\mathrm{Area}(G)/\pi$, show that for any closed subset X of the complex plane $\mathbb{C}$, the map $\gamma_X : \mathrm{Ext}(X) \to \mathrm{Hom}(\pi^1(X), \mathbb{Z})$ is surjective.

Hint: If the complement of X has infinitely many bounded components, say, $X_1, X_2, \ldots$, then $\mathrm{Area}(X_n) \to 0$ as $n \to \infty$, One may now imitate the proof of Theorem 5.1.1.

Exercise 5.6.2 Show that there are no compact circular (respectively homogeneous) operators.

Hint: The spectrum of a compact operator is countable.

Exercise 5.6.3 Show that the spectrum of a circular operator in $\mathcal{L}(\mathcal{H})$ is the union of a closed disc centered at the origin or annuli centered at the origin.

Exercise 5.6.4 ([79]) Let $\mathcal{H}$ be a separable Hilbert with orthonormal basis $\{e_n\}_{n \geqslant 0}$. Define $S : \mathcal{H} \to \mathcal{H}$ by

$$S e_n = \begin{cases} e_0 + e_1 & \text{if } n = 0, \\ e_{n+1} & \text{otherwise.} \end{cases}$$

Show that S is a circular operator.

Exercise 5.6.5 Show that any isometry is hyponormal. What are all the hyponormal m-isometries?

Hint: Note that isometries are the only hyponormal m-isometries. One may apply Lemma 5.2.1 and Remark 5.2.5.

Exercise 5.6.6 ([5, 30]) Let $T \in \mathcal{L}(\mathcal{H})$ be *concave*, that is, $I - 2T^*T + T^{*2}T^2 \leqslant 0$. Show that if T is Fredholm, then T is essentially unitary.

Hint: If T is Fredholm, then $\pi(T)$ is invertible. Now use the fact that any invertible concave operator is necessarily unitary.

Exercise 5.6.7 ([5])   Consider the operator T given by

$$T =: \bigoplus_{n=1}^{\infty} \begin{pmatrix} \alpha_n & c \\ 0 & \alpha_n \end{pmatrix} \quad \text{on} \quad \mathcal{H} =: \bigoplus_{n=1}^{\infty} \mathbb{C}^2$$

where c is a positive real and $\{\alpha_n\}_{n \geqslant 1}$ is a sequence in the unit circle which does not contain any of its accumulation point. Show that T is a cyclic 3-isometry which is not essentially normal.

Hint: Verify that

$$T^*T - TT^* = \bigoplus_{n=1}^{\infty} \begin{pmatrix} -c^2 & 0 \\ 0 & c^2 \end{pmatrix}.$$

Exercise 5.6.8   For $f : \mathbb{N} \to \mathbb{C}$, define $(\nabla f)(n) = f(n) - f(n+1)$, $n \in \mathbb{N}$. Show that f is a polynomial of degree at most $m - 1$ if and only if $\nabla^m f = 0$. Conclude that $T \in \mathcal{L}(\mathcal{H})$ is an m-isometry if and only for every $h \in \mathcal{H}$, $f_h(k) = \|T^k h\|^2$ is a polynomial of degree at most $m - 1$.

Exercise 5.6.9   For T in $\mathcal{L}(\mathcal{H})$, verify the following:

(i) If $T^*T \leqslant I$, then there exists a positive operator $A \leqslant I$ in $\mathcal{B}(\mathcal{H})$ such that

$$T^{*n}T^n \to A \ (\text{wot}) \ \text{as } n \to \infty.$$

(ii) If $T^*T \geqslant I$ and $\sup_{n \geqslant 0} \|T^n\| < \infty$, then there exists a positive operator $A \geqslant I$ in $\mathcal{B}(\mathcal{H})$ such that

$$T^{*n}T^n \to A \ (\text{wot}) \ \text{as } n \to \infty.$$

Hint: For (i), note that as $T^*T \leqslant I$, $\{T^{*n}T^n\}_{n \geqslant 0}$ is a monotonically decreasing sequence bounded below by 0. Note that $\{\|T^n x\|\}_{n \geqslant 0}$ is a decreasing sequence bounded above by $\|x\|$ for every $x \in \mathcal{H}$. Thus, if for $y \in \mathcal{H}$, we define a linear functional $\phi_y(x) = \lim_{n \to \infty} \langle T^n x, T^n y \rangle$ (which exists in view of the polarization identity; see Exercise 1.7.2), then $|\phi_y(x)| \leq \|x\| \|y\|$. Now apply Lemma B.1.8 to define A.

Exercise 5.6.10   Show that the Bergman kernel of the unit disc $\mathbb{D}$, that is, the vector $k_w \in L_a^2(\mathbb{D}, dA)$ with the reproducing property

$$\langle f, k_w \rangle = f(w), \quad f \in L_a^2(\mathbb{D}, dA), w \in \mathbb{D},$$

is of the form $k_w(z) = (1 - \bar{w}z)^{-2}$, $z, w \in \mathbb{D}$.

Hint: Use the Fourier expansion of f with respect to the orthonormal basis $\{\sqrt{n+1}z^n\}_{n\geq 0}$ in $L^2_a(\mathbb{D}, dA)$.

Exercise 5.6.11 Let μ be a rotation-invariant probability measure on $\mathbb{T}$. Show that μ is the arc-length measure.

Hint: Let σ be the arc-length. By Fubini's theorem and rotation-invariance of μ and σ, for any $f \in C(\mathbb{T})$,

$$
\begin{aligned}
\int_{\mathbb{T}} f(z)d\mu(z) &= \int_{\mathbb{T}}\left(\int_{\mathbb{T}} f(z)d\mu(z)\right)d\sigma(w) = \int_{\mathbb{T}}\left(\int_{\mathbb{T}} f(wz)d\mu(z)\right)d\sigma(w) \\
&= \int_{\mathbb{T}}\left(\int_{\mathbb{T}} f(zw)d\sigma(w)\right)d\mu(z) = \int_{\mathbb{T}} f(w)d\sigma(w).
\end{aligned}
$$

Exercise 5.6.12 Show that the multiplication operator M_z on $H^2(\mathbb{T})$ is homogeneous.

Hint: Verify that $U_\phi f = \sqrt{\phi'} \cdot (f \circ \phi)$ is the desired unitary.

Epilogue

So far, we have exclusively focussed on the original proof [26, Theorem 11.1] of Brown, Douglas, and Fillmore classifying essentially normal operators. In what follows, we will refer to this proof as the "BDF proof". We have therefore left out other proofs that simplify parts of the BDF proof. Indeed, there is a proof of the BDF theorem proposed by O'Donovan [104] that separates the techniques obtained from homological algebra and algebraic topology used in the proof of BDF from that of techniques obtained from operator theory. Although, it might appear surprising at first, it turns out that the BDF theorem is actually equivalent to what may appear to be a much simpler statement: If the index of an essentially normal operator T is trivial, then T must be of the form $N + K$ for some normal operator N and a compact operator K. This is Exercise 4.6.4.

As we have seen, it is not difficult to show that $\mathrm{Ext}(X)$ is an abelian semi-group. With a little more effort, the existence of a unique element that serves as the identity in $\mathrm{Ext}(X)$ is established. However, the proof in [23] of the existence of the inverse in $\mathrm{Ext}(X)$ is intimidating. A simpler proof due to Arveson [6] (see also [40]) appeared soon afterwards. Secondly, the BDF theorem established that Ext is a covariant functor from the category of compact metric spaces to abelian groups (Corollary 2.7.1) naturally leading to the question of its connection with other known functors from topology. This question was investigated vigorously and its connections with K-theory was eventually established on a firm footing. We discuss some of these developments in the following sections.

Finally, we conclude this short chapter with the discussion of several open problems, which include the Arveson–Douglas conjecture for semi-invariant modules of Hilbert modules over function algebras and the problem of classifying commuting "homogeneous" essentially normal operators.

Other Proofs

Here we briefly summarize the simplification due to Arveson of the proof that $\mathrm{Ext}(X)$ is a group. There were two other papers, one by Davie and the other by O'Donovan, that provided simplifications to parts of the BDF proof. We describe them in this section.

Let X be a compact metric space. The proof that $\mathrm{Ext}(X)$ is an abelian semigroup with a trivial element is not very hard (see Theorems 2.6.5 and 2.7.1). However, as we have seen, a lot of the theory developed in [26] was needed to prove the existence of an inverse in $\mathrm{Ext}(X)$ making it a group. In particular, the second splitting lemma (Theorem 4.1.3) was an essential ingredient in the proof. An alternative proof of the existence of the inverse, using a lifting theorem (cf. [9, Theorem B]) due to Andersen and Vesterstrøm, was given by Arveson [9, Theorem A].

Properties of $\mathrm{Ext}(X)$

In [42], Davie, isolated seven properties of $\mathrm{Ext}(X)$ listed below. For the proof of the BDF theorem (Theorem 4.5.1), it is enough to take X to be a planar set. Therefore, Davie verifies these properties of $\mathrm{Ext}(X)$ for $X \subseteq \mathbb{C}$ and points out that some of the proofs of [26] become a little simpler in this case. Then using these seven properties of $\mathrm{Ext}(X)$, he proves the BDF theorem.

- Corollary 4.3.1 – $\mathrm{Ext}(X)$ is an abelian group.
- Corollary 2.7.1 – A continuous map $f : X \to Y$ induces a homomorphism $f_* : \mathrm{Ext}(X) \to \mathrm{Ext}(Y)$, and $(fg)_* = f_* g_*$, $I_* = I$.
- Theorem 3.5.1 – Suppose $A \subseteq X$ is closed, $\iota : A \to X$ is the inclusion map, $p : X \to X/A$ the quotient map. Let $\sigma \in \mathrm{Ext}(X)$ with $p_* \sigma = 0$. Then, $\sigma = \iota_* \tau$ for some $\tau \in \mathrm{Ext}(A)$.
- Theorem 4.2.4 – Suppose X is the projective limit of the system (X_n, ρ_n), $n \in \mathbb{N}$, where ρ_n is onto for each n. Suppose $\sigma_n \in \mathrm{Ext}(X_n)$ for each n, and $(\rho_n)_* \sigma_{n+1} = \sigma_n$. Then there exists $\sigma \in \mathrm{Ext}(X)$ with $(\pi_n)_* \sigma = \sigma_n$, $n \in \mathbb{N}$.
- Example 2.3.4 – If $X \subseteq \mathbb{R}$, then $\mathrm{Ext}(X) = 0$.
- Remark 4.4.2 – There is a homomorphism $\iota : \mathrm{Ext}(X) \to G(X)$, which is natural in the sense that if $f : X \to Y$ is continuous, then $f_+ \iota(\sigma) = \iota(f_* \sigma)$, $\sigma \in \mathrm{Ext}(X)$.

  Here, the group $G(X)$ is $\mathrm{Hom}(\pi^1(X), \mathbb{Z})$ and $f_+ : G(X) \to G(Y)$ is the natural homomorphism induced by a continuous map $f : X \to Y$.
- Theorem 2.3.5, Corollary 4.4.1, and Remark 1.3.12– $\iota : \mathrm{Ext}(\mathbb{T}) \to G(\mathbb{T}) \approx \mathbb{Z}$ is an isomorphism.

The short exact sequence

There is a complete proof of the BDF theorem in [104] except for the proof that $\mathrm{Ext}(X)$ is a group. O'Donovan first proves that if X is any compact metric space and $A \subseteq X$ is a closed subset, then

$$\mathrm{Ext}(A) \xrightarrow{\iota_*} \mathrm{Ext}(X) \xrightarrow{q_*} \mathrm{Ext}(X/A), \tag{E.1.1}$$

where ι is the inclusion map and q is the quotient map, is a short exact sequence. The original proof from [26] is reproduced in these notes as Theorem 3.5.1. He then proves that any essentially normal operator T such that index $(T - \lambda) = 0$ for all $\lambda \notin X := \sigma_e(T)$, $T = N + K$ for some normal operator N and a compact operator K. He notes that the proof of the BDF theorem is equivalent to this statement, see Exercise 4.7.4. In [104], there is a direct proof of this equivalence using the Berger–Shaw theorem.

Arveson's proof of "Ext(X) is a group"

We now briefly outline the proof that Ext(X), is a group, following Arveson [9], for any compact metric space X. As we have already mentioned, his proof was based on an extension theorem of Andersen and Vesterstrøm. We reproduce here an unpublished result due to Davie stated in [25], which serves equally well for the proof of Arveson as we will see now. The advantage is that the proof of this version of the extension theorem is less demanding. The details are in [27, Theorem 1.19] and not included here.

Theorem E.1.1 ([27], Theorem 1.19)

Let X be compact and metrizable, let $\mathcal{A}$ be a C^*-algebra with unit, let $\mathcal{I}$ be a closed two-sided $*$-ideal of $\mathcal{A}$ and $\pi : \mathcal{A} \to \mathcal{A}/\mathcal{I}$ be the quotient map, and let $\tau : C(X) - \mathcal{A}/\mathcal{I}$ be a unital positive linear map. Then there is a unital positive linear map $\sigma : C(X) \to \mathcal{A}$ such that $\pi \circ \sigma = \tau$.

The proof of the existence of an inverse in Ext(X), which is from [27, Theorem 1.23], is a consequence of the extension theorem combined with the dilation theorem of Naimark. It first appeared as Proposition 4.4 in [42]. Here we set $\mathcal{A} := \mathcal{L}(\mathcal{H})$, $\mathcal{I} = C(\mathcal{H})$ and $Q := \mathcal{L}(\mathcal{H})/C(\mathcal{H})$.

Let $\sigma : C(X) \to Q(\mathcal{H})$ be an extension. Owing to the extension theorem, there is a unital positive linear map $\sigma : C(X) \to \mathcal{L}(\mathcal{H})$ such that $\tau = \pi \circ \sigma$. By Naimark's dilation theorem, there is a Hilbert space $\mathcal{H}' \supseteq \mathcal{H}$ and a unital $*$-homomorphism $\phi : C(X) \to \mathcal{L}(\mathcal{H}')$ such that σ is the compression of ϕ to $\mathcal{H}$. In other words,

$$\phi(f) = \begin{pmatrix} \sigma(f) & K_f \\ L_f & M_f \end{pmatrix}, \ f \in C(X).$$

Note first that K_f and L_f are compact. As ϕ is multiplicative, we have

$$\sigma(fg) = \sigma(f)\sigma(g) + K_f L_g, \ f, g \in C(X).$$

Applying π, it follows that $K_f L_g$ is compact. However, $K_f^* = L_{\bar{f}}$. In conclusion, ϕ being a $*$ map ensures that K_f and L_g are compact. The verification that $\tau'(f) := \pi(M_f)$ is a $*$-homomorphism is not hard. Moreover, $\tau + \tau' = \pi \circ \phi$. Thus, if τ_0 is the trivial element and if we set $\tau_1(f) = \tau'(f) \oplus \tau_0(f)$, $f \in C(X)$, then τ_1 is an extension such that $\tau + \tau_1$ is equivalent to $\pi \circ \phi$. Therefore, $\tau + \tau_1$ is trivial. Hence, τ_1 is the inverse of τ in Ext(X).

Related Developments

The reader going through these notes may have noticed that some of the notions could have been introduced in a more general setting. We have only considered extensions by $C(X)$ of the ideal of compact operators $C(\mathcal{H}) \subseteq \mathcal{L}(\mathcal{H})$, namely, the short exact sequence

$$0 \to C(\mathcal{H}) \to S \to C(X) \to 0,$$

S is an extension of $C(X)$ by the compact operators. The main goal has been to classify these extensions. This is sufficient for applications to operator theory.

Ext(A, B)

Recalling that any unital commutative C^*-algebra is of the form $C(X)$, one might wish to replace it by a more general C^*-algebra and even use an arbitrary C^*-algebra B rather than the ideal of compact operators C. For instance, let

$$0 \to B \xrightarrow{\iota} E \xrightarrow{q} A \to 0 \tag{E.1.2}$$

be a short exact sequence of C^*-algebras. Thus, B is an ideal in E. Let $M(B)$ be the two-sided multipliers of the algebra B. If B is unital, then $M(B) = B$. As $M(B)$ is always unital, it follows that $M(B) \neq B$ if B is nonunital. Let $Q(B)$ denote the algebra $M(B)/B$. An ideal B in E is said to be *essential* if each non-zero closed ideal in E has non-zero intersection with B. There is an injective $*$-homomorphism σ of E into $M(B)$ if and only if B is an essential ideal in E (see [111, Proposition 3.12.8]). The composition of $\sigma : E \to M(B)$ with the quotient map $\pi : M(B) \to Q(B)$ clearly gives rise to a $*$-homomorphism

$$\tau := \hat{\pi} \circ \hat{\sigma} : A \equiv E/B \xrightarrow{\hat{\sigma}} M(B)/B \xrightarrow{\hat{\pi}} Q(B), \tag{E.1.3}$$

which is the *Busby invariant* of the extension (E.1.2). Note that τ is injective if and only if B is essential in A.

There are many possible notions of equivalence for the two short exact sequences of the form (E.1.2), see [24, Section 15.4]. One of these is the strong equivalence of extensions: There exists a $*$-homomorphism γ such that the diagram

$$
\begin{array}{ccccccccc}
0 & \longrightarrow & B & \xrightarrow{j_1} & E_1 & \xrightarrow{q_1} & A & \longrightarrow & 0 \\
 & & \| & & \downarrow{\gamma} & & \| & & \\
0 & \longrightarrow & B & \xrightarrow{j_1} & E_2 & \xrightarrow{q_1} & A & \longrightarrow & 0
\end{array}
$$

commute. Now, the short exact sequence (E.1.2) can be recovered from its Busby invariant, that is, if two extensions are strongly isomorphic, then their Busby invariant coincide. This is not hard to verify, see [24, pp. 146]. The situation is completely analogous to the correspondence between "extensions of $C(X)$ by the compact operators $C(\mathcal{H})$" and "*-monomorphisms τ : $C(X) \to Q(\mathcal{H})$".

An extension is said to be *trivial* if it is split, that is, if there is a cross-section that is a *-homomorphism $s : A \to E$ such that $q \circ s = I_A$. In general, there are many trivial extensions.

To define the additive structure, assume that B is stable, that is, B is *-isomorphic to $B \otimes C$, where C is the algebra of compact operators on some separable Hilbert space $\mathcal{H}$. As $C \otimes C \equiv C$, it follows that an algebra of the form $B \otimes C$ is stable. Now, following [24, 15.6], we define the sum of two extensions.

First, fix an isomorphism of C in the 2×2 matrices $\mathcal{M}_2(C)$. This induces a natural isomorphism B in $\mathcal{M}_2(B)$. Hence, we obtain isomorphisms: $M(B) \equiv \mathcal{M}_2(MB))$ and $Q(B) \equiv \mathcal{M}_2(Q(B))$.

A sum of two extensions is defined by taking the direct sum of their Busby invariants as follows. If τ_1, τ_2 are the Busby invariants of two extensions of the form (E.1.2), then their sum is the extension determined by the Busby invariant $\tau_1 \oplus \tau_2 \to Q(B) \oplus Q(B) \subseteq \mathcal{M}_2(Q(B)) \equiv Q(B)$. It is shown in [24, Proposition 15.6.2] that the addition on the set of strong equivalence classes **Ext(A, B)** of A by B is an associative and commutative semi-group. Define Ext(A, B) to be the quotient of **Ext(A, B)** by the trivial extensions. This means that if τ_1 and τ_2 are two extensions of A by B, then $[\tau_1] = [\tau_2]$ in Ext(A, B) if there are trivial extensions τ_1', τ_2' such that $\tau_1 \oplus \tau_1'$ and $\tau_2 \oplus \tau_2'$ are strongly equivalent. This equivalence relation is called *stable equivalence*. If B is not stable, then Ext(A, B) is defined by taking B; an alternative would be to consider Ext$(A, B \otimes C)$. There are several different choices for which Ext(A, B) is a group. We will not go into the details here, except for one special case, where $B = C$ and A is a separable nuclear C^*-algebra. In this case, Ext$(A) := $ Ext(C, A) is a group. First, Voiculescu proves that there is a trivial extension, see [138], and also [10]. Seocond, Choi and Effros prove the lifting theorem for any spearable nuclear C^*-algebra in [32]. Finally, Arveson [10] puts these two results together to show that Ext(C, A) is a group. Salinas, with a slightly different hypothesis on the C^*-algebra A, proves that Ext(A) is a group, see [121, Theorem 1].

A covariant version of Ext is studied in [107], where connections with the work of Kasparov on Ext(A, B) are indicated. Moreover, essentially n-normal operators have been studied in a series of papers [106, 121, 123]. An overview of this topic along with that of the BDF theorem appears in [109].

Homotopy invariance

We have already seen that Ext(X) is homotopy invariant for planar compact sets. However, in [27], this is proved for any compact metric space X, see Proposition 4.5.1.

> **Theorem E.1.2**
>
> *The correspondence $X \mapsto Ext(X)$ is a homotopy functor, that is, if $f, g : X \to Y$ are homotopic, then $f_*, g_* : Ext(X) \to Ext(Y)$ are equal.*

The outline of the proof below is taken from [47]. To show that Ext is homotopy invariant, using the exactness of (E.1.1), it is enough to check that Ext(CX), where CX is the cone

$(X \times [0,1])/(X \times \{0\})$ over X, is trivial. Once $\mathrm{Ext}(CX)$ is assumed to be trivial, the homotopy invariance of Ext follows easily as outlined in the following.

Outline of the proof of Theorem 6.2.1 assuming that $\mathrm{Ext}(CX)$ is trivial. Recall that $X \mapsto \mathrm{Ext}(X)$ is a covariant functor, which is Corollary 2.7.1. Now, apply Theorem 3.5.1 to the compact metrizable space $X \times I$ and its closed subset $A = X \times \{0\}$ to obtain the short exact sequence

$$\mathrm{Ext}(X \times \{0\}) \xrightarrow{i_*} \mathrm{Ext}(X \times I) \xrightarrow{q_*} \mathrm{Ext}(CX),$$

where $i : X \times \{0\} \to X \times I$ is the inclusion map and $q : X \times I \to CX$ is the quotient map. Now as $\mathrm{Ext}(CX)$ is trivial, $\ker q_* = \mathrm{Ext}(X \times I)$, and hence i_* is surjective. Similar argument shows that for the inclusion map $j : X \times I \to X \times \{1\}$, the induced map j_* is surjective (although we do not need this fact). To complete the proof of Theorem E.1.2, assume that $f, g : X \to Y$ are homotopic. Thus there exists a continuous map $H : X \times [0,1] \to Y$ such that $H(x,0) = f(x)$ and $H(x,1) = g(x)$ for all $x \in X$. Note that

$$f = H \circ i \circ h, \quad g = H \circ j \circ k, \tag{E.1.4}$$

where $h : X \to X \times \{0\}$ and $k : X \to X \times \{1\}$ are natural homeomorphisms. As i_* is surjective, so is $i_* \circ h_*$, and hence, for any $r \in \mathrm{Ext}(X)$, there exists $v \in \mathrm{Ext}(X)$ such that

$$j_* \circ k_*(r) = i_* \circ h_*(v). \tag{E.1.5}$$

However, $i_* \circ h_*$ and $j_* \circ k_*$ are injective. Indeed, if $p : X \times I \to X$ denotes the projection map, then $p \circ i \circ h = p \circ j \circ k$ is the identity on X. It now follows from (E.1.5) that $r = v$, and hence, $i_* \circ h_* = j_* \circ k_*$. Now apply (E.1.4) to conclude that $f_* = g_*$. □

We record an immediate Corollary of the homotopy invariance of Ext, which is [27, Corollary 2.15].

Corollary E.1.1

If X is contractible, then $\mathrm{Ext}(X) = 0$.

There is an analytic proof in [103] of the verification that $\mathrm{Ext}(CX)$ is trivial (see also [122]). For any extension τ of $\mathrm{Ext}(CX)$, suppose there is an extension τ' of $\mathrm{Ext}(CX)$ such that $[\tau] + [\tau'] = [\tau']$. Then as $\mathrm{Ext}(CX)$ is a group, $[\tau] = 0$. The proof begins by showing that $\mathcal{E} = \pi^{-1}(\mathrm{im}\,\tau)$ is quasi-diagonal (Recall that an algebra $\mathcal{E} \subseteq \mathcal{L}(\mathcal{H})$ is said to be *quasi-diagonal* if there exists a sequence $\{P_n\}_{n \geqslant 0}$ of finite dimensional projections such that $\sum_{n=0}^{\infty} P_n = I$ and $T - P_n T P_n$ is compact for every integer $n \geqslant 0$ and every $T \in \mathcal{E}$). To complete the proof, a τ' with the desired property is exhibited showing that $\mathrm{Ext}(CX)$ is trivial. This is enough, as before, to establish that Ext is homotopy invariant.

Having established that $\mathrm{Ext}(X)$ is a covariant homotopy functor with the exact sequence property (E.1.1), it is shown in [27, Section 7] that Ext actually defines a generalized homology theory which is an operator theoretic realization of K-homology. Rather than going into further details here, we refer the reader to the expositions [24, 37, 47, 40, 77, 95] for these and many more exciting developments around this topic.

K-theory

There are at least two instances, where connections between K-theory and operator theory occur in a natural way. In the first instance, there is a pairing induced by the index from homotopy classes of maps from X to Fredholm operators on some separable Hilbert space $\mathcal{H}$ to the group $K^0(X)$. The other is obtained by generalizing the map $\gamma_X : \text{Ext}(X) \to \text{Hom}(\pi^1(X), \mathbb{Z})$ to the case where $\pi^1(X)$ is replaced by $K^1(X)$.

As the connected components of the collection $\mathcal{F}(\mathcal{H})$ of Fredholm operators in $\mathcal{L}(\mathcal{H})$ are precisely $\mathcal{F}_n(\mathcal{H})$, $n \in \mathbb{Z}$, we conclude that there is a bijection from the collection of components of $\mathcal{F}(\mathcal{H})$ to $\mathbb{Z}$ induced by the map $\text{ind} : \mathcal{F}(\mathcal{H}) \to \mathbb{Z}$. This is proved in Corollary 1.3.3. It is an indication of a possibly deeper connection between homotopy classes of Fredholm operators and some other abelian group. To explore this connection, consider a continuous family of Fredholm operators $F : X \to \mathcal{F}$ over some compact set X, where $\mathcal{F} := \mathcal{F}(\mathcal{H})$ for some fixed Hilbert space $\mathcal{H}$. Let $\{e_n\}_{n \geqslant 0}$ be an orthonormal basis for $\mathcal{H}$. Set H_n to be the closed subspace spanned by the vectors $\{e_i : i \geq n\}$, P_n to be the projection onto subspace H_n and $F_n(x) = (P_n \circ F)(x)$, $x \in X$. For large n, this has the effect of fixing the dimension of the spaces $\ker F(x)$. Clearly, $F_n(x)$ is Fredholm and $\text{ind} F_n(x) = \text{ind} F(x)$, $n \in \mathbb{N}$ (see Equation (1.3.9)). The following Lemma and brief discussion are from [16]. However, for a leisurely account with complete details of the proof of the Atiyah–Jänich theorem, see [95].

Lemma E.1.1

Let $F : X \to \mathcal{F}$ be a continuous map and $x_0 \in X$ be fixed but arbitrary. Choose an integer $n \geqslant 0$ such that $F_n(x_0)(\mathcal{H}) = H_n$. Then there exists a neighborhood U of x_0 such that

$$F_n(x)(\mathcal{H}) = H_n, \ x \in U.$$

Moreover, $\dim \ker F_n(x)$ is constant for $x \in U$ and there are d continuous functions $s_i : X \to \mathcal{H}$, $1 \leq i \leq d$ such that $\{s_1(x), \dots, s_d(x)\}$ is a basis for $\ker F_n(x)$, $x \in U$.

Proof Define the operator $G(x) : \mathcal{H} \to H_n \oplus \ker F_n(x_0)$ by setting $G(x)u = (F_n(x)u, P_0u)$, where P_0 is the projection onto the subspace $\ker F_n(x_0)$. By definition, $G(x_0)$ is bijective. Consequently, by the closed graph theorem (see Theorem B.1.10), $G(x_0)$ is a bounded invertible operator. As the set of invertible maps is open and G is continuous, it follows that it is an isomorphism in some neighborhood of x_0. Hence, $F_n(x)(\mathcal{H}) = H_n$, $x \in U$. If $e_1, \dots, e_d$ is a basis of $\ker F_n(x_0)$, the vectors $s_i(x) = G(x)^{-1}(e_i)$, $1 \leq i \leq d$, form a basis for $F_n(x)$, $x \in U$ completing the proof. $\qquad \Box$

Combining Corollary 1.3.1 with the previous Lemma, we see that $\text{ind} F(x) = d - n$ is a locally constant function. Now, suppose X is compact. Then there is a nonnegative integer n such that $F_n(x)(\mathcal{H}) = H_n$ for all $x \in X$. Set $\ker F_n := \cup_{x \in X} \ker F_n(x)$ and identify $\ker F_n$ as a subspace of $X \times \mathcal{H}$ via the map $\ker F_n \ni y \to (x, \ker F_n(x))$. Lemma E.1.1 shows that $\ker F_n$ is a locally trivial family of d-dimensional vector spaces. In other words, $\pi : E \to X$, $\pi^{-1}(x) =$

$\ker F_n(x)$, is a vector bundle of rank d. This means, we must find d continuous maps $s_1, \ldots, s_d :$ $X \to E$ such that $\pi \circ s_i = \mathrm{id}$, $1 \le i \le d$. However, Lemma E.1.1 shows how to do this. Moreover, $\ker F(x)^* = H_n^\perp$ for all $x \in X$. Hence, $\ker F_n^*$ is the trivial bundle on $X \times H_n^\perp$.

Let $\mathrm{Vect}(X)$ be the set of isomorphism classes of all vector bundles over X. If X is a singleton, then a vector bundle over X is just a vector space and the equivalence class of these is determined by its dimension. Thus in this case, $\mathrm{Vect}(X)$ is the set of non-negative numbers $\mathbb{Z}_+$. However, in general, there exist simple examples of non-trivial vector bundles. Therefore the isomorphism class of a vector bundle is not necessarily determined by dimension alone.

Given two vector bundles E, F over the same space X, their direct sum $E \oplus F$ is defined by taking the point-wise direct sum, i.e., $(E \oplus F)_x := \pi_E^{-1}(x) \oplus \pi_F^{-1}(x)$ $x \in X$. This induces an abelian semi-group structure on $\mathrm{Vect}(X)$. Recall that the *Grothendieck group* of any semi-group Z is defined to be the group $G(Z)$ of formal differences. Thus, $G(Z)$ is defined to be the set of all pairs in $Z \times Z$ modulo the equivalence: $(a,b) \sim (a'+z, b'+z)$, $z \in Z$. The abelian group $K(X)$ (same as $K^0(X)$) is obtained by applying this to the semi-group $\mathrm{Vect}(X)$.

In view of Corollary 1.3.1 and Lemma E.1.1, a candidate for the definition of an index of $F : X \to \mathcal{F}$ presents itself:

$$\mathrm{ind}\, F := [\ker F_n] - [\ker F_n^*] = [\ker F_n] - [X \times H_n^\perp], \tag{E.1.6}$$

where $[\cdot]$ denotes the equivalence class in $\mathrm{Vect}(X)$. Thus, we see that the index of $F : X \to \mathcal{F}$ takes values in $K(X)$.

To define the index of $F : X \to \mathcal{F}$, a large enough n was chosen to ensure that $\dim \ker F(x)$, $x \in X$, is constant. Moreover, for this definition, a fixed orthonormal basis was chosen in the Hilbert space $\mathcal{H}$. However, it is pointed out in [16] that the definition is independent of both these choices.

Theorem E.1.3 (Atiyah–Jänich)

If X is a compact space and $[X; \mathcal{F}]$ denotes the set of homotopy classes of continuous maps $F : X \to \mathcal{F}$, then $F \to \mathrm{ind}\, F$ induces an isomorphism $\mathrm{ind} : [X; \mathcal{F}] \to K(X)$.

In [16], using the index of the continuous family of the map $F : X \to \mathcal{F}$, a new proof for the Bott periodicity theorem is given.

The second relationship with K-theory comes from the pairing of the group $K^1(X)$ with $\mathrm{Ext}(X)$, that is, a map $\mathrm{Ext}(X) \to \mathrm{Hom}(K^1(X), \mathbb{Z})$. This pairing is described in [47]. Let $[X, \mathrm{GL}_n]$ be the group of homotopy classes of maps from X to $\mathrm{GL}_n(\mathbb{C})$. Identifying $C(X) \otimes M_n$ with the algebra $M_n(X)$, we see that $(C(X) \otimes M_n)^{-1}$ is the algebra of continuous functions taking values in $\mathrm{GL}_n(\mathbb{C})$. Let j be the map taking a function f to its homotopy class. Finally, identifying $Q(\mathcal{H}) \otimes M_n$ with $Q(\mathcal{H} \otimes \mathbb{C}^n)$ and setting $(Q(\mathcal{H} \otimes \mathbb{C}^n))^{-1}$ to be the invertible elements in $Q(\mathcal{H} \otimes \mathbb{C}^n)$, we get the index map $\mathrm{ind} : (Q(\mathcal{H} \otimes \mathbb{C}^n))^{-1} \to \mathbb{Z}$. Now, we can define the map $\gamma_n(\tau) : [X, \mathrm{GL}_n] \to \mathbb{Z}$ requiring that the diagram

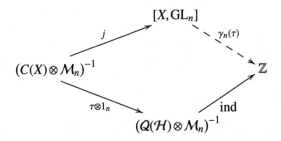

commute. That the map $\gamma_n(\tau)$ is well defined follows from the properties of the index map. The inclusion of GL_n in GL_{n+1} via the map $\lambda(A) = \begin{pmatrix} A & 0 \\ 0 & 1 \end{pmatrix}$ ensures that the diagram

$$[X, \text{GL}_n] \xrightarrow{\lambda_*} [X, \text{GL}_{n+1}]$$

with $\gamma_n(\tau)$ and $\gamma_{n+1}(\tau)$ mapping to $\mathbb{Z}$

commutes. As the direct limit of the groups $[X, \text{GL}_n]$ defines the group $K^1(X)$, we obtain a homomorphism $\gamma_\infty(\tau) : K^1(X) \to \mathbb{Z}$ and $\gamma_\infty : \text{Ext}(X) \to \text{Hom}(K^1(X), \mathbb{Z})$. The BDF theorem shows that for $X \subseteq \mathbb{C}$, the map γ_∞ is an isomorphism, but in general, it is not. However, the planar case, as we have seen, is enough for applications to operator theory.

Levy [90, 91] has proposed a spectral picture for commuting tuples of bounded linear operators with deep connections to an analytic K-theory class after introducing an appropriate notion of the index map.

Open Problems

In this concluding section, we discuss some open problems. Each of these problems has to do with concrete realization of $\text{Ext}(X)$, when X is not necessarily planar.

Essentially normal tuples

An m-tuple $T = (T_1, \ldots, T_m)$ of operators $T_1, \ldots, T_m \in \mathcal{L}(\mathcal{H})$ is said to be *essentially commuting* if $[T_i, T_j] \in C(\mathcal{H})$ for every $1 \leqslant i, j \leqslant m$. An essentially commuting m-tuple T is said to be *essentially normal* if $[T_i^*, T_j] \in C(\mathcal{H})$ for every $1 \leqslant i, j \leqslant m$.

In a manner completely analogus to the case of $X \subseteq \mathbb{C}$, we can obtain a $*$-monomorphism of $C(X)$ into $Q(\mathcal{H})$ from essentially normal m-tuples $T = (T_1, \ldots, T_m)$ of operators on $\mathcal{H}$. Fix the essential spectrum of T to be a compact subset X of $\mathbb{C}^m$. It is easy to see that the m-tuple T of operators, as earlier, determines a $*$-monomorphism τ from $C(X)$ to $Q(\mathcal{H})$ by mapping the co-ordinate functions $z_1, \ldots, z_m$ to $\pi(T_1), \ldots, \pi(T_m)$. Two such essentially normal operator tuples $T = (T_1, \ldots, T_m)$ and $T' = (T'_1, \ldots, T'_m)$ are said to be *essentially equivalent* if there is an essentially unitary operator $V \in \mathcal{L}(\mathcal{H})$ such that $VTV^* - T'$ is compact (that is, $VT_jV^* - T'_j$ is compact for every $j = 1,\ldots,m$). The equivalence class $[T]$ determines an element of $\mathrm{Ext}(X)$ (This is a weak equivalence; it coincides with strong equivalence, Proposition 2.5.1).

The C^*-algebra $\mathcal{S}_T$ generated by the essentially commuting tuple T, the identity operator and the compact operators $C(\mathcal{H})$ on some Hilbert space $\mathcal{H}$ determines an extension

$$0 \to C(\mathcal{H}) \to \mathcal{S}_T \to C(X) \to 0,$$

where the maximal ideal space X of the quotient $\mathcal{S}_T/C$ is the essential spectrum of T. Now, the classification problem for the essentially normal m-tuples T of operators, as before, is equivalent to the classification problem of the extension of $C(X)$ by the compacts, or equivalently, that of $*$-monomorphisms $\tau : \mathcal{S}_T \to C(X)$.

The homomorphism $\gamma_X : \mathrm{Ext}(X) \to \mathrm{Hom}(\pi^1(X), \mathbb{Z})$ is an isomorphism if $X \subset \mathbb{C}$ but it is not in general. For the unit sphere in $\mathbb{C}^2$, in Example 4.4.5, we exhibit a non-trivial element of $\mathrm{Ext}(\mathbb{S})$ and point out that $\gamma_\mathbb{S}$ cannot be injective as $\pi^1(\mathbb{S}) = 0$. None the less, using Bott periodicity, Ext of the unit sphere $\mathbb{S}^{n-1} \subset \mathbb{R}^n$ is shown to be $\mathbb{Z}$ or 0 according as n is even or odd (see [27, Corollary 7.1]). Finally, a complete set of representatives of $\mathrm{Ext}(\mathbb{S}_{2n-1})$ is given in [67, Lamma 3.1].

If two commuting essentially normal n-tuples T and T' are essentially equivalent, then the joint essential spectrum of T and T' are same, say equal to X, and $\mathrm{ind}(T - \lambda) = \mathrm{ind}(T' - \lambda)$ for every $\lambda \in \mathbb{C}^n \setminus X$ (see [36, Corollary 3.9 and Theorem 3]). Here the *Taylor essential spectrum* of an essentially normal n-tuple may be defined as the maximal ideal space of the abelian C^*-algebra generated commuting n-tuple $\pi(T)$ (image in the Calkin algebra). The *Fredholm index* is defined as the alternating sum of the dimensions of homology modules appearing in the *Koszul complex* associated with T (refer to [136] and [38] for details). Thus, Taylor essential spectra and index data are unitary invariants for the essential equivalence of essentially normal tuples. However, these are not complete unitary invariants. This may be derived from the fact that the set of trivial extensions for higher dimensional cubes is not closed (refer to [25, Section 4]). In operator theoretic terms, this says that the collection $\mathcal{N} + C$ given by

$$\{(N_1 + K_1, N_2 + K_2) : N_1, N_2 \text{ are commuting normal and } K_1, K_2 \text{ are compact}\}$$

is not norm-closed in $\mathcal{L}(\mathcal{H}) \times \mathcal{L}(\mathcal{H})$. Assuming this fact, it is easy to see that Taylor essential spectra and index data are not complete unitary invariants for the essential equivalence of essentially normal tuples. Indeed, if there exists a sequence $\{(N_1^{(m)} + K_1^{(m)}, N_2^{(m)} + K_2^{(m)})\}_{m \in \mathbb{N}}$ in $\mathcal{N} + C$ converging to $S = (S_1, S_2)$, then by the continuity of the Fredholm index (see [38, Theorem 3]), $\mathrm{ind}\, S = 0$. Moreover, as $m \to \infty$,

$$[N_j^{*(m)} + K_j^{*(m)}, N_j^{(m)} + K_j^{(m)}] \to [S_j^*, S_j], \quad j = 1, 2,$$

and hence, S is essentially normal. If we assume that the Taylor essential spectra and index data form a complete set of unitary invariants, then $S = (S_1, S_2)$ must belong to $\mathcal{N} + C$, which is a contradiction as $\mathcal{N} + C$ is not norm closed. The problem of finding all complete invariants for equivalence of essentially normal tuples appears to be intractable at the moment (the reader is referred to [57], [67] for some special instances of this problem).

We conclude this subsection with another open problem related to the BDF theory. Let $\mathcal{F}(\mathcal{H})$ denote the collection of essentially commuting Fredholm n-tuple of operators on $\mathcal{H}$. If $T \in \mathcal{F}(\mathcal{H})$, ind T is a well-defined integer which is an invariant for the path-components of Fredholm essentially commuting tuples (refer to [38]). In [46], R. G. Douglas raised the following question: is it the only invariant? In other words, given two essentially commuting Fredholm n-tuples T and T' with the same Fredholm index, is it always possible to find a continuous path $\gamma : [0,1] \to \mathcal{F}(\mathcal{H})$ such that $\gamma(0) = T$ and $\gamma((1) = T'$? This is the *deformation problem*. For $n = 1$, the answer is known to be yes (see Corollary 1.3.3). If T, T' are essentially normal, then an affirmative answer is provided in [46] using techniques from extension theory [26, 25, 27] (refer to [38] for a detailed exposition of this fact). An affirmative answer to the deformation problem for the class of essentially doubly commuting Fredholm pairs with a semi-Fredholm coordinate is due to Curto (see [38, Section 14]).

Essentially homogeneous tuples

We have discussed essentially normal essentially homogeneous operators in Section 5.4, where Corollary 5.4.1 provides a complete list of such operators. The classification of essentially homogeneous tuples appears to be not so simple. We describe the problem in some detail.

Let $\mathbb{B}_m \subset \mathbb{C}^m$ be the unit ball and $\mathbb{S}_m$ be its Silov boundary consisting of unit vectors in $\mathbb{C}^m$. Let G be the bi-holomorphic automorphism group of $\mathbb{B}_m$. The survey paper [9] has an explicit description of the bi-holomorphic automorphism groups of classical bounded symmetric domains. The ball $\mathbb{B}_m$ is a bounded symmetric domain of rank 1. The unitary group $U(m)$ is the maximal compact subgroup of G, which acts on $\mathbb{B}_m$ as linear transformations.

As G consists of functions holomorphic in a neighborhood of $\overline{\mathbb{B}}_m$, $g \in G$ acts holomorphically on $\overline{\mathbb{B}}_m$, the operator $g \cdot T$ is defined via the usual functional calculus for any m-tuple T of operators with $\sigma_{\text{ess}}(T) \subseteq \overline{\mathbb{B}}_m$. The unitary group $U(m)$ also acts on such m-tuple of operators, see (5.5.14).

An essentially commuting m-tuple $T := (T_1, \ldots, T_m)$ of operators acting on some Hilbert space $\mathcal{H}$ is said to be *essentially homogeneous* if $\sigma_{\text{ess}}(T) \subseteq \overline{\mathbb{B}}_m$, and there is a unitary representation U of G on $\mathcal{H}$ such that

$$U_g^* T U_g - g \cdot T \in C(\mathcal{H}), \quad g \in G, \tag{E.1.7}$$

where $U_g^* T U_g := (U_g^* T_1 U_g, \ldots, U_g^* T_m U_g)$. The only possibilities for the essential spectrum of an essentially normal essentially homogeneous m-tuple T, using the spectral mapping theorem, is either $\overline{\mathbb{B}}_m$ or its Silov boundary $\mathbb{S}_m$. As the unit ball $\overline{\mathbb{B}}_m$ is contractible, by homotopy invariance, we have that $\text{Ext}(\overline{\mathbb{B}}_m) = 0$. Hence, the only essentially normal essentially homogeneous m-tuple T of operators with $\sigma_{\text{ess}}(T) = \overline{\mathbb{B}}_m$ must be the m-tuple of multiplication by the coordinate functions on the Hilbert space $L^2(\mathbb{B}_m, dV)$.

Finding a model for the essentially normal essentially homogeneous operators T when its essential spectrum is $\mathbb{S}_m$ would be more interesting. Moreover, the question of essentially homogeneous operators makes perfectly good sense even if we start with an arbitrary classical bounded symmetric domain Ω. Again, finding models for these predictably must be even more interesting. We conclude with a problem of this kind that might be accessible.

A similar question arises if we replace the bi-holomorphic automorphism group by the unitary group $U(m)$. Now, say that an m-tuple T of operators acting on a Hilbert space $\mathcal{H}$ with $\sigma_{\text{ess}}(T) \subseteq \overline{\mathbb{B}}_m$ is *essentially spherical*, or equivalently, $U(m)$-essentially homogeneous, if there is a unitary representation U of the unitary group $U(m)$ on the Hilbert space $\mathcal{H}$ such that

$$U_\chi^* T U_\chi \text{ is essentially equivalent to } \chi \cdot T \text{ for every } \chi \in U(m).$$

We ask what are the essentially normal essentially spherical operators (see [Theorems 2.5 and 4.2, 29] for the case in which T is cyclic with cyclic vector belonging to the cokernel)?

Arveson–Douglas conjecture

At the outset, we must admit that what follows is not meant to be a survey of the many exciting results involving the Arveson–Douglas conjecture. For the general background, one may consult [126] along with the discussion in Section 5.5. A recent survey of the analytic aspects of the Arveson–Douglas conjecture has appeared in [59]. The recent papers [23, 51, 58, 56] is not an exhaustive list, never the less, they are expected to serve as a pointer to the large body of work in this area.

In the concluding paragraph of Section 5.5, we have shown that the quotient module $H^2(\mathbb{D}^2) \ominus [(z-w)^2]$ is essentially reductive. To state yet another variant of the Arveson–Douglas conjecture, we discuss quotient modules in slightly greater generality. This is taken from the recent paper [63]. The detailed proofs that are omitted here can be found in [63].

Let K_1 and K_2 be two scalar valued non-negative definite kernels defined on $\Omega \times \Omega$, $\Omega \subseteq \mathbb{C}^m$. The Hilbert space tensor product $(\mathcal{H}, K_1) \otimes (\mathcal{H}, K_2)$ is again a reproducing kernel Hilbert space with the reproducing kernel $K_1 \otimes K_2$, where $K_1 \otimes K_2 : (\Omega \times \Omega) \times (\Omega \times \Omega) \to \mathbb{C}$ is given by

$$(K_1 \otimes K_2)(z, \zeta; w, \rho) = K_1(z, w) K_2(\zeta, \rho), \ z, \zeta, w, \rho \in \Omega.$$

Assume that the multiplication operators M_{z_i}, $i = 1, \dots, m$, are bounded on $(\mathcal{H}, K_1)$ as well as on $(\mathcal{H}, K_2)$. Then, $(\mathcal{H}, K_1) \otimes (\mathcal{H}, K_2)$ may be realized as a Hilbert module over $\mathbb{C}[z_1, \dots, z_{2m}]$ with the module action defined by

$$\mathbf{m}_p(h) = ph, \ h \in (\mathcal{H}, K_1) \otimes (\mathcal{H}, K_2), \ p \in \mathbb{C}[z_1, \dots, z_{2m}].$$

The module $(\mathcal{H}, K_1) \otimes (\mathcal{H}, K_2)$ admits a natural direct sum decomposition as follows.

For a non-negative integer k, let $\mathcal{A}_k$ be the subspace of $(\mathcal{H}, K_1) \otimes (\mathcal{H}, K_2)$ defined by

$$\mathcal{A}_k := \{f \in (\mathcal{H}, K_1) \otimes (\mathcal{H}, K_2) : ((\tfrac{\partial}{\partial \zeta})^i f(z, \zeta))_{|\Delta} = 0, \ |i| \le k\}, \tag{E.1.8}$$

where $i \in \mathbb{Z}_+^m$, $|i| = i_1 + \cdots + i_m$, $(\tfrac{\partial}{\partial \zeta})^i = \tfrac{\partial^{|i|}}{\partial \zeta_1^{i_1} \cdots \partial \zeta_m^{i_m}}$, and $((\tfrac{\partial}{\partial \zeta})^i f(z, \zeta))_{|\Delta}$ is the restriction of $(\tfrac{\partial}{\partial \zeta})^i f(z, \zeta)$ to the diagonal set $\Delta := \{(z, z) : z \in \Omega\}$. It is easily verified that each of the subspaces $\mathcal{A}_k$ is closed and invariant under multiplication by any polynomial in $\mathbb{C}[z_1,\ldots,z_{2m}]$ and therefore, they are sub-modules of $(\mathcal{H},K_1)\otimes(\mathcal{H},K_2)$. Setting $\mathcal{S}_0 = \mathcal{A}_0^\perp$, $\mathcal{S}_k := \mathcal{A}_{k-1}\ominus\mathcal{A}_k$, $k = 1,2,\ldots$, we obtain a direct sum decomposition of the Hilbert space

$$(\mathcal{H},K_1)\otimes(\mathcal{H},K_2) = \bigoplus_{k=0}^{\infty}\mathcal{S}_k.$$

In this decomposition, the subspaces $\mathcal{S}_k \subseteq (\mathcal{H},K_1)\otimes(\mathcal{H},K_2)$ are not necessarily sub-modules. Indeed, one may say that they are *semi-sub-modules* following the terminology commonly used in Sz.–Nagy–Foias model theory for contractions.

Now, fix any non-negative definite kernel K on Ω. Pick $\alpha > 0$ and $\beta > 0$ such that $K_1 = K^\alpha$ and $K_2 = K^\beta$ are non-negative definite on Ω. In this case, $\mathcal{S}_0$ is the Hilbert space $(\mathcal{H},K^{\alpha+\beta})$. We describe the semi-sub-module $\mathcal{S}_1$ in some detail.

Recall that $K^\alpha\otimes K^\beta$ is the reproducing kernel for the Hilbert space $(\mathcal{H},K^\alpha)\otimes(\mathcal{H},K^\beta)$, where the kernel $K^\alpha\otimes K^\beta$ on $(\Omega\times\Omega)\times(\Omega\times\Omega)$ is given by

$$K^\alpha\otimes K^\beta(z,\zeta;z',\zeta') = K^\alpha(z,z')K^\beta(\zeta,\zeta'),$$

$z = (z_1,\ldots,z_m)$, $\zeta = (\zeta_1,\ldots,\zeta_m)$, $z' = (z_{m+1},\ldots,z_{2m})$, $\zeta' = (\zeta_{m+1},\ldots,\zeta_{2m})$ are in Ω. We realize the Hilbert space $(\mathcal{H},K^\alpha)\otimes(\mathcal{H},K^\beta)$ as a space consisting of holomorphic functions on $\Omega\times\Omega$. Let $\mathcal{A}_0$ and $\mathcal{A}_1$ be the subspaces defined by

$$\mathcal{A}_0 = \{f \in (\mathcal{H},K^\alpha)\otimes(\mathcal{H},K^\beta) : f_{|\Delta} = 0\}$$

and

$$\mathcal{A}_1 = \{f \in (\mathcal{H},K^\alpha)\otimes(\mathcal{H},K^\beta) : f_{|\Delta} = (\partial_{m+1}f)_{|\Delta} = \cdots = (\partial_{2m}f)_{|\Delta} = 0\},$$

where Δ is the diagonal set $\{(z,z) \in \Omega\times\Omega : z\in\Omega\}$, $\partial_i f$ is the derivative of f with respect to the ith variable, and $f_{|\Delta}$, $(\partial_i f)_{|\Delta}$ denote the restrictions to the set Δ of the functions f, $\partial_i f$, respectively. It is easy to see that both $\mathcal{A}_0$ and $\mathcal{A}_1$ are closed subspaces of the Hilbert space $(\mathcal{H},K^\alpha)\otimes(\mathcal{H},K^\beta)$ and $\mathcal{A}_1$ is a closed subspace of $\mathcal{A}_0$. The quotient modules $\mathcal{S}_0$, $(\mathcal{S}_0\oplus\mathcal{S}_1),\ldots$ occur naturally in [54, 112, 61]. Here we wish to determine the semi-sub-module $\mathcal{A}_0\ominus\mathcal{A}_1$. To describe these, first observe that

$$\begin{aligned}\bar{\partial}_i(K^\alpha\otimes K^\beta)(\cdot,(z',\zeta')) &= \bar{\partial}_i K^\alpha(\cdot,z')\otimes K^\beta(\cdot,\zeta'), \ z',\zeta'\in\Omega \\ \bar{\partial}_{m+i}(K^\alpha\otimes K^\beta)(\cdot,(z',\zeta')) &= K^\alpha(\cdot,z')\otimes\bar{\partial}_i K^\beta(\cdot,\zeta'), \ z',\zeta'\in\Omega,\end{aligned} \tag{E.1.9}$$

$1 \le i \le m$. Put $z' = \zeta' = w \in \Omega$ and set, for $1 \le i \le m$,

$$\begin{aligned}\phi_i(w) &:= (\bar{\partial}_i(K^\alpha\otimes K^\beta)(\cdot,(z',\zeta')) - \bar{\partial}_{m+i}(K^\alpha\otimes K^\beta)(\cdot,(z',\zeta')))_{|z'=\zeta'=w} \\ &= \beta\bar{\partial}_i(K^\alpha\otimes K^\beta)(\cdot,(w,w)) - \alpha\bar{\partial}_{m+i}(K^\alpha\otimes K^\beta)(\cdot,(w,w)).\end{aligned} \tag{E.1.10}$$

Proposition E.1.1 (Proposition 3.2, [63])

The quotient module $\mathcal{A}_0 \ominus \mathcal{A}_1$ is the closed linear span of the vectors $\{\phi_i(w) : w \in \Omega, 1 \le i \le m\}$ in $(\mathcal{H}, K^\alpha) \otimes (\mathcal{H}, K^\beta)$.

The Hilbert module $\mathcal{S}_1 := \mathcal{A}_0 \ominus \mathcal{A}_1$ is the semi-sub-module in the Hilbert module $\mathcal{H}^{(\alpha,\beta)} := (\mathcal{H}, K^\alpha) \otimes (\mathcal{H}, K^\beta)$. It can be identified as a reproducing kernel Hilbert space consisting of holomorphic functions on Ω with the reproducing kernel

$$\mathbb{K}^{(\alpha,\beta)}(z,w) := K^{\alpha+\beta}(z,w) \Big((\partial_i \bar{\partial}_j \log K)(z,w) \Big)_{i,j=1}^m.$$

This is Theorem 3.3 of [63]. Finding a concrete realization of the other semi-sub-modules $\mathcal{S}_k$, $k > 2$ appears to be a challenging problem. We conclude by working out the example of the weighted Bergman spaces on the bi-disc.

For $\alpha > 0$, let $\mathcal{H}^{(\alpha)}$ be the Hilbert space which is determined by requiring that $\{e_n^{(\alpha)}(z) = c_n^{-1/2} z^n : n \ge 0\}$ is a complete orthonormal set in it, where c_n is the coefficient of x^n in the expansion of $(1-x)^{-\alpha}$ or c_n is the set of binomial coefficients: $\begin{pmatrix} -\alpha \\ n \end{pmatrix} = \dfrac{\alpha(\alpha+1)\cdots(\alpha+n-1)}{n!}$.

It follows that $\mathcal{H}^{(\alpha)}$ possesses a reproducing kernel $K^{(\alpha)} : \mathbb{D} \times \mathbb{D} \to \mathbb{C}$, which is given by the formula

$$
\begin{aligned}
K^{(\alpha)}(z,w) &= \sum_{n=0}^{\infty} e_n^{(\alpha)}(z) \overline{e_n^{(\alpha)}(w)} \\
&= (1 - z\bar{w})^{-\alpha},
\end{aligned}
$$

where $\mathbb{D}$ is the open unit disc. Thus, $\mathcal{H}^{(\alpha)}$ consists of holomorphic functions on the open unit disc $\mathbb{D}$. For $\alpha, \beta > 0$, let $\mathcal{H}^{(\alpha,\beta)}$ be the tensor product $\mathcal{H}^{(\alpha)} \otimes \mathcal{H}^{(\beta)}$. The Hilbert space $\mathcal{H}^{(\alpha,\beta)}$ is then a space of holomorphic functions on the bi-disc $\mathbb{D}^2$ via the identification of the elementary tensor $e_m^{(\alpha)} \otimes e_n^{(\beta)}$ with the function of two variables $z_1^m z_2^n$ on the bi-disc $\mathbb{D} \times \mathbb{D}$. It naturally possesses the reproducing kernel $K^{(\alpha,\beta)}(\mathbf{z}, \mathbf{w}) = (1 - z_1 \bar{w}_1)^{-\alpha}(1 - z_2 \bar{w}_2)^{-\beta}$, where $\mathbf{z} = (z_1, z_2)$ and $\mathbf{w} = (w_1, w_2)$ are both in $\mathbb{D} \times \mathbb{D}$. We see that the kernel $\mathbb{K}^{(\alpha,\beta)} = \dfrac{1}{(1-z\bar{w})^{\alpha+\beta+2}}$, $z, w \in \mathbb{D}$, and determine a weighted Bergman space on the disc. Hence, all the semi-sub-modules $\mathcal{S}_1^{(\alpha,\beta)}$ are essentially reductive.

This prompts a natural question, modeled after the Arveson–Douglas conjecture, namely, which semi-sub modules of a Hilbert module are essentially reductive.

The calculations for the bidisc overlaps with the work of Ferguson and Rochberg on higher order Hankel forms [61]. Moreover, some of the results in the paper [112] for bounded symmetric domains are of a similar nature. The module multiplication for the quotient module $(\mathcal{S}_0^{(\alpha,\beta)} \oplus \mathcal{S}_1^{(\alpha,\beta)})$ is in [53] for the weighted Bergman spaces on the bidisc, and in general, in [63]. The quotient module, in the first case, is again essentially reductive. It is the direct sum of two shifts modulo compact. This phenomenon appears to be a general feature of homogeneous operators (see [86, Theorem 4.1]).

Appendix A

Point Set Topology

In this appendix, we present some topological ingredients used in the main body of this book. In particular, the results stated are confined mostly to the metric space set up and most of the time without the proofs.

Urysohn's Lemma and Tietze Extension Theorem

Let (X, d) be a metric space with metric d. Let A be a non-empty subset of X, and for $x \in X$, let $d(x, A) = \inf\{d(x, a) : a \in A\}$. Then, $d(x, A)$ is a continuous function of x. Let A and B be disjoint non-empty closed subsets of X. For $x \in X$, define

$$f(x) = \frac{d(x, A)}{d(x, A) + d(x, B)}.$$

Clearly, $f : X \to [0, 1]$ is a continuous function. Note that $f(a) = 0$ and $f(b) = 1$ for every $a \in A$ and $b \in B$. In particular, the disjoint non-empty closed subsets A and B of X are separated by the continuous function f. Thus, we obtain the following special case of Urysohn's lemma (refer to [96]).

Theorem A.1.1 (Urysohn's Lemma)

Let X be a metric space. Given closed non-empty disjoint subsets A and B of X, there exists a continuous function $f : X \to [0, 1]$ such that $f|_A = 0$ and $f|_B = 1$.

The following particular consequence of Urysohn's lemma is invoked in Chapter 3.

Corollary A.1.1

Let X be a compact Hausdorff space. Let $g \in C(X)$ be such that g vanishes on a non-empty closed subset K of X. For any $\epsilon > 0$, there exists a continuous function G on X vanishing in a neighborhood of K such that

$$\sup_{x \in X} |g(x) - G(x)| < \epsilon.$$

Proof   Note that there exists a neighborhood V of K such that $\sup_{x \in V} |g(x)| < \epsilon$. Let U be an open subset of X such that $K \subseteq U \subseteq \overline{U} \subseteq V$. Apply Urysohn's lemma to $\overline{U}$ and $X \backslash V$ to get $f \in C(X)$ such that $0 \leqslant f \leqslant 1$, $f = 0$ on U and $f = 1$ on $X \backslash V$. Let $G = gf$. $\square$

We also need the following extension of Urysohn's lemma in the main body.

Theorem A.1.2 (Tietze Extension Theorem)

Any continuous real-valued function on a closed subspace of a metric space may be extended to a continuous real-valued function on the entire space.

Product Topology and and Tychonoff's Theorem

Consider an arbitrary family $\{X_\alpha\}_{\alpha \in I}$ of topological spaces indexed by a set I. The *product topology* on $\prod_{\alpha \in I} X_\alpha$ is the topology generated by the basis

$$\left\{ \prod_{\alpha \in I} U_\alpha : U_\alpha \text{ is open in } X_\alpha \text{ and } U_\alpha \neq X_\alpha \text{ for only finitely many values of } \alpha \right\}.$$

For $\beta \in I$, one may define the *projection map* $\pi_\beta : \prod_{\alpha \in I} X_\alpha \to X_\beta$ by $\pi_\alpha(x) = x_\beta$, where $x = (x_\alpha)_{\alpha \in I}$. It is worth mentioning that the product topology is the smallest topology that makes all projections π_β continuous. The following simple fact enables us to understand the product topology using the convergence of sequences.

Proposition A.1.1

Let $\{x_n = (x_{n\alpha})_{\alpha \in I}\}_{n \geqslant 0}$ be a sequence in the product space $\prod_{\alpha \in I} X_\alpha$ with product topology. The sequence $\{x_n\}_{n \geqslant 0}$ converges to $x = (x_\alpha)_{\alpha \in I} \in \prod_{\alpha \in I} X_\alpha$ iff for every $\beta \in I$, $\{\pi_\beta(x_n) = x_{n\beta}\}$ converges to x_β.

Proof   Suppose that for every $\beta \in I$, $\{\pi_\beta(x_n) = x_{n\beta}\}$ converges to x_β. Let $\prod_{\alpha \in I} U_\alpha$ be an open neighborhood of x in the product topology. Thus, there exists finitely many indices $\alpha_1, \cdots, \alpha_k \in I$ such that $U_\alpha = X_\alpha$ for every $\alpha \neq \alpha_1, \cdots, \alpha_k$. Fix $i = 1, \cdots, k$. As $\{\pi_{\alpha_i}(x_n) = x_{n\alpha_i}\}$ converges to x_{α_i}, there exists positive integer N_i such that $x_{n\alpha_i} \in U_{\alpha_i}$ for all $n \geq N_i$. Check that $x_n \in \prod_{\alpha \in I} U_\alpha$ for all $n \geq \max\{N_1, \cdots, N_k\}$. $\square$

Here is another important property of the product topology.

Proposition A.1.2

Let $f : A \to \prod_{\alpha \in I} X_\alpha$ be given by $f(a) = (f_\alpha(a))_{\alpha \in I}$, where the functions $f_\alpha : A \to X_\alpha$, $\alpha \in I$ are given. Suppose $\prod_{\alpha \in I} X_\alpha$ carries the product topology. Then, f is continuous iff each f_α is continuous.

Proof   Suppose $U_\alpha = X_\alpha$ for finitely many values of α, say $\alpha_1, \cdots, \alpha_m$. Then,

$$
\begin{aligned}
f^{-1}\left(\prod_{\alpha \in I} U_\alpha \right) &= \left\{ a \in A : f(a) \in \prod_{\alpha \in I} U_\alpha \right\} \\
&= \{a \in A : f_{\alpha_i}(a) \in U_\alpha \text{ for } i = 1, \cdots, m\} = \cap_{i=1}^{m} (f_{\alpha_i})^{-1}(U_{\alpha_i}).
\end{aligned}
$$

The desired equivalence is now immediate. □

We conclude this section with Tychonoff's theorem required in Chapter 4.

Theorem A.1.3 (Tychonoff's Theorem)

The product space $\prod_{\alpha \in I} X_\alpha$ carrying the product topology is compact if each X_α is compact.

Alexandroff–Hausdorff Theorem and Totally Disconnected Metric Spaces

For two nonempty sets A and B, the notation A^B stands for the collection of all functions from B into A. With this notation, we claim that any compact metric space can be embedded into the product space $[0,1]^{\mathbb{N}}$ endowed with the product topology, where $\mathbb{N}$ denotes the set of nonnegative integers. We closely follow [117, Lemma 1]. To see this, let (X,d) be a compact metric space with metric d. For every integer $n \geqslant 0$, cover X by balls of radius $\frac{1}{n+1}$ to obtain, by the compactness argument, finitely many balls covering X. It is easy to see that the collection of all centres of these balls forms a countable dense subset of X. Let us denote this countable dense set by $\{x_n\}_{n \geqslant 0}$. This shows that X is separable. In addition, assume that X consists of at least two elements. Clearly, the diameter $\text{diam}\, X$ of X is positive. For every integer $n \geqslant 0$, define $f_n : X \to [0,1]$ by

$$f_n(x) = d(x, x_n)/\text{diam}\, X. \tag{A.1.1}$$

As the distance function is continuous, so is f_n. Define $f : X \to [0,1]^{\mathbb{N}}$ by

$$f(x) = (f_0(x), f_1(x), \ldots,), \quad x \in X.$$

By Proposition A.1.2, f is continuous. Moreover, f is one-to-one. Indeed, if $f(x) = f(y)$, then $d(x, x_n) = d(y, x_n)$ for all integers $n \geqslant 0$. As $\{x_n\}_{n \geqslant 0}$ is dense, there exists a subsequence $\{x_{n_k}\}_{k \geqslant 0}$ converging to x. However, then $d(y, x_{n_k}) = d(x, x_{n_k}) \to 0$ as $k \to \infty$, and hence, by the uniqueness of the limit, $x = y$. Thus, we have a continuous injection from a compact space into the Hausdorff space $[0,1]^{\mathbb{N}}$. Hence, f must be a homeomorphism onto the image. This proves the following:

Proposition A.1.3

Any compact metric space is homeomorphic to a closed subset of $[0,1]^{\mathbb{N}}$, where $[0,1]^{\mathbb{N}}$ is endowed with the product topology.

The following theorem revealing the structure of a compact metric space is not needed in the main text.

Theorem A.1.4

Every compact metric space X is a continuous image of the (compact metric) space $2^{\mathbb{N}}$, the Cartesian product of countably infinitely many copies of a two point space.

Proof We closely follow [134]. As X is totally bounded, we may find finitely many closed sets $\{B_k : 1 \leqslant k \leqslant n_1\}$ of diameter less than 1 with union X. Since each B_j is totally bounded, we may find finitely many closed sets $\{B_{j,k} : 1 \leqslant k \leqslant m_j\}$ of diameter less than $1/2$ with union B_j. Let $n_2 = \max_{j=1}^{n_1} m_j$ and set $B_{j,k} = B_{j,m_j}$ for $k = m_j, \ldots, n_2$. Thus, for every $j = 1, \ldots, n_1$, we have a collection $\{B_{j,k} : 1 \leqslant k \leqslant n_2\}$ of closed sets of diameter less than $1/2$ with union B_j. By induction on j, for every integer $j \geqslant 1$, we have a collection $\{B_{i_1,i_2,\ldots,i_j} : 1 \leqslant i_j \leqslant n_j\}$ of closed sets of diameter less than $1/j$ such that

$$B_{i_1,i_2,\ldots,i_{j-1}} = \bigcup_{i_j=1}^{n_j} B_{i_1,i_2,\ldots,i_j}, \quad i_l, \ldots, i_l \geqslant 1, \, l = 1, \ldots, j-1. \tag{A.1.2}$$

By Cantor's intersection theorem, $\cap_{j \geqslant 1} B_{i_1,i_2,\ldots,i_j}$ is single-ton, say, $\{f(i_1, i_2, \ldots,)\}$. This allows us to define $f : \prod_{j=1}^{\infty} \{1, 2, \ldots, n_j\} \to X$. Further, if i and j agree in the first q co-ordinates, then $f(i)$ and $f(j)$ are at a distance of at most $2/q$ from one another. This shows that the function f is continuous. As $X = \cup_{j=1}^{n_1} B_j$, by (A.1.2), f is surjective. Clearly, the aforementioned argument holds for the choice $n_j = 2^{m_j}$ (by extending the indices of closed sets trivially). As there is a natural continuous surjection from $2^{\mathbb{N}}$ onto $\prod_{j=1}^{\infty} \{1, 2, \ldots, 2^{m_j}\}$, we get the conclusion. $\quad\square$

Recall that the *Cantor set* is obtained by removing 2^{k-1} centrally situated open subintervals of $[0, 1]$ each of length $1/3^k$ at the kth step, where $k = 1, 2, \ldots$. A part of the proof of the following Lemma is used to complete the proof of Lemma 2.6.2.

Theorem A.1.5 (Alexandroff–Hausdorff Theorem)

If X is a compact metric space, then there is a continuous map from the Cantor set onto X.

Proof   In view of Proposition A.1.3, it suffices to show that there is a continuous map from the Cantor set onto $[0, 1]^{\mathbb{N}}$. Consider the map

$$\varphi : \{0, 1\}^{\mathbb{N}} \to [0, 1], \quad \varphi(\{a_n\}_{n \geqslant 0}) = \sum_{n=1}^{\infty} \frac{2a_{n-1}}{3^n}.$$

We claim that φ is a one-to-one map from $\{0, 1\}^{\mathbb{N}}$ onto the Cantor set. To see injectivity of φ, let $\{a_n\}_{n \geqslant 0}, \{b_n\}_{n \geqslant 0} \in \{0, 1\}^{\mathbb{N}}$ and let $k \in \mathbb{N}$ denote the smallest integer such that $a_k \neq b_k$. As seen in the proof of Lemma 2.2.3, one may verify that $|\varphi(\{a_n\}_{n \geqslant 0}) - \varphi(\{b_n\}_{n \geqslant 0})| \geqslant \frac{1}{3^{k+1}}$. This shows that φ is one-to-one. To verify surjectivity, note that every number in $[0, 1]$ has a ternary expansion $x = \sum_{n=1}^{\infty} \frac{x_n}{3^n}$, where $x_n \in \{0, 1, 2\}$. The claim stands verified from the fact that if x belongs to the Cantor set, then x_n does not take the value 1 for any $n \in \mathbb{N}$ (this in turn follows from the fact that x belongs to the set obtained in the nth step in the Cantor set construction if and only if it admits a ternary expansion with a_k belonging to $\{0, 2\}$ for $k = 1, \ldots, n$). Finally, as φ is a homeomorphism, by Theorem A.1.4, there exists a continuous surjection from $2^{\mathbb{N}}$ onto X. $\quad\square$

Recall that a topological space X is *totally disconnected* if the only connected subsets of X are single points. Here are some examples of such sets:

(a) The rationals in $[0, 1]$ with the subspace topology inherited from $[0, 1]$ is totally disconnected.

(b) The Cantor set.

The Cantor set is a compact metric space without any isolated points. An application of the result of Alexandroff and Hausdorff shows that these properties determine totally disconnected metric spaces up to homeomorphism.

Corollary A.1.2

A totally disconnected, compact metric space without any isolated points is homeomorphic to the Cantor set.

For a proof of the last corollary, the reader is referred to [130].

Quotient Topology

Let X be a set with an equivalence relation $\sim$. Given $x \in X$, let

$$[x] = \{y \in X : y \sim x\}$$

be the equivalence class containing x. Let $X/\sim$ denote the set of all equivalence classes of elements in X.

Our primary aim of this subsection is to introduce a "best possible" topology (of course inherited from X) on $X/\sim$, which will make $X/\sim$ a topological space. To do that, suppose (X,Ω) is a topological space, and consider the *quotient map* $q : X \to X/\sim$ given by $q(x) = [x]$. As q is surjective, we endow with $X/\sim$, the *quotient topology* Ω_q:

$$\Omega_q = \{U \subseteq X/\sim : q^{-1}(U) \in \Omega\}.$$

Let us check that Ω_q is indeed a topology. Clearly,

$$q^{-1}(\emptyset) = \emptyset \in \Omega \text{ and } q^{-1}(X/\sim) = X \in \Omega.$$

If $\{U_\alpha\} \subseteq \Omega_q$, then $q^{-1}(U_\alpha) \in \Omega$, and hence, $q^{-1}(\cup_\alpha U_\alpha) = \cup_\alpha q^{-1}(U_\alpha) \in \Omega$. It follows that Ω_q is closed under arbitrary union. Similarly, one may check that Ω_q is closed under finite intersection.

Remark A.1.6 Note that the quotient map $q : (X,\Omega) \to (X/\sim, \Omega_q)$ is a continuous surjection. In particular, if X is compact, then so is $X/\sim$.

Theorem A.1.7

Let X and Y be compact spaces. Assume further that Y is Hausdorff. Let $f : X \to Y$ be a continuous surjection. Define the equivalence relation $\sim$ on X by $x_1 \sim x_2$ iff $f(x_1) = f(x_2)$. Then, $g : X/\sim \to Y$ given by $g([x]) = f(x)$ is a well-defined homeomorphism.

Proof By Remark A.1.6, $q(X) = X/\sim$ is compact. Recall that if $f : X \to Y$ is a continuous bijection and if X is compact and Y is Hausdorff, then f is a homeomorphism. Thus, it suffices to check that $g : X/\sim \to Y$ is a continuous bijection.

Note that $g \circ q = f$. As f is a surjection, so is g. Moreover, if $g([x_1]) = g([x_2])$, then $f(x_1) = f(x_2)$, and hence, $x_1 \sim x_2$, so that $[x_1] = [x_2]$. Thus, g is injective.

We know that $g \circ q = f$ is continuous. For an open set V in Y, note that $f^{-1}(V) = (g \circ q)^{-1}(V) = q^{-1}(g^{-1}(V))$ is an open subset of X. By the definition of quotient topology, $g^{-1}(V)$ is open in $X/\sim$. $\qquad\square$

Let X be a compact Hausdorff space and let A be a closed subset of X. Consider the topological space $Y = (X\backslash A) \cup \{\infty\}$ with the topology Ω_Y

$$\Omega_Y = \{U \subseteq X\backslash A : U \text{ is open}\} \cup \{Y\backslash C : C \text{ is compact in } X\backslash A\}.$$

It is easy to see that Ω_Y defines a topology on Y.

Lemma A.1.1

If X is a compact Hausdorff space, then so is Y.

Proof   To verify that Y is compact, let $\{U_\alpha\} \cup \{V_\beta\}$ be an open cover of Y, where $U_\alpha \subseteq X\backslash A$ is open in X, and $V_\beta = Y\backslash C_\beta$ for some compact set C_β in $X\backslash A$. Fix β_0. Note that the compact set C_{β_0} has the open cover $\{U_\alpha\} \cup \{V_\beta \cap (X\backslash A)\}$, and hence, there exists finite subcover $\{U_{\alpha_i}\} \cup \{V_{\beta_j} \cap (X\backslash A)\}$ of C_{β_0}. It follows that Y has the finite subcover $\{U_{\alpha_i}\} \cup \{V_{\beta_j}\} \cup \{V_{\beta_0}\}$.

To see that Y is Hausdorff, let $x, y \in Y$ be two distinct points. As X is Hausdorff, we may assume without loss of generality that $x \in X\backslash A$ and $y = \infty$. One may choose a neighborhood U of x such that $\overline{U} \subseteq X\backslash A$. Since $\overline{U}$ is a subset of $X\backslash A$ and as it is compact in X, it is also compact in $X\backslash A$. Then, U and $Y\backslash\overline{U}$ are disjoint neighborhoods of x and y. $\qquad\square$

If X is a compact space, any space homeomorphic to Y is known as the *one-point compactification* of $X\backslash A$ (cf. [93, Theorem 2]).

Corollary A.1.3

(One-point Compactification) Let X be a compact Hausdorff space and let A be a closed subset of X. Let $\sim$ be the equivalence relation defined on X as follows: If $x_1, x_2 \in A$, then $x_1 \sim x_2$. If $x_1, x_2 \in X\backslash A$, then $x_1 \sim x_2$ iff $x_1 = x_2$. Consider the topological space $Y = (X\backslash A) \cup \{\infty\}$ with topology Ω_Y as discussed earlier. Then, Y is homeomorphic to $X/\sim$. Moreover, if X is a metric space with metric $d : X\times X \to [0,\infty)$, then (Y, Ω_Y) is metrizable with metric $d_Y : Y\times Y \to [0,\infty)$ given by

$$d_Y(x, x') = d(x, x'), \quad d_Y(x, \infty) = \inf_{a\in A} d(x, a),$$

$$d_Y(\infty, x) = d_Y(x, \infty) \quad d_Y(\infty, \infty) = 0, \quad x, x' \in X\backslash A.$$

Proof   Define $f : X \to Y$ by $f(x) = x$ if $x \in X\backslash A$, and $f(a) = \infty$ if $a \in A$. Clearly, f is a surjection with levels set precisely the equivalence classes in $X/\sim$. We check that f is continuous. Let $U \in \Omega_Y$. If $U \subseteq X\backslash A$, then $f^{-1}(U) = U$, which is open in X. If $U = Y\backslash C$ for some compact C in $X\backslash A$, then $f^{-1}(U) = f^{-1}(Y)\backslash f^{-1}(C) = X\backslash C$, which is open as C is closed in X. By Theorem A.1.7, the space $X/\sim$ is homeomorphic to Y.

We now check the 'moreover' part. Clearly, d_Y is symmetric and identically zero on the diagonal of $Y\times Y$. As A and $\{x\}$, $x \in X\backslash A$ are disjoint, the distance between $\{x\}$ and A is positive, and hence, $d_Y(x, \infty) > 0$. Further, for any $x, x' \in X\backslash A$ and $a \in A$,

$$d_Y(\infty, x) \leqslant d(a, x) \leqslant d(a, x') + d(x', x) = d(a, x') + d_Y(x', x).$$

After taking infimum on the right-hand side over A, we obtain

$$d_Y(\infty, x) \leqslant d_Y(\infty, x') + d_Y(x', x).$$

It is now easy to see that d_Y defines a metric on Y. To see that the topology generated by the metric d_Y is same as the quotient topology on Y, note first that any open subset of $X \backslash A$ is open in the d_Y-metric topology. If C is a compact subset of $X \backslash A$, then there exists $\epsilon > 0$ such that the distance between C and A is at least ϵ. It follows that $\{x \in Y : d_Y(\infty, x) < \epsilon/2\} \subseteq Y \backslash C$. This shows that every open set in Ω_Y is open in the d_Y-metric topology. Finally, note that any ball in d_Y-metric is open in Ω_Y, and hence, d_Y-metric topology coincides with Ω_Y. This completes the proof. $\qquad\square$

Appendix B

Linear Analysis

In this appendix, we collect some miscellaneous topics from linear analysis referred throughout the main text.

Stone–Weierstrass Theorem

In this section, we present a proof of the Stone–Weiersrass theorem, which does not rely on the Weierstrass theorem. We closely follow the treatment of [119].

Throughout this section, K denotes a compact Hausdorff space.

Theorem B.1.1

Let $\mathscr{A}$ be an algebra of continuous functions $f : K \to \mathbb{R}$ with the following properties:

(1) If $x \neq y \in K$, then there exists $f \in \mathscr{A}$ such that $f(x) \neq f(y)$.

(2) For every $x \in K$, there exists $f \in \mathscr{A}$ such that $f(x) \neq 0$.

Then, $\mathscr{A}$ is dense in the algebra $C_{\mathbb{R}}(K)$ of continuous real-valued functions on K endowed with the uniform norm.

We start the proof with a lemma, which shows that under some modest assumption, pointwise convergence yields uniform convergence.

Lemma B.1.1

Let $\{f_n\}$ be a sequence in $C[a,b]$ converging pointwise to a continuous function f. If $\{f_n(x)\}$ is decreasing for all $x \in [a,b]$, then $\{f_n\}$ converges uniformly to f.

Proof Let $g_n = f_n - f \geq 0$. For $\epsilon > 0$, consider the closed subset

$$K_n = \{x \in [a,b] : g_n(x) \geq \epsilon\}$$

201

of $[a,b]$. As $g_n \geqslant g_{n+1}$, $K_{n+1} \subseteq K_n$. In particular, finite intersection of sets from $\{K_n\}$ is non-empty if every K_n is non-empty. If each K_n is non-empty, then by Cantor's intersection theorem, $\cap_{n=1}^{\infty} K_n \neq \emptyset$. However, if $x \in [a,b]$, then as $g_n(x) \to 0$, $x \notin K_n$ for sufficiently large n. Hence, K_N is empty for some N, that is, $0 \leqslant g_n(x) < \epsilon$ for every $x \in [a,b]$ and for every $n \geqslant N$. $\qquad \square$

Here is an important special case of Weierstrass' theorem.

Lemma B.1.2

Define a sequence $\{p_n\}_{n\geqslant 0}$ of polynomials by $p_0(x) = 0$, and

$$p_{n+1}(x) = p_n(x) + (x^2 - p_n(x)^2)/2, \quad n \geqslant 0.$$

If $q_n(x) = p_n(x^2)$, then $\{q_n\}_{n\geqslant 0}$ converges uniformly to $f(x) = |x|$ on $[-1,1]$.

Proof A routine calculation shows that

$$|x| - P_{n+1}(x) = (|x| - P_n(x))(1 - (|x| + P_n(x))/2), \quad n \geqslant 0.$$

One may now verify inductively that

$$0 \leqslant p_n(x) \leqslant p_{n+1}(x) \leqslant |x|, \quad x \in [-1,1], n \geqslant 0.$$

In particular, $\{p_n(x)\}_{n\geqslant 0}$ converges pointwise to $|x|$. Now apply Lemma B.1.1 to $f_n = -p_n, n \geqslant 0$. $\qquad \square$

The last lemma yields some basic properties of closed subalgebras of $C_{\mathbb{R}}(K)$.

Lemma B.1.3

Let $\mathscr{A}$ be a subalgebra of $C_{\mathbb{R}}(K)$ and let $\overline{\mathscr{A}}$ denote the uniform closure of $\mathscr{A}$ in $C_{\mathbb{R}}(K)$. If $f_1,\ldots,f_k \in \overline{\mathscr{A}}$, then so are $\max\{f_1,\ldots,f_k\}$ and $\min\{f_1,\ldots,f_k\}$.

Proof Recall that $|f|(x) = |f(x)|$, $x \in K$. It may be concluded from Lemma B.1.2 that $|f| \in \overline{\mathscr{A}}$ for every $f \in \overline{\mathscr{A}}$. The first part now is immediate from

$$\max\{f_1,f_2\} = \frac{f_1 + f_2}{2} + \frac{|f_1 - f_2|}{2}$$

and finite induction, whereas the remaining part follows from $\min\{f_1,\ldots,f_k\} = -\max\{-f_1,\ldots,-f_k\}$. $\qquad \square$

We need one more lemma in the proof of Theorem B.1.1.

Lemma B.1.4

Under the assumptions of Theorem B.1.1, for distinct points x_1, x_2 of K and real scalars c_1, c_2, there exists $f \in \mathscr{A}$ such that

$$f(x_1) = c_1 \text{ and } f(x_2) = c_2.$$

Proof By assumptions (1) and (2) of Theorem B.1.1, there exist $g, h, k \in \mathscr{A}$ such that $g(x_1) \neq g(x_2)$, $h(x_1) \neq 0$, and $k(x_2) \neq 0$. Then, $u = (g - g(x_1))k$ and $v = (g - g(x_2))h$ both belong to $\mathscr{A}$, and $f = \frac{c_1 v}{v(x_1)} + \frac{c_2 u}{u(x_2)}$ satisfies the desired properties and the proof is over. □

Proof [Proof of Theorem B.1.1] Fix $x \in K$ and $\epsilon > 0$. For every $y \in K$, by Lemma B.1.4, there exists $f_y \in \mathscr{A}$ such that $f_y(x) = f(x)$ and $f_y(y) = f(y)$. Note that $f_y - f$ is continuous such that $(f_y - f)(x) = 0$. Thus, there exists an open neighborhood U_y of y such that

$$f_y(t) - f(t) > -\epsilon, \quad t \in U_y.$$

Note that $\{U_y\}$ is an open cover of the compact set K. Thus, for some $y_1, \ldots, y_k \in K$, $K \subseteq \cup_{i=1}^{k} U_{y_i}$. Moreover, for $g_x = \max\{f_{y_1}, \ldots, f_{y_k}\} \in \mathscr{A}$ (see Lemma B.1.3),

$$g_x(t) > f(t) - \epsilon, \quad t \in K. \tag{B.1.1}$$

Now we vary x over K. Note that $g_x(x) = \max\{f_{y_1}(x), \ldots, f_{y_k}(x)\} = f(x)$. Hence, by the continuity of g_x, there exists an open neighborhood V_x of x such that

$$g_x(t) - f(t) < \epsilon, \quad t \in V_x.$$

Now $\{V_x\}$ is an open cover of the compact set K. Thus, for some $x_1, \ldots, x_l \in K$, $K \subseteq \cup_{i=1}^{l} V_{x_i}$. Moreover, for $h = \min\{g_{x_1}, \ldots, g_{x_l}\} \in \mathscr{A}$,

$$h(t) < f(t) + \epsilon, \quad t \in K.$$

Moreover, by (B.1.1), $h(t) > f(t) + \epsilon$ for every $t \in K$. That is, $\|h - f\|_\infty < \epsilon$. □

In the remaining part of this appendix, we discuss some applications of the Stone–Weierstrass theorem used in the main text.

For $f : K \to \mathbb{C}$, define $\overline{f} : K \to \mathbb{C}$ by $\overline{f}(x) = \overline{f(x)}$, $x \in K$.

Corollary B.1.1

Let $\mathscr{A}$ be an algebra of continuous functions $f : K \to \mathbb{C}$ with the following properties:

(1) If $x \neq y \in K$, then there exists $f \in \mathscr{A}$ such that $f(x) \neq f(y)$.

(2) For every $x \in K$, there exists $f \in \mathscr{A}$ such that $f(x) \neq 0$.

(3) For every $f \in \mathscr{A}$, $\overline{f} \in \mathscr{A}$.

Then, $\overline{\mathscr{A}} = C(K)$.

Proof Let $\mathscr{A}_\mathbb{R}$ denote the algebra functions in $\mathscr{A}$ that are real-valued, and let $\mathfrak{R}\mathscr{A}$ be the set of real parts of functions in $\mathscr{A}$. As $\mathfrak{R}f = (f + \overline{f})/2 \in \mathscr{A}$ for every $f \in \mathscr{A}$, $\mathfrak{R}\mathscr{A} = \mathscr{A}_\mathbb{R}$. Check that $\mathscr{A}_\mathbb{R}$ satisfies (1) and (2) of Theorem B.1.1. Thus, $\mathfrak{R}\mathscr{A}$ is dense in the algebra $C_\mathbb{R}(K)$, and hence, $\overline{\mathscr{A}} = C(K)$. □

Corollary B.1.2

Let $\mathscr{A}$ be an algebra of continuous functions $f : K \to \mathbb{C}$ with the following properties:

(1). If $x \neq y \in K$, then there exists $f \in \mathscr{A}$ such that $f(x) \neq f(y)$.

(2) There exists $a \in K$ such that $f(a) = 0$ for every $f \in \mathscr{A}$.

(3) For every $f \in \mathscr{A}, \bar{f} \in \mathscr{A}$.

Then, $\overline{\mathscr{A}} = \{f \in C(K) : f(a) = 0\}$.

Proof   As there exists $a \in K$ such that $f(a) = 0$ for every $f \in \mathscr{A}$,

$$\mathscr{A} \subseteq \{f \in C(K) : f(a) = 0\}.$$

Moreover, as $\{f \in C(K) : f(a) = 0\}$ is closed in $C(K)$, $\overline{\mathscr{A}} \subseteq \{f \in C(K) : f(a) = 0\}$. Now let $f \in C(K)$ be such that $f(a) = 0$. By Stone's theorem, the algebra $\mathscr{B}$ generated by $\mathscr{A}$ and 1 is dense in $C(K)$. Thus, there exists a sequence $\{f_n\}$ in $\mathscr{B}$ such that $\|f_n - f\|_\infty \to 0$ as $n \to \infty$. Note that $f_n = g_n + \alpha_n$ for $g_n \in \mathscr{A}$. Moreover, $f_n(a) \to f(a) = 0$, and hence, $\alpha_n \to 0$. It follows that $\|g_n - f\|_\infty \leqslant \|f_n - f\|_\infty + |\alpha_n|$, which converges to 0 as $n \to \infty$.   □

Corollary B.1.3

Let $f \in C_\mathbb{R}(K)$ be such that $f(x) > 0$ for all $x \in K$. Let $\mathscr{A}$ be a subalgebra of $C_\mathbb{R}(K)$ that contains f. If f is injective, then $\overline{\mathscr{A}} = C_\mathbb{R}(K)$.

Proof   Note that $f \in \mathscr{A}$ satisfies (1) and (2) of Theorem B.1.1.   □

Remark B.1.2   Let $X = [a, b]$. Consider the algebra $\mathscr{A}$ of functions of the form $p(f)$, where p is a polynomial such that $p(0) = 0$. The last corollary is applicable to $f(x) = x^\alpha$ for $\alpha > 0$, $f(x) = e^x$, $f(x) = 1/x$ if $a > 0$.

Hilbert spaces and Parseval's identity

A *Hilbert space* $\mathcal{H}$ is a complete inner-product space. The identity appearing in part (iii), commonly known as *Parseval's identity*, provides a formula for the norm of an element in a Hilbert space.

Proposition B.1.1

If $\mathcal{H}$ is a separable Hilbert space with orthonormal basis $\{e_n\}_{n \geqslant 0}$, then the following statements hold:

(i) Every $h \in H$ takes the form $\sum_{n=0}^\infty \langle h, e_n \rangle e_n$.

(ii) The co-ordinate functionals corresponding to the orthonormal basis $\{e_n\}$ are continuous.

(iii) $\|h\|^2 = \sum_{n=0}^\infty |\langle h, e_n \rangle|^2$.

Proof   If $h = \sum_{n=0}^\infty \alpha_n e_n$ for some complex numbers α_n, then by the orthonormality of $\{e_n\}_{n \geqslant 0}$, $\alpha_n = \langle h, e_n \rangle$. The first part follows from this whereas the second part follows from the first part and the continuity of the inner-product. The last part also follows from the first part.   □

Lemma B.1.5

(Maximal Orthonormal Set) Let $\{e_n\}_{n\geqslant 0}$ be an orthonormal set in a Hilbert space H with the property: If $x \in H$ such that $\langle x, e_n \rangle = 0$ for all $n \geqslant 0$, then $x = 0$. Then, $\{e_n\}_{n\geqslant 0}$ is an orthonormal basis.

Proof Note that $\langle x - \sum_{n=0}^{\infty} \langle x, e_n \rangle e_n, e_m \rangle = 0$. By hypothesis,

$$x - \sum_{n=0}^{\infty} \langle x, e_n \rangle e_n = 0,$$

and hence, $\{e_n\}$ is an orthonormal basis. $\square$

Theorem B.1.3

For $n \in \mathbb{Z}$, let $E_n(t) := e^{int}$. Then, $\{E_n\}_{n\in\mathbb{Z}}$ is an orthonormal basis for $L^2[0, 2\pi]$.

Proof Let $f \in L^1[0, 2\pi]$ be such that $\langle f, E_n \rangle_2 = 0$ for every $n \in \mathbb{Z}$. Define $F(t) := \int_{[0,t]} f(s)ds$. We verify the following:

- There is a scalar $a \in \mathbb{C}$ such that $\langle F - a, E_n \rangle_2 = 0$ for every $n \in \mathbb{Z}$. This follows from an application of integration by parts.

- There exists a sequence of trigonometric polynomials p_n such that

$$\|F - a - p_n\|_\infty \to 0 \text{ as } n \to \infty.$$

  This may be concluded from the Stone–Weierstrass theorem.

- $F = a$ almost everywhere. This follows from $\lim_{n\to\infty} \|F - a - p_n\|_2 = 0$ and

$$\|F - a\|_2^2 = \lim_{n\to\infty} \langle F - a, p_n \rangle_2 = 0.$$

- f is zero almost everywhere.

By Lemma B.1.5, $\{E_n\}_{n\in\mathbb{Z}}$ is an orthonormal basis for $L^2[0, 2\pi]$. $\square$

As an application of the last theorem, we obtain an approximation result:

Corollary B.1.4

For every $f \in L^2[0, 2\pi]$, we have

$$\int_{[0,2\pi]} |f(s) - \sum_{n=-k}^{k} \langle f, e^{int} \rangle e^{-ins}|^2 ds \to 0 \text{ as } k \to \infty.$$

Quotient spaces and C^*-algebras

Let X be a normed linear space and let Y be a subspace of X. Let X/Y denote the quotient space with natural vector space structure. Define

$$\|x+Y\|_Q = \inf_{y \in Y} \|x+y\|, \quad x+Y \in X/Y. \tag{B.1.2}$$

We leave the verifications of the following elementary facts to the reader:

Lemma B.1.6

Let X be a normed linear space and let Y be a subspace of X. We have the following statements:

(1) If $x+Y = 0$, then $\|x+Y\|_Q = 0$.

(2) If Y is closed in X, then $\|x+Y\|_Q$ defines a norm on X/Y.

(3) X/Y is complete if so are X and Y.

(4) The quotient map $q : X \to X/Y$ given by $q(x) = x + Y$ is an open map with norm at most 1.

A *Banach algebra* B is an algebra over $\mathbb{C}$ with identity 1, which has a norm making it into a Banach space and satisfying $\|1\| = 1$ and the inequality $\|fg\| \leqslant \|f\|\|g\|$ for all $f, g \in B$. A *C^*-algebra* B is a Banach algebra together with the linear map $x \to x^*$ from B into itself (to be referred to as an *involution*) such that for all $x, y \in B$ and $\lambda \in \mathbb{C}$,

$$x^{**} = (x^*)^* = x, \quad (x+y)^* = x^* + y^*,$$
$$(xy)^* = y^*x^*, \quad (\lambda x)^* = \bar{\lambda}x^*,$$
$$\|x^*\| = \|x\|, \quad \|x^*x\| = \|x\|\|x^*\|.$$

The identity $\|x^*x\| = \|x\|^2$, $x \in B$ is commonly known as the *C^*-identity*. It is easy to see that the C^*-identity implies $\|x^*\| = \|x\|$ for every $x \in B$. A subset of B is said to be *self-adjoint* if it is closed under the involution. An ideal in B is said to be a *∗-ideal* if it is self-adjoint.

In this text, we will be interested in two basic examples of C^*-algebras described in the following theorems.

Theorem B.1.4

For every compact Hausdorff space X, the space $C(X)$ of continuous functions from X into $\mathbb{C}$ forms a C^*-algebra, where the constant function 1 is the unit, pointwise product of functions is the multiplication and the conjugate of a continuous function on X is the involution.

Theorem B.1.5

For a complex Hilbert space $\mathcal{H}$, the algebra $\mathcal{L}(\mathcal{H})$ of bounded linear operators from $\mathcal{H}$ into $\mathcal{H}$ is a unital C^*-algebra, where the identity operator is the unit, composition of operators is the multiplication and the uniquely defined adjoint of a bounded linear operator on a Hilbert space is the involution.

We leave the verifications of the two foregoing theorems to the reader. Recall that for any infinite dimensional space $\mathcal{H}$, $\mathcal{L}(\mathcal{H})$ contains a copy of $\ell^\infty(\mathbb{N})$. Unlike $\mathcal{L}(\mathcal{H})$, $C(X)$ is separable for a large family of compact Hausdorff spaces X (see [82, pp. 221]).

Theorem B.1.6

For any compact metric space X, $C(X)$ is norm separable, that is, $C(X)$ is countably generated as a Banach algebra.

Proof Without loss of generality, we may assume that $\operatorname{diam} X > 0$. For an integer $n \geqslant 0$, let $\{x_n\}_{n \geqslant 0}$ be a dense subset of X and f_n be as defined in (A.1.1). For $p, q \in X$, if $f_n(p) = f_n(q)$ for every $n \geqslant 0$, then $d(p, x_n) = d(q, x_n)$ and density of $\{x_n\}_{n \geqslant 0}$ in X implies that $p = q$. Thus, $\{f_n\}_{n \geqslant 0}$ separates points of X. It follows from Corollary B.1.1 that the complex algebra generated by $\{f_n\}_{n \geqslant 0}$ is dense in $C(X)$. If we allow finite complex-rational linear combinations of finite products of functions in $\{f_n\}_{n \geqslant 0}$, then we get the norm separability of $C(X)$. $\quad\square$

Open Mapping Theorem

Let X and Y be normed linear spaces. A mapping $T : X \to Y$ is said to be *open* if $T(U)$ is open in Y for every open subset U of X.

***Example* B.1.7** Let X be a normed linear space and let $a \in X$ and $\alpha \in \mathbb{C} \setminus \{0\}$. Define the translation τ_a and dilation d_α on X by

$$\tau_a(x) = x + a, \ d_\alpha(x) = \alpha x.$$

Then, τ_a and d_α are open mappings. This follows from the fact that τ and d are homeomorphisms with inverses τ_{-a} and $d_{\alpha^{-1}}$ respectively. $\quad\blacksquare$

If T has continuous inverse then clearly T is open. What is surprising is that this is true even if T is not invertible. This is the content of the open mapping theorem. Before we state it, let us see a handy characterization of linear open mappings.

Lemma B.1.7

If $T : X \to Y$ is a linear transformation, then the following statements are equivalent:

(1) T sends the open unit ball to an open subset of Y.

(2) T is an open mapping.

(3) There exists $c > 0$ such that for each $y \in Y$ there corresponds $x \in X$ with the properties $\|x\| \leqslant c\|y\|$ and $Tx = y$.

Proof (1) $\Rightarrow$ (2): Let U be an open subset of X. Let $y = Tx \in T(U)$. Then, $\mathbb{B}(x, R) \subseteq U$ for some $R > 0$, where $\mathbb{B}(x, R)$ denotes the ball centered at x and of radius R in X. Further,

$$\mathbb{B}(0, 1) \subseteq \frac{U - x}{R} = \left\{ \frac{y - x}{R} : y \in U \right\}.$$

As $T(\mathbb{B}(0,1))$ is open in Y, so is $R \cdot T(\mathbb{B}(0,1)) + Tx = T(\mathbb{B}(x,R))$ in $T(U)$.

(2) $\Rightarrow$ (3): Note that $0 \in T(\mathbb{B}(0,1))$ is open, and hence, $\mathbb{B}(0,R) \subseteq T(\mathbb{B}(0,1))$ for some $R > 0$. Thus, for every $y \in Y$, $r\frac{y}{\|y\|} \in B(0,R)$ for $r = R/2$. Then, there exists $x_0 \in \mathbb{B}(0,1)$ such that $Tx_0 = r\frac{y}{\|y\|}$. Check that $Tx = y$, where $x = x_0\|y\|/r$. Check that $\|x\| \leqslant r^{-1}\|y\|$, so that (3) holds with $c = r^{-1}$.

(3) $\Rightarrow$ (1): For $r > 0$, note that $\mathbb{B}(y_0, rc^{-1}) \subseteq T(\mathbb{B}(x_0, r))$, where $y_0 = Tx_0$. $\quad\square$

***Remark* B.1.8** Every linear open map is surjective.

Theorem B.1.9 (Open Mapping Theorem)

Every bounded linear transformation from a Banach space onto a Banach space is open.

Proof Let X, Y be Banach spaces and let $T : X \to Y$ be a surjective bounded linear transformation. For a unit vector $y \in Y$, we verify the following:

- $Y = \cup_{n=1}^{\infty} \overline{T(\mathbb{B}(0,n))}$.

- There exists a positive integer k such that $\overline{T(\mathbb{B}(0,k))}$ contains a non-empty open set W.

- Let $y_0 \in W$ be such that $\mathbb{B}(y_0,R) \subseteq W$ for some $R > 0$. If $z \in \overline{\mathbb{B}(y_0,R)}$, then there exists $\{u_n\} \subseteq \overline{\mathbb{B}(0,2k)}$ such that $Tu_n \to z - y_0$.

- There exists $x_1 \in X$ such that $\|x_1\| \leqslant \frac{2k}{R}$ and $\|y - Tx_1\| < 1/2$.

- There exists a sequence $\{x_n\}$ such that $\|x_n\| \leq \frac{2k}{R}\frac{1}{2^{n-1}}$ and
$$\|y - (Tx_1 + \cdots + Tx_n)\| \leqslant 1/2^n.$$

- The sequence $\{x_1 + \cdots + x_n\}$ converges to some $x \in X$. Moreover, $\|x\| \leqslant 4k/R$ and $Tx = y$.

The desired conclusion may now be deduced from Lemma B.1.7 . $\quad\square$

The following is often known as the bounded inverse theorem.

Corollary B.1.5

A bijective bounded linear transformation between two Banach spaces is a homeomorphism.

A linear transformation $T : X \to Y$ is said to be *closed* if $x_n \to x$ in X and $Tx_n \to y$ in Y, then $Tx = y$.

Theorem B.1.10 (Closed Graph Theorem)

Let X and Y be Banach spaces and let $T : X \to Y$ be a linear operator. If T is closed, then T is continuous.

Proof Define the norm $|x|$ on X by $|x| := \|x\| + \|Tx\|$, $x \in X$. Then, $(X, |\cdot|)$ is complete. Now, apply Corollary B.1.5 to the identity mapping from X onto the Banach space $(X, |\cdot|)$. $\quad\square$

Dual spaces and weak convergence

Let X be a normed linear space. The *dual space* X' of X is defined as the normed linear space of all bounded linear functionals $f : X \to \mathbb{C}$. Then, X' is a Banach space with norm $\|f\| := \sup\{|f(x)| : \|x\| \leqslant 1\}$. We compute here the dual space of $\ell^p(\mathbb{N})$, $1 \leqslant p < \infty$.

Example B.1.11 Let $1 \leqslant p < \infty$ and let q be such that $1/p + 1/q = 1$. By Hölder's inequality, ϕ_y given by $\phi_y((x_n)) = \sum_n x_n \bar{y}_n$ defines a bounded linear functional on $\ell^p(\mathbb{N})$ with $\|\phi_y\| \leqslant \|y\|_q$. Thus, for any $y = (y_n)_{n \geqslant 0} \in \ell^q$, one may define $F : \ell^q(\mathbb{N}) \to (\ell^p(\mathbb{N}))'$ by $F((y_n)_{n \geqslant 0}) = \phi_y$. Moreover, if $\phi \in (\ell^p(\mathbb{N}))'$, then $\phi = \phi_y$ with $y = (\phi(e_0), \phi(e_1), \cdots,)$, where $\{e_k\}_{k \geqslant 0}$ is the standard orthonormal basis of $\ell^q(\mathbb{N})$. We claim that $\|\phi_y\| = \|y\|_q$ if $y \in \ell^q(\mathbb{N})$. If $p = 1$, then $\|y\|_\infty \leqslant \sup_{n \geqslant 0} \|\phi_y\| \|e_n\|_1 = \|\phi_y\|$. Otherwise, define $x \in \ell^p(\mathbb{N})$ by letting $x_n = 0$ if $y_n = 0$ and $x_n = \frac{|y_n|^q}{y_n}$ otherwise, and note that $\|x\|_p^p = \|y\|_q^q$ and $\phi_y(x) = \|y\|_q^q$. Hence, $\phi_y((x_n / \|x\|_p)_{n \geqslant 0}) = \|y\|_q$. ∎

We need to identify the dual of a Hilbert space in obtaining a variant of the Bolzano–Weierstrass property for Hilbert spaces.

Lemma B.1.8 (Riesz Representation Lemma)

Let $\mathcal{H}$ be a complex Hilbert space. If $f : \mathcal{H} \to \mathbb{C}$ is a bounded linear functional, then there exists a unique $y \in \mathcal{H}$ such that $f(x) = \langle x, y \rangle$ for all $x \in H$.

Proof Let $\{e_n\}_{n \in I}$ be an orthonormal basis for $\mathcal{H}$ (see Exercise 1.7.1). For a finite subset F of I, consider $y_F = \sum_{n \in F} \overline{f(e_n)} e_n \in \mathcal{H}$. Note that

$$f(y_F) = \sum_{n \in F} |f(e_n)|^2 = \|y_F\|^2.$$

Moreover, there is $M > 0$ such that $|f(y_F)| \leqslant M\|y_F\|$. This yields $\|y_F\| \leqslant M$ for every finite subset F of I. In particular, $y = \sum_{n \in I} \overline{f(e_n)} e_n \in \mathcal{H}$. Check that $f(x) = \langle x, y \rangle$. To see the uniqueness part, note that $\langle x, y_1 \rangle = \langle x, y_2 \rangle$ implies $\langle x, y_1 - y_2 \rangle$ for every $x \in H$, and hence, for $x = y_1 - y_2$. This gives $y_1 = y_2$. □

We say that a sequence $\{x_n\}$ in a Hilbert space $\mathcal{H}$ *converges weakly* to $x \in \mathcal{H}$ if $\langle x_n, y \rangle \to \langle x, y \rangle$ for every $y \in \mathcal{H}$.

Remark B.1.12 A weakly convergent sequence in a Hilbert space is necessarily bounded. Indeed, for a positive integer m, consider the open set $V_m = \{y \in \mathcal{H} : |\langle x_n, y \rangle| > m \text{ for some } n\}$ and note that by the Baire category theorem, V_{m_0} is not dense for some integer m_0, that is, $|\langle x_n, y \rangle| \leqslant m_0$ for every n and every y in some ball in $\mathcal{H}$.

Theorem B.1.13

Every bounded sequence in a Hilbert space admits a weakly convergent subsequence.

Proof Let $\{h_n\}_{n \in \mathbb{N}}$ be a bounded sequence in the Hilbert space $\mathcal{H}$ and let $\mathcal{K}$ denote the closed linear span of $\{h_n\}_{n \in \mathbb{N}}$. Clearly, $\mathcal{K}$ is separable; hence, it has a countable dense subset, say, $\{x_k\}_{k \in \mathbb{N}}$. Let $M = \sup_{n \in \mathbb{N}} \|h_n\|$ and set $h'(x) = \langle h, x \rangle$ for $h, x \in \mathcal{H}$. As $|h'_n(x_1)| \leqslant M\|x_1\|$,

$\{h'_n(x_1)\}$ is a bounded sequence in $\mathbb{C}$. By the Bolzano–Weierstrass theorem, $\{h'_n(x_1)\}$ has a convergent subsequence $\{h'_{n1}(x_1)\}$. As $|h'_{n1}(x_2)| \leq M\|x_2\|$, $\{h'_{n1}(x_2)\}$ is a bounded sequence in $\mathbb{C}$. Again by the Bolzano–Weierstrass theorem, $\{h'_{n1}(x_2)\}$ has a convergent subsequence $\{h'_{n2}(x_2)\}$. Inductively, for $k \geq 1$, $\{h'_{nk-1}(x_k)\}$ has a convergent subsequence $\{h'_{nk}(x_k)\}$.

For fixed $k \geqslant 1$, consider the sequence $\{h'_{nn}(x_k)\}$. Note that $h'_{nn}(x_k)$ belongs to the convergent sequence $\{h'_{nk}(x_k)\}$ for $n \geqslant k$. Hence, $\{h'_{nn}(x_k)\}$ is also convergent. To complete the proof, let $x \in \mathcal{K}$. Given $\epsilon > 0$, choose k large enough so that $\|x - x_k\| < \epsilon$. For $m, n \geqslant 1$, note that

$$|h'_{mm}(x) - h'_{nn}(x)| \;\leq\; |h'_{mm}(x) - h'_{mm}(x_k)| + |h'_{mm}(x_k) - h'_{nn}(x_k)| + |h'_{nn}(x_k) - h'_{nn}(x)|$$
$$\leqslant\; 2M\epsilon + |h'_{mm}(x_k) - h'_{nn}(x_k)|.$$

Thus, $\{h'_{nn}(x)\}$ is a Cauchy sequence. It follows that $h'_{nn}(x)$ is convergent, and as $\{h_n\}$ is orthogonal to $\mathcal{K}^\perp$, the same conclusion holds for any $x \in \mathcal{H}$. Finally, as $\sup_n \|h_{nn}\| \leqslant M$, one may conclude from Lemma B.1.8 that $\{h_{nn}\}$ converges weakly to some element in $\mathcal{H}$. □

We conclude this section by describing the dual of the space $C(X)$ of complex-valued continuous functions on a compact space X.

Theorem B.1.14 (Riesz Representation Theorem)

Let X be a compact space and $M(X)$ denote the space of $\mathbb{C}$-valued regular Borel measures on X with the total variation norm. For $\mu \in M(X)$, define

$$\phi_\mu(f) = \int_X f\,d\mu, \quad f \in C(X).$$

Then, $\phi_\mu \in C(X)'$ and the map $\mu \mapsto \phi_\mu$ defines an isometric isomorphism of $M(X)$ onto $C(X)'$.

Spectrum of a Bounded Linear Operator

In this section, we prove several basic facts pertaining to spectrum of a bounded linear operator on a Hilbert space. These include in particular the fact that the spectrum is always upper semi-continuous [102], [75, Problem 86] in the following sense:

Theorem B.1.15 (Newburgh)

If $A \in \mathcal{L}(\mathcal{H})$, then for every open set U containing $\sigma(A)$, there exists $\epsilon > 0$ with the property

$$\sigma(B) \subseteq U \text{ whenever } B \in \mathcal{L}(\mathcal{H}) \text{ and } \|A - B\| < \epsilon.$$

Before we present a proof of Theorem B.1.15, let us recall that the spectrum $\sigma(T)$ of $T \in \mathcal{L}(\mathcal{H})$ is defined as the subset

$$\{\lambda \in \mathbb{C} : T - \lambda I \text{ is not invertible in } \mathcal{L}(\mathcal{H})\}$$

of the complex plane $\mathbb{C}$. It turns out that $\sigma(T)$ is a bounded and non-empty subset of $\mathbb{C}$. In fact,

$$\sigma(T) \subseteq \{z \in \mathbb{C} : |z| \leqslant \|T\|\}, \tag{B.1.3}$$

where $\|T\| = \sup\{\|Tx\| : x \in \mathcal{H}, \|x\| = 1\}$ is the operator norm of T. The inclusion in (B.1.3) may be derived from the following fact essential in the proof of Theorem B.1.15.

Proposition B.1.2

If $C \in \mathcal{L}(\mathcal{H})$ has norm less than 1, then $I - C$ is invertible. Moreover,

$$\lim_{n \to \infty} \left\| \sum_{k=0}^{n} C^k - (I - C)^{-1} \right\| = 0.$$

Proof Note that for integers $m < n$, one has

$$\left\| \sum_{k=m}^{n} C^k \right\| \leq \sum_{k=m}^{n} \|C^k\| \leq \sum_{k=m}^{n} \|C\|^k,$$

where we used the fact that $\|C^k\| \leq \|C\|^k$ for every non-negative integer k. As $\|C\| < 1$, by the convergence of the geometric series for real numbers [119, Theorem 3.26],

$$P(m,n) = \sum_{k=m}^{n} C^k \to 0 \text{ in the operator norm as } m, n \to \infty.$$

Thus, $\{P(0,n)\}_{n \geq 0}$ is a Cauchy sequence in $\mathcal{L}(\mathcal{H})$. As $\mathcal{L}(\mathcal{H})$ is complete in the operator norm, there exists $T \in \mathcal{L}(\mathcal{H})$ such that $\{P(0,n)\}_{n \geq 0}$ converges to T in the operator norm. Since $\|C\| < 1$, $\{C^k\}_{k \geq 0}$ converges to 0 in the operator norm. To conclude the proof, note that

$$(I - C)P(0,n) = I - C^{n+1} = P(0,n)(I - C)$$

converges to $(I - C)T$, I, $T(I - C)$ simultaneously in the operator norm. As the limit is unique, we must have $(I - C)T = I = T(I - C)$. □

Corollary B.1.6

The group of invertible operators in $\mathcal{L}(H)$ is open in $\mathcal{L}(H)$.

Proof Let $A \in \mathcal{L}(H)$ be an invertible operator and let $B \in \mathcal{L}(H)$. If $\|A - B\| < \|A\|$ then by Proposition B.1.2, $A^{-1}B$ is invertible, and hence, so is B. □

In the proof of Theorem B.1.15, we also need the so-called *resolvent function* $R_T : \mathbb{C} \backslash \sigma(T) \to \mathcal{L}(\mathcal{H})$ given by

$$R_T(z) = (T - zI)^{-1}, \quad z \in \mathbb{C} \backslash \sigma(T).$$

In the following lemma, we record some observations pertaining to the spectrum and the resolvent function.

Lemma B.1.9

For $T \in \mathcal{L}(H)$, we have the following statements:

(1) $\sigma(T)$ *is a non-empty compact subset of* $\mathbb{C}$,

(2) *the resolvent function* $R_T : \mathbb{C} \backslash \sigma(T) \to \mathcal{L}(\mathcal{H})$ *is continuous,*

(3) *the resolvent function* R_T *satisfies*

$$\|R_T(z)\| \leqslant \frac{1}{|z| - \|T\|}, \quad z \in \mathbb{C}, \ |z| > \|T\|.$$

(4) *the spectral radius* $r(T) = \sup\{|z| : z \in \sigma(T)\}$ *of* T *is given by*

$$r(T) = \lim_{n \to \infty} \|T^n\|^{1/n}.$$

(5) *If* $\lambda \in \partial\sigma(T)$, *then* $\inf_{\|f\|=1} \|(T - \lambda)f\| = 0.$

Proof We closely follow [131] and [35]. For $z_0 \in \mathbb{C} \backslash \sigma(T)$, let $r = \|R_T(z_0)\|^{-1}$ and let $z \in \mathbb{C}$ be such that $|z - z_0| < r$. Observe that

$$T - zI = (T - z_0I)(I - R_T(z_0)(z - z_0)).$$

By the previous lemma, $I - R_T(z_0)(z - z_0)$ and hence, $T - zI$ is invertible. This together with (B.1.3) shows that $\sigma(T)$ is a compact subset of $\mathbb{C}$. To verify (2), observe first that $R_T(z) = R_T(z_0)(I - R_T(z_0)(z - z_0))^{-1}$. The continuity of R_T now follows from

$$R_T(z) = \lim_{n \to \infty} \sum_{k=0}^{n} R_T(z_0)^{k+1}(z - z_0)^k \text{ in the operator norm.}$$

To prove (3), notice that $T - wI = -w(I - w^{-1}T)$ for any nonzero $w \in \mathbb{C}$, and then use Proposition B.1.2. We now complete the verification of (1). For $x, y \in \mathcal{H}$, note that the series

$$\langle R_T(z)x, y \rangle = \sum_{k=0}^{\infty} (z - z_0)^k \langle (R_T(z_0))^{k+1}x, y \rangle$$

is uniformly convergent on $\{z \in \mathbb{C} : |z - z_0| < r\}$. Thus, $\langle R_T(\cdot)x, y \rangle$ is analytic in $\mathbb{C} \backslash \sigma(T)$. Moreover, $\lim_{|z| \to \infty} \langle R_T(z)x, y \rangle = 0$. Indeed, by (3),

$$|\langle R_T(z)x, y \rangle| \leqslant \|R_T(z)\| \|x\| \|y\| \leqslant \frac{\|x\| \|y\|}{|z| - \|T\|}, \quad z \in \mathbb{C}, \ |z| > \|T\|.$$

Suppose now $\sigma(T) = \emptyset$. Then, $\langle R_T(\cdot)x, y \rangle$ becomes an entire function which is bounded. Hence, by Liouville's theorem [120, Theorem 10.23], $\langle R_T(\cdot)x, y \rangle$ is identically 0 for all $x, y \in \mathcal{H}$, which is absurd.

We next verify (4). Without loss of generality, we may assume that $T^n \neq 0$ for every positive integer n (in this case, $\sigma(T) = \{0\}$ and $r(T) = 0$). Note that $\alpha_n = \log \|T^n\|$, $n \geqslant 0$ satisfies $\alpha_{n+m} \leqslant \alpha_n + \alpha_m$, $m \geqslant 0$. Thus, for any fixed integer k and $n = mk + r$ with $0 \leqslant r \leqslant k - 1$,

$$\alpha_n \leqslant m\alpha_k + \sup_{0 \leqslant r \leqslant k-1} \alpha_r.$$

It is now easy to see that

$$\limsup \frac{\alpha_n}{n} \leq \inf \frac{\alpha_k}{k} \leq \liminf \frac{\alpha_n}{n}.$$

Thus, the limit $\mathrm{spr}(T) := \lim_{n\to\infty} \|T^n\|^{1/n}$ exists.

Let $\mu \in \mathbb{C}$ be such that $|\mu|\mathrm{spr}(T) < 1$ and let $\epsilon = \frac{1-|\mu|\mathrm{spr}(T)}{2} > 0$. Then, for integer $N \geq 1$,

$$|\mu|\left| \|T^n\|^{1/n} - \mathrm{spr}(T) \right| < \epsilon, \quad n \geq N.$$

It follows that

$$|\mu|^n \|T^n\| \leq \frac{(1 + |\mu|\mathrm{spr}(T))^n}{2^n}, \quad n \geq N.$$

Thus, $I - \mu T$ is invertible, and hence, $r(T) \leq \mathrm{spr}(T)$.

To see that $r(T) \geq \mathrm{spr}(T)$, let $r > r(T)$. Then, for any unit vector $x \in \mathcal{H}$, $\langle (I - \mu T)^{-1}x, x \rangle$ is analytic in a neighborhood of $\mathbb{D}_{r^{-1}}$. By Cauchy estimates, for some constant C_x,

$$\|T^n\| \leq C_x r^n, \quad n \geq 1,$$

so that $\mathrm{spr}(T) \leq r$. As $r > r(T)$ is arbitrary, we obtain $\mathrm{spr}(T) \leq r(T)$. This completes the verification of (4).

To see (5), let $\lambda \in \partial\sigma(T)$ and let $\{\lambda_n\}_{n\geq 1}$ be a sequence in $\mathbb{C}\backslash\sigma(T)$ such that $\lambda_n \to \lambda$ as $n \to \infty$. We claim that

$$\|(T - \lambda_n)^{-1}\| \to \infty \text{ as } n \to \infty. \tag{B.1.4}$$

Indeed, if $M = \sup_{n\geq 1} \|(T - \lambda_n)^{-1}\| < \infty$, then

$$\|(T - \lambda) - (T - \lambda_n)\| \leq M^{-1} \leq \|(T - \lambda_n)^{-1}\|^{-1}$$

for sufficiently large n, and hence, by Proposition B.1.2, we arrive at the contradiction that $\lambda \notin \sigma(T)$. This proves (B.1.4). We now complete the verification of (5). For every integer $n \geq 1$, choose a unit vector $x_n \in \mathcal{H}$ such that $y_n = (T - \lambda_n)^{-1}x_n$ satisfies

$$\|y_n\| > \|(T - \lambda_n)^{-1}\| - \frac{1}{n}.$$

By (B.1.4), $\|y_n\| \to \infty$ as $n \to \infty$. However,

$$\|(T - \lambda)(y_n/\|y_n\|)\| \leq \|y_n\|^{-1} + |\lambda - \lambda_n|.$$

It follows that $\|(T - \lambda)(y_n/\|y_n\|)\| \to 0$ as $n \to \infty$. $\qquad\square$

Remark B.1.16 The *approximate-point spectrum* of $T \in \mathcal{L}(\mathcal{H})$ is defined as

$$\sigma_{\mathrm{ap}}(T) = \{\lambda \in \mathbb{C} : T - \lambda I \text{ is not bounded below }\}.$$

Part (5) says that $\partial\sigma(T) \subseteq \sigma_{\mathrm{ap}}(T)$.

We now complete the proof of the upper semi-continuity of the spectrum.

Proof   [Proof of Theorem B.1.15] Let U be an open subset of $\mathbb{C}$ that contains $\sigma(A)$. As $\lim_{|z| \to \infty} \|R_A(z)\| = 0$ (Proposition B.1.9(3)) and R_A is continuous, there exists $M > 0$ such that

$$\|R_A(z)\| < M \text{ for all } z \in \mathbb{C} \backslash U.$$

Let $\epsilon = \frac{1}{M}$ and let $B \in \mathcal{L}(\mathcal{H})$ be such that $\|A - B\| < \epsilon$. Now it suffices to check that $B - \lambda I$ is invertible if $\lambda \notin U$. Suppose $\lambda \notin U$, note that

$$
\begin{aligned}
\|I - (B - \lambda I)R_A(\lambda)\| &= \|I - (B - A + A - \lambda I)R_A(\lambda)\| \\
&= \|(A - B)R_A(\lambda)\| \leqslant \|A - B\| \|R_A(\lambda)\| \\
&< \epsilon M = 1.
\end{aligned}
$$

By Proposition B.1.2, $(B - \lambda I)R_A(\lambda)$, and hence, $B - \lambda I$ is invertible. $\square$

Remark B.1.17   Newburgh's theorem extends to any unital Banach algebra, and in particular, applying it to Calkin algebra, together with Atkinson's theorem, we conclude that the essential spectrum is upper semicontinuous.

Functional Calculus for Bounded Linear Operators

Let $p(z) = \sum_{k=0}^{n} a_k z^k$ be a complex polynomial of degree n in the variable z. If $T \in \mathcal{L}(\mathcal{H})$, then it is natural to define $p(T)$ by

$$p(T) = \sum_{k=0}^{n} a_k T^k,$$

which is obtained by replacing z by T in the expression for $p(z)$. Notice that $p(T)$ belongs to $\mathcal{L}(\mathcal{H})$, as $\mathcal{L}(\mathcal{H})$ is an algebra. The association $p \mapsto p(T)$ is an *algebra homomorphism*: For $\alpha \in \mathbb{C}$, and for complex polynomials p, q,

$$(p + \alpha q)(T) = p(T) + \alpha(q(T)) \text{ and } (pq)(T) = p(T)q(T).$$

One may refer to this algebra homomorphism as the *polynomial functional calculus* for T. The following is known as the spectral mapping property for polynomials.

Lemma B.1.10

For any polynomial in the complex variable and for any A in $\mathcal{L}(\mathcal{H})$, we have

$$\sigma(p(A)) = p(\sigma(A)).$$

Proof   Replacing p by $p - \lambda$, it suffices to verify that $p(T)$ is invertible if and only if p is nowhere-vanishing on $\sigma(T)$. By the fundamental theorem of algebra,

$$p(z) = \alpha_0(z - \alpha_1)(z - \alpha_2) \cdots (z - \alpha_n), \ \{\alpha_k\}_{k=0}^{n} \subseteq \mathbb{C}.$$

Suppose that p is nowhere-vanishing on $\sigma(T)$. It follows that $\alpha_k \notin \sigma(T)$ for $k = 1, \ldots, n$. Hence,

$$p(T) = \alpha_0(T - \alpha_1 I)(T - \alpha_2 I) \cdots (T - \alpha_n I), \ \{\alpha_k\}_{k=1}^n \subseteq \mathbb{C} \backslash \sigma(T). \tag{B.1.5}$$

It is now clear that $p(T)$ is invertible. Conversely, assume that $p(T)$ is invertible. Then there exists $S \in \mathcal{L}(\mathcal{H})$ such that $p(T)S = I = S\,p(T)$. It follows from (B.1.5) that $\alpha_k \notin \sigma(T)$ for any $k = 1, \ldots, n$. Thus, $p(z)$ is nonzero for any $z \in \sigma(T)$. $\qquad\square$

Next, consider the rational function $r(z) = \frac{p(z)}{q(z)}$ in one variable z for nonzero polynomials $p(z)$ and $q(z)$. As in the polynomial case, one may be tempted to define $r(T)$ as $q(T)^{-1}p(T)$. However, one needs to be a little careful this time, as r may have a *pole*. In general, it is not possible to define $r(T)$, as a member of $\mathcal{L}(\mathcal{H})$ for $T \in \mathcal{L}(\mathcal{H})$. For example, if $r(z) = \frac{1}{z}$, then $r(T)$ is not defined for any non-invertible operator T. However, in case the rational function r does not have any pole in $\sigma(T)$, $r(T)$ can be defined as a bounded linear operator in $\mathcal{L}(\mathcal{H})$. To see this, assume that $q(z)$ is non-vanishing on $\sigma(T)$. Then, by Lemma B.1.10, $q(T)$ is invertible, and hence, $r(T) = q(T)^{-1}p(T)$ is a member of $\mathcal{L}(\mathcal{H})$. Note that $r(T)$ is independent of the representation of r: If $r(z) = \frac{p_1(z)}{q_1(z)} = \frac{p_2(z)}{q_2(z)}$, then

$$q_1(T)^{-1}p_1(T) = q_2(T)^{-1}p_2(T).$$

The algebra homomorphism $r \mapsto r(T)$ will be referred to as the *rational functional calculus* for T. Notice that the rational functional calculus extends the polynomial functional calculus: If $r(z)$ is a polynomial, that is, $r(z) = \frac{p(z)}{q(z)}$ with $q(z) = 1$, then $r(T) = q(T)^{-1}p(T) = I^{-1}p(T) = p(T)$.

If $T \in \mathcal{L}(\mathcal{H})$, then the question arises: does there exist any functional calculus for T that extends the rational functional calculus? The answer to this question is affirmative. In fact, there exists an analytic functional calculus $f \mapsto f(T)$ commonly known as the *Riesz–Dunford functional calculus*, which extends the rational functional calculus. Roughly, this is obtained by integrating $f(z)(z - T)^{-1}$ over certain chains surrounding the spectrum of T (the reader is referred to [129] for details). In the remaining part of this appendix, we will focus on a functional calculus for a special class of so-called normal operators. Recall that $N \in \mathcal{L}(\mathcal{H})$ is *normal* if its *self-commutator* $[N^*, N] = N^*N - NN^*$ is the zero operator. We say that $A \in \mathcal{L}(\mathcal{H})$ is *self-adjoint* if $A^* = A$. Further, we say that $P \in \mathcal{L}(\mathcal{H})$ is an *orthogonal projection* (or simply a *projection*) if P is a self-adjoint idempotent, that is, $P^* = P$ and $P^2 = P$. A normal operator U is said to be *unitary* if $U^*U = I$.

Let us see basic examples of all these subclasses of normal operators.

***Example* B.1.18** Let (Ω, Σ, μ) be a measure space and let $\phi : \Omega \to \mathbb{C}$ be an essentially bounded function, that is,

$$\|\phi\|_{\mu,\infty} = \inf\{M \in \mathbb{R}_+ : |\phi(z)| \leqslant M \text{ for } z \text{ a.e.} - \mu\} < \infty.$$

Consider the Hilbert space $L^2(\mu)$ of square-integrable functions on Ω with respect to μ. Define the *multiplication operator* $M_\phi : L^2(\mu) \to L^2(\mu)$ by

$$M_\phi(f)(z) \equiv \phi(z)f(z), \quad z \text{ a.e.} - \mu, \ f \in L^2(\mu).$$

Note that M_ϕ satisfies $\|M_\phi f\| \leqslant \|\phi\|_{\mu,\infty}\|f\|$ for every $f \in L^2(\mu)$. Thus, M_ϕ is a bounded linear operator on $L^2(\mu)$ with $\|M_\phi\| \leqslant \|\phi\|_{\mu,\infty}$. Let us compute the Hilbert space adjoint M_ϕ^* of M_ϕ. Note that for all $f, g \in L^2(\mu)$,

$$
\begin{aligned}
\langle M_\phi^* f, g \rangle &= \langle f, M_\phi g \rangle = \int_\Omega f(z)\overline{\phi(z)g(z)}d\mu(z) \\
&= \int_\Omega \overline{\phi(z)}f(z)\overline{g(z)}d\mu(z) = \langle M_{\bar\phi}f, g \rangle,
\end{aligned}
$$

where $\bar\phi : \Omega \to \mathbb{C}$ given by $\bar\phi(z) = \overline{\phi(z)}$ ($z \in \Omega$) is essentially bounded with $\|\bar\phi\|_{\mu,\infty} = \|\phi\|_{\mu,\infty}$. Thus, $M_\phi^* = M_{\bar\phi}$. It is now clear that M_ϕ is normal. It is easy to verify the following statements:

(1) M_ϕ is self-adjoint if and only if ϕ takes real values μ-a.e.

(2) M_ϕ is an orthogonal projection if and only if ϕ is an indicator function of a μ-measurable subset of Ω.

(3) M_ϕ is unitary if and only if ϕ takes values in the unit circle μ-a.e.   ∎

Let $p(z,\bar z) = \sum_{p,q=0}^m a_{pq}z^p\bar z^q$ and let N be a normal operator. We obtain the operator $p(N, N^*)$ by replacing z and $\bar z$ by N and N^* respectively. As N is normal, $p(N, N^*)$ is well-defined. It is easy to see that $p \mapsto p(N, N^*)$ is an algebra homomorphism. In view of the Stone–Weierstrass theorem, one may be tempted to extend this functional calculus to all continuous functions on the spectrum of N. In Appendix C, we will see that the so-called spectral theorem provides even larger functional calculus for a normal operator.

Appendix C

The Spectral Theorem

This appendix is devoted to the various forms of spectral theorems for normal operators used in this book. We also present several applications of the spectral theorem.

Spectral Theorem

Let Σ be a σ-algebra over a set Ω. A *spectral measure* is an orthogonal projection-valued mapping $E : \Sigma \to \mathcal{L}(\mathcal{H})$ with $E(\Omega) = I$ such that

(1) $E(\omega \cap \omega') = E(\omega)E(\omega')$ and $E((\omega\backslash\omega')\cup(\omega'\backslash\omega)) = E(\omega\backslash\omega')+E(\omega'\backslash\omega)$ for every $\omega, \omega' \in \Sigma$,

(2) $E_{x,y}(\omega) = \langle E(\omega)x, y \rangle$, $\omega \in \Sigma$ is a complex measure for every $x, y \in \mathcal{H}$.

***Example* C.1.1 (Multiplication by Characteristic Function as Spectral Measure)**
Consider the measure space (Ω, Σ, μ). Note that for each $\omega \in \Sigma$, χ_ω is essentially bounded with $\|\chi_\omega\|_{\mu,\infty} \leqslant 1$. Let M_{χ_ω} denote the operator of multiplication by χ_ω on $L^2(\Omega, \mu)$. We check that the mapping $E(\omega) = M_{\chi_\omega}$ from Σ into $\mathcal{B}(L^2(\Omega, \mu))$ defines a spectral measure:

(1) Notice that $E(\Omega) = M_{\chi_\Omega} = M_1 = I$. As χ_ω is real-valued, it follows that M_{χ_ω} is self-adjoint. Moreover, as $\chi_\omega^2 = \chi_\omega$, we have $E(\omega)^2 = E(\omega)$.

(2) Since $\chi_{\omega'}\chi_{\omega''} = \chi_{\omega'\cap\omega''}$, we have $E(\omega' \cap \omega'') = E(\omega')E(\omega'')$. Further, $\chi_{\omega\cup\omega'} = \chi_\omega + \chi_{\omega'}$ if $\omega \cap \omega' = \emptyset$, and hence, $E(\omega \cup \omega') = E(\omega) + E(\omega')$ in this case.

(3) Finally, note that for any $f, g \in L^2(\Omega, \mu)$,

$$E_{f,g}(\omega) = \langle M_{\chi_\omega}f, g \rangle = \int_\Omega \chi_\omega(z)f(z)\overline{g(z)}d\mu(z) = \int_\omega f(z)\overline{g(z)}d\mu(z).$$

As $f\overline{g}$ is integrable, $E_{f,g}$ defines a complex measure. ∎

Example C.1.2 **(Projection as Spectral Measures)**   Let Ω be any set, Σ be its power set $P(\Omega)$, and $\mathcal{H}$ be any separable Hilbert space. Fix a sequence $\{x_n\}_{n\in\mathbb{N}}$ in Ω, and an orthonormal basis $\{e_n\}_{n\in\mathbb{N}}$ of $\mathcal{H}$. By Parseval's identity, one may rewrite $h \in \mathcal{H}$ as $\sum_{n=1}^{\infty}\langle h, e_n\rangle e_n$. For $\omega \in \Sigma$, let $\mathbb{N}_\omega = \{n \in \mathbb{N} : x_n \in \omega\}$. Define $E : P(\Omega) \to \mathcal{B}(\mathcal{H})$ by

$$E(\omega)h = \sum_{n\in\mathbb{N}_\omega} \langle h, e_n\rangle e_n, \quad \omega \in \Sigma.$$

Clearly, $E(\Omega) = I$. As for every $\omega, \omega' \in \Sigma$, we have $\mathcal{N}_{\omega\cap\omega'} = \mathcal{N}_\omega \cap \mathcal{N}_{\omega'}$ and $\mathcal{N}_{(\omega\backslash\omega')\cup(\omega'\backslash\omega)} = \mathcal{N}_{\omega\backslash\omega'} \cup \mathcal{N}_{\omega'\backslash\omega}$, $E(\cdot)$ is easily seen to be a spectral measure. ∎

Here is the first version of the spectral theorem for normal operators.

Theorem C.1.3 (Spectral Theorem)

If $N \in \mathcal{L}(\mathcal{H})$ is a normal operator, then there exists a unique spectral measure E on the Borel σ-algebra $B(\sigma(N))$ which satisfies for all $x, y \in \mathcal{H}$,

$$\langle Nx, y\rangle = \int_{\sigma(N)} \lambda \, dE_{x,y}(\lambda).$$

Moreover, $E(\omega)S = S E(\omega)$, $\omega \in B(\sigma(N))$ whenever $SN = NS$, $S \in \mathcal{L}(\mathcal{H})$.

Let K be a compact subset of the complex plane. Let $B_\infty(K)$ denote the normed algebra of complex-valued bounded Borel-measurable functions on K endowed with the sup norm $\|\cdot\|_\infty$.

Theorem C.1.4 (Functional Calculus)

Suppose $N \in \mathcal{L}(\mathcal{H})$ is normal with the spectral measure E as guaranteed by Theorem C.1.3. Then, for every $f \in B_\infty(\sigma(N))$, there exists a unique normal operator $f(N) \in \mathcal{L}(\mathcal{H})$ such that for all $x, y \in \mathcal{H}$,

$$\langle f(N)x, y\rangle = \int_{\sigma(N)} f(\lambda) \, dE_{x,y}(\lambda).$$

*Further, the map $f \mapsto f(N)$ defines a contractive algebra *-homomorphism, which is isometric on the algebra $C(\sigma(N))$ of continuous functions on $\sigma(N)$.*

An outline of the proof of the last two theorems will be presented later in this appendix. In the remaining part of this section, we discuss several applications of spectral theorem (see also Exercises 1.23–1.26). The following says that a *-cyclic normal operator can be realized as a multiplication operator M_z on an L^2 space.

Corollary C.1.1

Let $N \in \mathcal{L}(H)$ be a normal operator. If

$$\mathcal{H} = \bigvee \{N^k N^{*l} h : k, l \geqslant 0\}$$

for some $h \in \mathcal{H}$, then there exists a compactly supported measure μ on $\sigma(N)$ such that N is unitarily equivalent to the operator M_z of multiplication by z on $L^2(\sigma(N), \mu)$.

Proof Let $E(\cdot)$ denote a spectral measure of N. For $\Delta \in B(\sigma(N))$, define a scalar-valued positive measure μ by

$$\mu(\Delta) = \langle E(\Delta)h, h \rangle.$$

For a polynomial $f(z, \bar{z})$ in z and $\bar{z}$, set $Uf = f(N, N^*)h$. One may conclude from Theorem C.1.4 that $\|Uf\| = \|f\|_{L^2(\sigma(N), \mu)}$. As the polynomials in z and $\bar{z}$ are dense in $L^2(\sigma(N), \mu)$ (by the Stone–Weierstrass theorem and the fact that continuous functions are dense in $L^2(\sigma(N), \mu)$), U extends uniquely to a linear isometry $U : L^2(\sigma(N), \mu) \to \mathcal{H}$ given by $Uf = f(N)h$. As U has dense range, it is unitary. Finally, note that $UM_z = NU$. $\square$

We exploit in the following the continuous functional calculus for normal operators (see Theorem C.1.4) to describe the C^*-algebra generated by a normal operator (see Definition 1.6.3).

Corollary C.1.2

If $N \in \mathcal{L}(\mathcal{H})$ is a normal operator, then there exists a unital $*$-isometric isomorphism $\phi : C^*(N) \to C(\sigma(N))$ such that $\phi(N) = z$, where z denotes the identity function on the spectrum $\sigma(N)$ of N.

Proof The argument is similar to one given in Corollary C.1.1. $\square$

***Remark* C.1.5** Assume that $C^*(N)$ contains a non-trivial orthogonal projection E. Then there exists $f \in C(\sigma(N))$ such that $E = f(N)$. As $f(N)$ is self-adjoint, f is real-valued. Moreover, since $f(N)$ is idempotent, $f^2 = f$. As f is continuous, either $f = \chi_\omega$ for some proper subset ω of $\sigma(N)$ or f is 0 or 1. In the first case $\sigma(N)$ must be disconnected, and in the last two cases E is either 0 or I. It follows that $\sigma(N)$ is disconnected if and only if E is a non-trivial orthogonal projection.

Corollary C.1.3

For every normal operator N, there exists a normal operator $M = N^{1/2}$ such that $N = M^2$.

Proof Consider the function $f : \sigma(N) \to \mathbb{C}$ given by

$$f(z) = \begin{cases} |z|^{1/2} e^{i\theta/2} & \text{if } z \neq 0, \ \frac{z}{|z|} = \exp(i\theta) \text{ with } \theta \in [0, 2\pi), \\ 0 & \text{if } z = 0. \end{cases}$$

Then f is a (possibly discontinuous) square root branch of the identity function on $\sigma(N)$. Clearly, $f(z)$ is a bounded, Borel measurable function. By the functional calculus of N, we obtain a bounded, normal operator $f(N)$. Clearly, $f(N)^2 = N$, and hence, $f(N)$ provides a square root of N. $\square$

***Remark* C.1.6** If N is positive, then the spectrum of N is contained in the non-negative real line. The identity function on $\sigma(N)$ admits a positive, continuous square-root, say $f(z)$. It follows that $f(N)$ is the desired positive square-root of N. This square root is indeed unique.

The spectrum of a normal operator can be described purely in terms of its spectral measure.

Corollary C.1.4

If $N \in \mathcal{L}(\mathcal{H})$ is a normal operator with spectral measure $E(\cdot)$, then the spectrum of N is given by

$$\sigma(N) = \{\lambda \in \mathbb{C} : E(\omega) \neq 0 \text{ for every neighborhood } \omega \text{ of } \lambda\}.$$

Proof If $E(\omega) = 0$ for some neighborhood ω of $\lambda \in \mathbb{C}$, then by the functional calculus for N, the operator $N - \lambda$ is invertible with bounded inverse given by

$$\int_{\sigma(N)\setminus\omega} (z - \lambda)^{-1} dE(z).$$

Conversely, assume that $N - \lambda$ is invertible for some $\lambda \in \mathbb{C}$. As $\sigma(N)$ is compact, $\|(N - \lambda)^{-1}\| = \sup_{z \in \sigma(N)} |z - \lambda|^{-1}$, $\sigma(N)$ is disjoint from some neighborhood, say ω, of λ. We then immediately have $E(\omega) = \int_{\sigma(N)} \chi_\omega(z) dE(z) = 0$. □

Corollary C.1.5 (Spectral Mapping Property)

Let $N \in \mathcal{L}(\mathcal{H})$ be normal. If $f : \sigma(N) \to \mathbb{C}$ is a continuous mapping, then $\sigma(f(N)) = f(\sigma(N))$.

Proof The idea of the inclusion $f(\sigma(N)) \subseteq \sigma(f(N))$ is simple. We first verify the inclusion for the family of polynomials in z and $\bar{z}$, and obtain it for arbitrary continuous functions with the help of the Stone–Weierstrass theorem (Theorem B.1.1) and upper semi-continuity of the spectrum (Theorem B.1.15). We begin with the claim that

$$\{p(z, \bar{z}) : z \in \sigma(N)\} \subseteq \sigma(p(N, N^*)),$$

where p is a complex polynomial in two variables. To see this, let $\lambda \in \sigma(N)$, and note that a normal operator is invertible if and only if it is bounded from below (see Exercise 1.7.21). Thus, there exists a sequence $\{x_n\}_{n \in \mathbb{N}}$ of unit vectors in $\mathcal{H}$ such that

$$\lim_{n \to \infty} \|(N - \lambda I)x_n\| = 0 = \lim_{n \to \infty} \|(N^* - \bar{\lambda}I)x_n\|.$$

As for any $T \in \mathcal{L}(\mathcal{H})$ and any $z \in \mathbb{C}$,

$$T^m - z^m I = \left(T^{m-1} + zT^{m-2} + \cdots + z^{m-1}I\right)(T - zI) \ (m \in \mathbb{N}),$$

it follows that for any $i \in \mathbb{N}$,

$$\lim_{n \to \infty} \left\|\left(N^i - \lambda^i I\right)x_n\right\| = 0 = \lim_{n \to \infty} \left\|\left(N^{*i} - \bar{\lambda}^i I\right)x_n\right\|.$$

Moreover, as $\left(p(N, N^*) - p(\lambda, \bar{\lambda})I\right)x_n$ is same as

$$\sum_{i,j=0}^{n} a_{i,j}\left(N^i N^{*j} - \lambda^i \bar{\lambda}^j I\right)x_n = \sum_{i,j=0}^{n} a_{i,j}\left(N^i(N^{*j} - \bar{\lambda}^j I)x_n + \bar{\lambda}^j\left(N^i - \lambda^i I\right)x_n\right),$$

we conclude that $p(\lambda, \bar{\lambda}) \in \sigma(p(N, N^*))$. Thus, the claim stands verified.

We now verify the inclusion $f(\sigma(N)) \subseteq \sigma(f(N))$. Let $X = \sigma(N)$ and $\mathcal{S}$ be the uniform closure of polynomials in z and $\bar{z}$. By the Stone–Weierstrass theorem, $\mathcal{S} = C(X)$. Thus, there exists a sequence $\{p_n\}_{n\in\mathbb{N}}$ of complex polynomials in z and $\bar{z}$ such that $\lim\limits_{n\to\infty} \|p_n - f\|_\infty = 0$. By the continuous functional calculus,

$$\lim_{n\to\infty} \|p_n(N, N^*) - f(N)\| = 0.$$

Let $\lambda \in f(\sigma(N))$. Thus, $\lambda = f(z)$ for some $z \in \sigma(N)$ and $\lim\limits_{n\to\infty} p_n(z, \bar{z}) = \lambda$. We contend that $\lambda \in \sigma(f(N))$. Notice that it suffices to check that λ belongs to any open set V that contains $\sigma(f(N))$. Let U be an open set such that $\sigma(N) \subseteq U \subseteq \overline{U} \subseteq V$. By the upper semi-continuity of the spectrum, there exists a positive real ϵ with the following property:

$$\sigma(B) \subseteq U \text{ whenever } B \in \mathcal{B}(\mathcal{H}) \text{ and } \|f(N) - B\| < \epsilon.$$

Choose n_0 sufficiently large so that $\|f(N) - p_n(N, N^*)\| < \epsilon$, $n \geq n_0$. It follows that $p_n(z, \bar{z}) \in \sigma(p_n(N, N^*)) \subseteq U$ for all $n \geq n_0$. Hence, $\lambda \in \overline{U} \subseteq V$ as desired.

To see the remaining part, suppose that f has no zero on $\sigma(N)$. Then, $g = \frac{1}{f}$ is a bounded Borel measurable function on $\sigma(N)$, and hence, $g(N)$ is a bounded linear operator on $\mathcal{H}$. By the functional calculus,

$$f(N)g(N) = I = g(N)f(N).$$

Thus, $f(N)$ is invertible. Now we come to the last part of the proof. Let $\beta \in \sigma(f(N))$. Then, $f(N) - \beta I = (f - \beta)(N)$ is not invertible. Hence, $f - \beta$ has a zero on $\sigma(N)$. That is, $f(\alpha) = \beta$ for some $\alpha \in \sigma(N)$. This shows that $\sigma(f(N)) \subseteq f(\sigma(N))$. This completes the proof of the corollary. □

Remark C.1.7   If $N = A + iB$ is a normal operator with real and imaginary parts A and B respectively, then

$$\sigma(A + iB) \subseteq \{x + iy : x \in \sigma(A), y \in \sigma(B)\}. \tag{C.1.1}$$

This may be obtained from the spectral mapping property applied to the real and imaginary parts $\mathfrak{R}(z)$ and $\mathfrak{I}(z)$ of the identity function $id(z) = z$.

Let $p(z, \bar{z}) = \sum_{p,q=0}^{m} a_{pq} z^p \bar{z}^q$ and let N be a normal operator. Then, $p(N, N^*) = 0$ if and only if $\sigma(N)$ is contained in the algebraic set

$$\{\lambda \in \mathbb{C} : p(\lambda, \bar{\lambda}) = 0\}.$$

In case $p(z, \bar{z})$ is given by $z - \bar{z}$, $z\bar{z} - 1$, $z^2 - z$, N is a self-adjoint operator, unitary operator, an orthogonal projection respectively.

Let us compute the spectra of multiplication operators with continuous symbols.

Example C.1.8   Let Ω be a compact subset of the complex plane. Consider a finite measure space (Ω, Σ, μ) and let ϕ be a continuous function in $L^2(\mu)$. We have already recorded in

Example B.1.18 that the multiplication operator M_ϕ on $L^2(\mu)$ is normal. As $M_\phi = \phi(M_z)$, by the spectral mapping property, $\sigma(M_\phi) = \phi(\Omega)$.

Consider the Hilbert space $L^2(\mathbb{T})$ of square-integrable functions on the unit circle with respect to the arc-length measure and consider the multiplication operator M_z on $L^2(\mathbb{T})$. Then, M_z is a normal operator with spectrum equal to the unit circle $\mathbb{T}$.   ∎

We conclude this section with an immediate consequence of the continuous functional calculus.

Corollary C.1.6

Let $A \in \mathcal{L}(\mathcal{H})$ be a normal operator on a Hilbert space $\mathcal{H}$. For every polynomial p in the complex variable z, we have

$$\|p(A)\| = \sup\{|p(z)| : z \in \sigma(A)\}.$$

In particular, norm and spectral radius of a normal operator are the same.

Positive operators as directed set

Let $\mathcal{H}$ be a complex Hilbert space. The real subspace of self-adjoint operators in $\mathcal{L}(\mathcal{H})$ admits a natural order structure. Given two self-adjoint operators $A, B \in \mathcal{L}(\mathcal{H})$, we say that $A \leqslant B$ if $B - A$ is a positive operator (an operator $P \in \mathcal{L}(\mathcal{H})$ is *positive* if P is a self-adjoint operator with spectrum contained in $[0, \infty)$).

We begin with the following elementary fact:

Lemma C.1.1

For any two invertible self-adjoint operators $A, B \in \mathcal{L}(\mathcal{H})$,

$$A \leqslant B \iff B^{-1} \leqslant A^{-1}. \tag{C.1.2}$$

Proof   To see the assertion (C.1.2), note first that $B^{1/2}$ is an invertible, self-adjoint operator. It follows that

$$B^{-1/2}AB^{-1/2} \leqslant I \implies (B^{-1/2}AB^{-1/2})^{-1} \geqslant I \implies B^{1/2}A^{-1}B^{1/2} \geqslant I.$$

Multiplying $B^{-1/2}$ from left and right, we obtain the desired inequality.   □

Theorem C.1.9

Consider the collection $\mathcal{E}$ of all positive operators A in $\mathcal{L}(\mathcal{H})$ of norm less than 1. Then, for any $A, B \in \mathcal{E}$, there exists $C \in \mathcal{E}$ such that $A \leqslant C$ and $B \leqslant C$. In particular, $\mathcal{E}$ is a directed set with respect to the ordering $\leqslant$.

Proof   Let $A, B \in \mathcal{E}$. Consider the functions

$$\phi(t) = t/(1-t), \ t \in [0, 1), \quad \psi(t) = t/(1+t), \ t \geqslant 0.$$

As A and B have norm less than 1, the operator $C = \psi(\phi(A) + \phi(B))$, as given by the continuous functional calculus (of A, B and $\phi(A) + \phi(B)$), defines a positive operator in $\mathcal{L}(\mathcal{H})$ of norm less than 1. To see that $A \leqslant C$ and $B \leqslant C$, note first that $\phi(A) \leqslant \phi(A) + \phi(B)$. It then follows from (C.1.2) that

$$I - A = (\phi(A) + I)^{-1} \geqslant (I + \phi(A) + \phi(B))^{-1} = I - C.$$

This gives the inequality $A \leqslant C$. Similarly, one can see that $B \leqslant C$. $\qquad\square$

Path-connectivity of the Group of Invertible Operators

If $S \in \mathcal{L}(\mathcal{H})$ is self-adjoint, then $U = \exp(iS)$, as given by the continuous functional calculus, is unitary: If $f(z) = \exp(iz)$, then obviously $|f|(z) = |f(z)| = 1$, $z \in \mathbb{C}$. It follows that

$$U^*U = f(S)^*f(S) = |f|^2(S) = 1(S) = I.$$

Similarly, one can check that $UU^* = I$. The converse is also true.

Lemma C.1.2

An operator $U \in \mathcal{L}(\mathcal{H})$ is unitary if and only if there exists a self-adjoint $S \in \mathcal{L}(\mathcal{H})$ such that $U = \exp(iS)$.

Proof For a unitary U, consider the function $\arg : \sigma(U) \to (0, 2\pi]$ given by

$$\arg(z) = \begin{cases} \theta & \text{if } z = \exp(i\theta) \text{ with } \theta \in (0, 2\pi), \\ 2\pi & \text{if } \theta = 2\pi. \end{cases}$$

Note that arg is a bounded Borel measurable function (with the only possible discontinuity at 1) and that $\exp(i\arg(z)) = z$, $z \in \sigma(U)$. As arg is non-negative-valued, by the spectral theorem, $\arg(U)$ is positive. We claim that $U = \exp(i\arg(U))$.

Caution. By the spectral theorem, $U = (\exp \circ\, i(\arg))(U)$. But what we need is $U = \exp(i\arg(U))$!

To see the claim, define

$$f_n(z) = \sum_{k=0}^{n} \frac{(i\arg(z))^k}{k!},$$

and note that $\{f_n(z)\}_{n \in \mathbb{N}}$ converges uniformly to z for every $z \in \sigma(U)$. Hence, $\lim_{n \to \infty} \|f_n(U) - U\| = 0$. Similarly, we obtain

$$\lim_{n \to \infty} \left\| \sum_{k=0}^{n} \frac{(i\arg(U))^k}{k!} - \exp(i\arg(U)) \right\| = 0.$$

Further, for every $n \in \mathbb{N}$,

$$\sum_{k=0}^{n} \frac{(i\arg(U))^k}{k!} = \sum_{k=0}^{n} \frac{(i\arg)^k(U)}{k!} = f_n(U).$$

Thus, $U = \exp(i\arg(U))$, as desired. □

***Remark* C.1.10**   If $E(\cdot)$ denotes the spectral measure for S, then the spectral measure $F(\cdot)$ for U is given by

$$F(\omega) = E(-i\log(\omega)), \quad \omega \in B(\sigma(U)),$$

where log is a (possibly multivalued) function governed by

$$\log(\exp(i\theta)) = \begin{cases} i\theta & \text{if } \theta \in (0, 2\pi], \\ 0 & \text{if } \theta = 0. \end{cases}$$

Let V be a normed linear space with the norm $\|\cdot\|$. A subset O of V is said to be *path-connected* if for any $v_1, v_2 \in O$ there exists a map (to be referred to as *path*) $\gamma : [0, 1] \to O$ such that

(1) $\gamma(0) = v_1$ and $\gamma(1) = v_2$, and

(2) $\lim_{n\to\infty} \|\gamma(t_n) - \gamma(t)\| = 0$ whenever $\lim_{n\to\infty} |t_n - t| = 0$ for any $t, t_n \in [0, 1]$.

Let $\mathcal{U}(\mathcal{H})$ denote the set of all unitary operators in $\mathcal{L}(\mathcal{H})$.

Theorem C.1.11

$\mathcal{U}(\mathcal{H})$ *is path-connected.*

Proof   Let $U_j \in U(\mathcal{H})$ and write $U_j = \exp(iS_j)$ for self-adjoint operator S_j, $j = 1, 2$ as ensured by Lemma C.1.2. Define $\gamma : [0, 1] \to \mathcal{U}(\mathcal{H})$ by

$$\gamma(t) = \exp(i(1 - t)S_1)\exp(itS_2), \quad t \in [0, 1].$$

Note that γ is well-defined. Obviously, $\gamma(0) = U_1$, $\gamma(1) = U_2$. It now suffices to check that for any $t \in [0, 1]$,

$$\lim_{n\to\infty} \|\exp(it_n\arg(U)) - \exp(it\arg(U))\| = 0$$

whenever $\{t_n\}_{n\in\mathbb{N}} \subset [0, 1]$ converges to $t \in [0, 1]$. By the continuous functional calculus for U,

$$
\begin{aligned}
\|\exp(it_n\arg(U)) - \exp(it\arg(U))\| &= \|\exp(it_n\arg) - \exp(it\arg)\|_\infty \\
&= |\exp(it_n\arg(z_0)) - \exp(it\arg(z_0))|
\end{aligned}
$$

for some $z_0 \in \sigma(U)$. Hence $\lim_{n\to\infty} \|\exp(it_n\arg(U)) - \exp(it\arg(U))\| = 0$. □

For an invertible $T \in \mathcal{L}(\mathcal{H})$, consider the operators P and U given by

$$P = (T^*T)^{\frac{1}{2}} \text{ and } U = TP^{-1}. \tag{C.1.3}$$

Note that P is a positive invertible operator in $\mathcal{L}(\mathcal{H})$. Clearly, U is invertible. Moreover, as

$$U^*U = P^{-1}T^*TP^{-1} = (T^*T)^{-\frac{1}{2}}T^*T(T^*T)^{-\frac{1}{2}} = I,$$

U is unitary. Thus, *every invertible $T \in \mathcal{L}(\mathcal{H})$ can be decomposed uniquely as $T = UP$ with positive invertible P and unitary U.* This is a special case of the polar decomposition of an operator as seen in Chapter 1.

Let $\mathcal{G}(\mathcal{H})$ denote the set of all invertible operators in $\mathcal{L}(\mathcal{H})$. Note that $\mathcal{G}(\mathcal{H})$ is a group under the operation of composition and that $\mathcal{U}(\mathcal{H}) \subsetneq \mathcal{G}(\mathcal{H}) \subsetneq \mathcal{L}(\mathcal{H})$.

Corollary C.1.7

$\mathcal{G}(\mathcal{H})$ *is path-connected.*

Proof   It suffices to show that there exists a path joining any $T \in \mathcal{G}(\mathcal{H})$ and the identity operator I. Let $T \in \mathcal{G}(\mathcal{H})$ and consider its polar decomposition $T = UP$ (see the discussion prior to Corollary C.1.7). Let $\gamma : [0,1] \to \mathcal{U}(\mathcal{H})$ be a path joining U and I (see Theorem C.1.11). Define $\delta : [0,1] \to \mathcal{G}(\mathcal{H})$ by

$$\delta(t) = \gamma(t)((1-t)P + tI), \quad t \in [0,1].$$

Note that δ is well-defined as $(1-t)P + tI$ is invertible in view of

$$\sigma((1-t)P_1 + tI) = \{(1-t)\lambda + t : \lambda \in \sigma(P)\}$$

and $\sigma(P) \subset (0,\infty)$. Clearly, the continuity of δ follows from that of γ. As $\delta(0) = T$ and $\delta(1) = I$, the proof is over.   □

Corollary C.1.8

For a nonzero Hilbert space $\mathcal{H}$, let $S_{\mathcal{H}}$ denote the unit sphere $\{h \in \mathcal{H} : \|h\| = 1\}$ of $\mathcal{H}$. Then, the unit sphere $S_{\mathcal{H}}$ of $\mathcal{H}$ is path-connected.

Proof   Let $h_1, h_2 \in S_{\mathcal{H}}$. By the Gram–Schmidt orthonormalization process, there exist orthonormal bases $\{h_1, e_1, \ldots\}$ and $\{h_2, f_1, \ldots\}$ of $\mathcal{H}$. Set

$$Uh = \begin{cases} h_2 & \text{if } h = h_1 \\ f_j & \text{if } h = e_j, \ j \geqslant 1. \end{cases}$$

Then, U extends to a bounded linear unitary operator on $\mathcal{H}$ such that $Uh_1 = h_2$. By Lemma C.1.2, one may write $U = \exp(iS)$ for some self-adjoint S. Now define $\gamma : [0,1] \to S_{\mathcal{H}}$ by $\gamma(t) = \exp(itS)h_1, t \in [0,1]$, and note that $\gamma(0) = h_1$ and $\gamma(1) = h_2$.   □

Putnam–Fuglede Theorem

In this section, we discuss the Putnam–Fuglede theorem as another application of the spectral theorem. The following measure-theoretic proof is taken from [135].

Theorem C.1.12

Let $N, T \in \mathcal{L}(\mathcal{H})$ be such that $TN = NT$. If N is normal, then $Tf(N) = f(N)T$ for every $f \in B_\infty(\sigma(N))$.

Proof   Assume that N is normal with spectral measure $E(\cdot)$ (extended trivially to the entire complex plane $\mathbb{C}$). It suffices to check that every Borel set ω of $\mathbb{C}$, $E(\omega)T = TE(\omega)$ (see the proof of Corollary C.1.5). Consider the collection

$$\mathscr{F} = \{\omega \in B(\mathbb{C}) : T(E(\omega)\mathcal{H}) \subseteq E(\omega)\mathcal{H}\},$$

where $B(\mathbb{C})$ is the σ-algebra of all Borel subsets of $\mathbb{C}$. Note that if ω and $\mathbb{C}\backslash\omega$ belong to $\mathscr{F}$, then $E(\omega)$ commutes with T. Hence, it enough to verify that $\mathscr{F} = B(\mathbb{C})$. We divide the proof into several steps.

Step I   *The closed unit disc* $\overline{\mathbb{D}} = \{z \in \mathbb{C} : |z| \leqslant 1\}$ *belongs to* $\mathscr{F}$:

Let $x \in \mathcal{H}$ and write $T(E(\overline{\mathbb{D}})x) = E(\overline{\mathbb{D}})y + E(\sigma(N)\backslash\overline{\mathbb{D}})z$ for some $y, z \in \mathcal{H}$. As N commutes with T and $E(\cdot)$, for any positive integer n,

$$TN^n(E(\overline{\mathbb{D}})x) = N^n(E(\overline{\mathbb{D}})y) + N^n E(\sigma(N)\backslash\overline{\mathbb{D}})z.$$

Since $\sup_{n\geqslant 1} \|N^n E(\overline{\mathbb{D}})\| < \infty$, we must have $\sup_{n\geqslant 1} \|N^n(E(\sigma(N)\backslash\overline{\mathbb{D}})z)\| < \infty$. However, for every positive integers m, n,

$$\begin{aligned}
\sup_{n\geqslant 1}\|N^n(E(\sigma(N)\backslash\overline{\mathbb{D}})z)\| &\geqslant \int_{|t|>1+1/m} |t^n|^2 \langle dE(t)z, z\rangle \\
&\geqslant (1+1/m)^n \langle E(\{z \in \sigma(N) : |z| > 1+1/m\})z, z\rangle,
\end{aligned}$$

which is possible only if $E(\sigma(N)\backslash\overline{\mathbb{D}})z = 0$. This shows that $T(E(\overline{\mathbb{D}})x) = E(\overline{\mathbb{D}})y$ belongs to $E(\overline{\mathbb{D}})\mathcal{H}$.

Step II   *Any closed disc* $\overline{\mathbb{D}}_r(z_0) = \{z \in \mathbb{C} : |z - z_0| \leqslant r\}$ *belongs to* $\mathscr{F}$:

We may apply Step I to the normal operator $f(N)$ with $f(t) = (t - z_0)/r$ to conclude that $T(F(\overline{\mathbb{D}})\mathcal{H}) \subseteq F(\overline{\mathbb{D}})\mathcal{H}$, where $F(\cdot) = E(f^{-1}(\cdot))$ is the spectral measure of $f(N)$. However, $f^{-1}(\overline{\mathbb{D}}) = \overline{\mathbb{D}}_r(z_0)$.

Step III   *If* $\omega_n \in \mathscr{F}$, $n \in \mathbb{N}$ *such that* $\omega_n \uparrow \omega$ (*resp.* $\omega_n \downarrow \omega$), *then* $\omega \in \mathscr{F}$:

By assumption,

$$T(E(\omega_n)\mathcal{H}) \subseteq E(\omega_n)\mathcal{H}, \quad n \in \mathbb{N}.$$

As $E(\omega_n) \uparrow E(\omega)$ (by the monotone convergence theorem), by the continuity of T, we obtain that $\omega \in \mathscr{F}$. Similarly, one can see the remaining part.

Step IV   *For any real numbers* a, b, c, d *such that* $0 \leqslant a < b$ *and* $0 \leqslant c < d$, *the rectangle* $R_{a,b,c,d} = \{z \in \mathbb{C} : a < \Re z \leqslant b, \ a < \Im z \leqslant b\}$ *belongs to* $\mathscr{F}$.

By Steps II and III, any open disc $\mathbb{D}_r(z_0)$ belongs to $\mathscr{F}$. Further, as $R_a = \{z \in \mathbb{C} : \Re z > a\}$ being the union $\cup_{n=1}^{\infty} \mathbb{D}_n(a + n)$, by Step III, R_a belongs to $\mathscr{F}$. Similarly, one can check that the half planes of the form

$$L_b = \{z \in \mathbb{C} : \Re z \leqslant b\}, \ U_c = \{z \in \mathbb{C} : \Im z > c\}, \ L_d = \{z \in \mathbb{C} : \Im z \leqslant d\}$$

belong to $\mathscr{F}$. By the multiplicativity of the spectral measure, $R_{a,b,c,d} = R_a \cap L_b \cap U_c \cap L_d$ belongs to $\mathscr{F}$.

To complete the proof, consider the collection $\mathscr{A}$ of all finite disjoint unions of rectangles of the form $R_{a,b,c,d}$. As $\mathscr{A}$ is an algebra of sets that generates $B(\mathbb{C})$ as a σ-algebra, and since the collection $\mathscr{F}$ is a monotone class containing $\mathscr{A}$, the desired conclusion is now immediate from the monotone class theorem (refer to [133]). □

Note that the last result generalizes the square root lemma: *If a positive operator commutes with a bounded linear operator, then so does its square root.* Here is another consequence.

Corollary C.1.9 (Putnam–Fuglede–Rosenblum Theorem)

Let $M, N, T \in \mathcal{L}(\mathcal{H})$ be such that M, N are normal operators. If $MT = TN$, then $M^*T = TN^*$. In particular, any commuting family $\mathcal{F}$ of normal operator in $\mathcal{L}(\mathcal{H})$ is doubly commuting in the sense that $NM^* = M^*N$ for every $M, N \in \mathcal{F}$.

Proof The following matrix-trick is due to Berberian [36, pp. 279]. Consider the operators on $\mathcal{H} \oplus \mathcal{H}$ given by

$$L = \begin{pmatrix} N & 0 \\ 0 & M \end{pmatrix} \quad \text{on} \quad \mathcal{K} = \mathcal{H} \oplus \mathcal{H},$$

$$T' = \begin{pmatrix} 0 & T \\ 0 & 0 \end{pmatrix} \quad \text{on} \quad \mathcal{K} = \mathcal{H} \oplus \mathcal{H}.$$

Check that $T'L = LT'$ and that L is normal. Now apply Theorem C.1.12. □

By Gelfand's theorem [35], any abelian unital C^*-algebra is isometrically *-isomorphic to $C(X)$ for some compact Hausdorff space X. In view of this, Corollary C.1.2 (with almost the same proof) actually extends to a commuting normal operator $N_1, \ldots, N_n$ in $\mathcal{L}(\mathcal{H})$ (even to an arbitrary family of commuting normal operators in $\mathcal{L}(\mathcal{H})$). The space X is referred to as the *maximal ideal space* X of $C^*(N) = C^*(\{N_1, \ldots, N_n\})$ (which is abelian by Corollary C.1.9). If $n \in \mathbb{N}$, then the unital *-isometric isomorphism $\phi : C^*(N) \to C(X)$ governs $\phi(N_j) = z_j$, where $X = \sigma(N_1) \times \cdots \times \sigma(N_n) \subseteq \mathbb{C}^n$ and z_j denotes the coordinate function on X, $j = 1, \ldots, d$.

Corollary C.1.10

Suppose $M, N, T \in \mathcal{L}(\mathcal{H})$, M, N are normal, T is invertible such that $M = TNT^{-1}$. If $T = U|T|$ is the polar decomposition of T (see (C.1.3)), then $M = UNU^{-1}$.

Proof Note that $M = U(|T|N|T|^{-1})U^{-1}$. Thus, it suffices to check that $|T|N = N|T|$. However, by the preceding corollary, $M^*T = TN^*$. It follows that $T^*TN = T^*MT = NT^*T$. Thus, N commutes with any polynomial in T^*T, and hence, by the continuous functional calculus, it commutes with $|T|$. □

Corollary C.1.11

If $N, M \in \mathcal{L}(\mathcal{H})$ are essentially normal commuting operators, then the cross-commutator $[N, M] = N^*M - MN^*$ is compact.

Proof If π denotes the Calkin map, then the desired conclusion may be derived from Corollary C.1.9 applied to the commuting normal operators $\pi(N)$ and $\pi(M)$ (see Remark 1.2.6). □

Outline of the Proof of the Spectral Theorem

The proof of the spectral theorem involves several facts pertaining to normal operators. We begin with the continuous functional calculus for self-adjoint operators.

Lemma C.1.3

If A is a self-adjoint operator in $\mathcal{L}(\mathcal{H})$, then there is a unique positive, isometric, algebraic *-homomorphism map $\phi : C(\sigma(A)) \to \mathcal{L}(\mathcal{H})$ such that $\phi(p) = p(A)$ for every polynomial $p \in C(\sigma(A))$.

Proof By the C^*-algebra identity and Lemma 5.2.1,

$$\|p(A)\|^2 = \|p(A)^* p(A)\| = \||p|^2(A)\| = r(|p|^2(A)).$$

By Lemma B.1.10, $\sigma(|p|^2(A)) = |p|^2(\sigma(A))$, and hence, $r(|p|^2(A)) = |p(z_0)|^2$ for some $z_0 \in \sigma(A)$. Thus, $\|p(A)\| = \|p\|_{\infty,\sigma(A)}$ showing that $\phi(p) = p(A)$ defines a positive, isometric, algebraic *-homomorphism. Now apply the Stone–Weierstrass Theorem (see Appendix A) to extend ϕ isometrically to $C(\sigma(A))$. □

Recall that a bounded linear operator T on a Hilbert space $\mathcal{H}$ is *cyclic* if there exists a vector $h \in \mathcal{H}$ (to be referred to as a *cyclic vector*) such that

$$\bigvee \{T^n h : n \text{ is a non-negative integer}\} = \mathcal{H}.$$

Lemma C.1.4

Let A be a self-adjoint operator in $\mathcal{L}(\mathcal{H})$. If A is cyclic with cyclic vector $f \in \mathcal{H}$, then there exists a finite positive Borel measure μ_f and a unitary operator $U : \mathcal{H} \to L^2(\sigma(A), d\mu_f)$ such that

$$(UAU^{-1}g)(\lambda) = \lambda g(\lambda), \quad g \in L^2(\sigma(A), \mu_f).$$

Proof Let ϕ be as given in Lemma C.1.3. Consider the bounded linear functional $\psi : C(\sigma(A)) \to \mathbb{C}$ by

$$\psi(g) = \langle \phi(g)f, f \rangle, \quad g \in C(\sigma(A)).$$

By the Riesz representation theorem (see Theorem B.1.14), there exists a finite positive Borel measure μ_f on $\sigma(A)$ such that

$$\psi(g) = \int_{\sigma(A)} g(t) d\mu_f(t), \quad g \in C(\sigma(A)).$$

Define U by $U\phi(g)f = g$ for $g \in C(\sigma(A))$, and note that

$$\|\phi(g)f\|^2 = \langle \phi(|g|^2)f, f \rangle = \int_{\sigma(A)} |g(t)|^2 d\mu_f(t) = \|g\|^2.$$

As f is cyclic for A, U extends isometrically from $\mathcal{H}$ into $L^2(\sigma(A), \mu_f)$. Since the range of U contains continuous functions, U is surjective. Note that for any $g \in L^2(\sigma(A), \mu_f)$ and $\lambda \in \sigma(A)$,

$$(UAU^{-1}g)(\lambda) = (UA\phi(g)f)(\lambda) = (U\phi(zg)f)(\lambda) = \lambda g(\lambda).$$

This completes the proof. □

Lemma C.1.5

Let $\mathcal{H}$ be a separable Hilbert space. If $A \in \mathcal{L}(\mathcal{H})$, then there exists an orthonormal family $\{g_j\}_{j=1}^N$ with $N \in \mathbb{N}$ or $N = \infty$, such that $\mathcal{H} = \oplus_{j=1}^N \mathcal{H}_j$, where

$$\mathcal{H}_j = \bigvee \{A^{*k}A^l g_j : k,l \in \mathbb{N}\}, \quad j = 1,\dots,N.$$

Proof Let $\{e_j\}_{j\in\mathbb{N}}$ be an orthonormal basis of $\mathcal{H}$. Let $g_1 = e_1$. If $\mathcal{H}_1 = \mathcal{H}$, then let $N = 1$. Otherwise, let k_1 be the smallest positive integer such that $P_{\mathcal{H}_1^\perp} e_{k_1} \neq 0$ and let $g_2 = \dfrac{P_{\mathcal{H}_1^\perp} e_{k_1}}{\|P_{\mathcal{H}_1^\perp} e_{k_1}\|}$. As $\langle g, g_2 \rangle = 0$ for every $g \in \mathcal{H}_1$, the spaces $\mathcal{H}_1$ and $\mathcal{H}_2$ are orthogonal. Further, $\{e_j\}_{j=1}^{k_1} \subseteq \mathcal{H}_1 \oplus \mathcal{H}_2$. Now proceed by induction. $\qquad\square$

Proposition C.1.1 (Spectral Theorem for Self-adjoint Operators)

Let $\mathcal{H}$ be a separable Hilbert space and let $A \in \mathcal{L}(\mathcal{H})$. If A is a self-adjoint operator, then there exist finite positive Borel measures $\mu_1,\dots,\mu_N$ ($N \in \mathbb{N}$ or $N = \infty$) and a unitary operator $U : \mathcal{H} \to \oplus_{n=1}^N L^2(\sigma(A),\mu_n)$ such that

$$(UAU^{-1}g)_n(\lambda) = \lambda g_n(\lambda), \quad g = (g_n)_{n=1}^N \in \oplus_{n=1}^N L^2(\sigma(A),\mu_n).$$

In particular, there exists a unique positive, contractive, algebraic *-homo-morphism $\phi : B_\infty(\sigma(A)) \to \mathcal{L}(\mathcal{H})$ such that

$$\phi(f) = f(A), \quad f \in C(\sigma(A)) \text{ is a polynomial.}$$

Moreover, ϕ is isometric on $C(\sigma(A))$. $\qquad\square$

Proof By Lemma C.1.5, there exist invariant subspaces $\mathcal{H}_1, \mathcal{H}_2, \dots, \mathcal{H}_N$ of A such that $\mathcal{H} = \oplus_{n=1}^N \mathcal{H}_n$ and $A|_{\mathcal{H}_n}$ is cyclic. The first part now follows from Lemma C.1.4. To see the second part, let $f_n \in L^2(\sigma(A),\mu_n)$ such that $\|f_n\| = 2^{-n}$, and let M denote the disjoint union of N copies of $\sigma(A)$. If μ be the restriction of μ_n to nth copy of $\sigma(A)$, then $\langle M, \mu \rangle$ is the desired finite measure space. Moreover, A is unitarily equivalent to the operator M_λ of multiplication by λ on $L^2(M,\mu)$. The Borel functional calculus of A now follows from that of M_λ. This completes the proof. $\qquad\square$

Lemma C.1.6

If A_1, A_2 are commuting self-adjoint operators in $\mathcal{L}(\mathcal{H})$, then for any bounded Borel measurable functions f, g on $\mathbb{R}$,

$$f(A_1)g(A_2) = g(A_2)f(A_1).$$

Proof If A_1 and A_2 commute then so do polynomials in A_1 and A_2. Now use the bounded Borel functional calculus for self-adjoint operators, as established in the previous proposition. $\qquad\square$

Theorem C.1.13 (Spectral Theorem for Normal Operators)

*If $A \in \mathcal{L}(\mathcal{H})$ is a normal operator on a separable Hilbert space $\mathcal{H}$, then there exists a unique, positive, contractive, algebraic *-homomorphism map $\phi : B_\infty(\sigma(A)) \to \mathcal{L}(\mathcal{H})$ such that $\phi(f) = f(A)$ for every polynomial $f \in C(\sigma(A))$. Moreover, ϕ is isometric on $C(\sigma(A))$.*

Proof   Write $N = A_1 + iA_2$ for commuting self-adjoint operators $A_1, A_2 \in \mathcal{L}(\mathcal{H})$. For a Borel subset Ω of $\mathbb{R}$, consider the orthogonal projection $P_\Omega(A_i) = \chi_\Omega(A_i)$ for $i = 1, 2$. By the preceding lemma, $P_\Omega(A_1)$ and $P_\Omega(A_2)$ are commuting.

Let $f : \mathbb{R}^2 \to \mathbb{C}$ be a linear combination of characteristic functions of measurable rectangles (that is, sets of form $\Omega = \Omega_1 \times \Omega_2$, where Ω_1 and Ω_2 are Borel subsets of $\mathbb{R}$). As union of measurable rectangles can be expressed as disjoint union of measurable rectangles, f can be rewritten as

$$f = \sum_{i=1}^{2} c_i \chi_{\Omega^{(i)}} \text{ with } \Omega^{(i)} \cap \Omega^{(j)} = \emptyset \text{ if } i \neq j,$$

where $\Omega^{(i)} = \Omega_1^{(i)} \times \Omega_2^{(i)}$, $i = 1, 2$, are measurable rectangles. Define $f(A_1, A_2)$ by

$$f(A_1, A_2) = \sum_{i=1}^{2} c_i P_{\Omega_1^{(i)}}(A_1) P_{\Omega_2^{(i)}}(A_2).$$

Note that $f(A_1, A_2)$ is well-defined. Indeed, $f(A_1, A_2) = 0$ if and only if $c_1 = 0, c_2 = 0$. Further, for any $h \in \mathcal{H}$,

$$\begin{aligned}
\|f(A_1, A_2)h\|^2 &= \sum_{i=1}^{2} |c_i|^2 \left\| P_{\Omega_1^{(i)}}(A_1) P_{\Omega_2^{(i)}}(A_2)h \right\|^2 \\
&\leqslant (\max_{i=1}^{2} |c_i|)^2 \sum_{i=1}^{2} \left\| P_{\Omega_1^{(i)}}(A_1) P_{\Omega_2^{(i)}}(A_2)h \right\|^2 \\
&= (\max_{i=1}^{2} |c_i|)^2 \left\| \sum_{i=1}^{2} P_{\Omega_1^{(i)}}(A_1) P_{\Omega_2^{(i)}}(A_2)h \right\|^2.
\end{aligned}$$

As $\max_{i=1}^{2} |c_i| = \|f\|_\infty$, it follows that

$$\|f(A_1, A_2)\| \leqslant \|f\|_\infty.$$

Recall that the co-ordinate function z can be uniformly approximated by the simple Borel measurable functions on any compact subset of $\mathbb{C}$ ([120, Proof of Theorem 1.17]). Hence, by the Stone–Weierstrass theorem, any continuous function on $[-\|A_1\|, \|A_1\|] \times [-\|A_2\|, \|A_2\|]$ can be approximated uniformly by simple functions, and hence, we can extend this functional calculus to a continuous functional calculus with the same bound. We can now argue as in the self-adjoint case to obtain spectral theorem for commuting self-adjoint operators A_1 and A_2. We leave the details to the reader.   □

The Multiplicity Function and the Spectral Theorem

Let (Λ, m) be a measure space and for $\lambda \in \Lambda$, let $\mathcal{H}_\lambda$ be a nonzero separable Hilbert space. A *section* is a map $s : \Lambda \to \bigcup_{\lambda \in \Lambda} \mathcal{H}_\lambda$ such that $s(\lambda) \in \mathcal{H}_\lambda$. We will denote the linear space of all sections by $\mathcal{S}$. We adopt the following definition from [55].

Definition C.1.14

The pair $(\{\mathcal{H}_\lambda\}_{\lambda \in \Lambda}, \Gamma)$ is said to be a *measurable field of Hilbert spaces* if Γ is a linear subspace of $\mathcal{S}$ such that

(i) for each $s \in \Gamma$, the function $\lambda \to \|s(\lambda)\|$ is measurable,

(ii) if s_0 is in $\mathcal{S}$ and for every $s \in \Gamma$, the function $\lambda \to \langle s_0(\lambda), s(\lambda) \rangle$ is measurable then s_0 is in Γ,

(iii) there exists a sequence $\{s_n\}_{n \geqslant 0} \subseteq \Gamma$ such that $\{s_n(\lambda)\}_{n \geqslant 0}$ spans $\mathcal{H}_\lambda$ for each $\lambda \in \Lambda$.

Mackey [92, pp. 91] calls a sequence $\{s_n\}_{n \geqslant 0}$ satisfying (iii) a *pervasive* sequence. It can be shown that the existence of a pervasive sequence is equivalent to the measurability of the extended integer valued function d on Λ defined by $d(\lambda) = \dim \mathcal{H}_\lambda$. The direct integral $\int_\Lambda \oplus \mathcal{H}_\lambda \, dm$ is the obvious Hilbert space of the set of sections s in Γ such that $\int_\Lambda \|s(\lambda)\|^2 \, dm$, where m is a Borel measure defined on Λ, is finite (two such sections are identified if they are almost everywhere (m) equal). We refer the reader to [55] for further details.

Suppose for each $\lambda \in \Lambda$, we have an operator $T(\lambda)$ on $\mathcal{H}_\lambda$ such that

- the function $\lambda \to \langle T(\lambda) s_1, s_2 \rangle$ is measurable for each pair of sections $s_1, s_2 \in \int_\Lambda \oplus \mathcal{H}_\lambda dm$,

- ess sup $\|T(\lambda)\| < \infty$.

We then define $\int_\Lambda \oplus T(\lambda)$, the direct integral of $\{T(\lambda)\}$ by the formula

$$\left(\left(\int_\Lambda \oplus T(\lambda) \right) s \right)(\lambda) = T(\lambda) s(\lambda), \quad s \in \int_\Lambda \oplus \mathcal{H}_\lambda dm; \quad \lambda \in \Lambda. \tag{C.1.4}$$

In particular, define the multiplication operator $M_\varphi : \int_\Lambda \oplus \mathcal{H}_\lambda \, dm \to \int_\Lambda \oplus \mathcal{H}_\lambda \, dm$, where φ is a bounded measurable function on Λ, by setting

$$(M_\varphi s)(\lambda) = \varphi(\lambda) s(\lambda), \quad s \in \int_\Lambda \oplus \mathcal{H}_\lambda \, dm.$$

Example C.1.15 Let (X_n, μ_n) be a measure space and $\mathcal{H}_n$ be a non-zero Hilbert space, $n \in \mathbb{N}$. Define $L^2(X_n, \mu_n; \mathcal{H}_n)$ to be the set of all weakly measurable functions $f : X_n \to H_n$ for which

$$\|f\|^2 = \int \|f(x)\|_{\mathcal{H}_n}^2 d\mu_n$$

is finite, Assume that $X_m \cap X_n = \emptyset$, $m \neq n$, and set $\Lambda := \bigcup \{X_n : n \in \mathbb{N}\}$. Let μ be the measure on Λ such that $\mu_{|X_n} = \mu_n$. The space

$$L^2(\Lambda, \mu; \mathcal{H}) := \bigoplus_{n \in \mathbb{N}} L^2(X_n, \mu_{|X_n}; \mathcal{H}_n)$$

is an example of a direct integral space. The operator M_φ of multiplication by any bounded measurable function φ on Λ defines a normal operator on $L^2(\Lambda, \mu; \mathcal{H})$.   ∎

Now, the various entities involved in defining the direct integral space actually determine it, modulo unitary equivalence, as a Hilbert module over the algebra $C(\Lambda)$. This is one form of the spectral theorem, which is stated as a remark at the bottom of page 846 in [1], and reproduced here.

Theorem C.1.16 (Invariants)

To each normal operator N in $\mathcal{L}(\mathcal{H})$, one can associate a compact set Λ in the plane, an equivalence class of Borel measures m on Λ and an m-measurable function $d : \Lambda \to \{0, 1, \ldots, \infty\}$ so that the triple $\{\Lambda, m, d\}$ forms a complete set of unitary invariants for the operator N.

Now any normal operator N is unitarily equivalent to a multiplication operator M_ϕ on the Hilbert space $L^2(X, \mu)$ for some ϕ in $L^\infty(X, \mu)$. Therefore to decide when two normal operators are unitarily equivalent, we just have to find its "direct integral" representation. This is [1, Theorem 3], which we reproduce as follows.

Set $\nu := \mu \circ \phi^{-1}$ to be the measure: $\nu(A) = \mu(\phi^{-1}(A))$ for any Borel subset A of $Y := \phi(X)$. The measure ν is the scalar spectral measure of the operator M_ϕ. Moreover, let $y \to \mu_y$ be a disintegration of the measure μ relative to the map $\phi : X \to Y$. In addition, if μ'_y is another disintegration of the measure μ relative to ϕ, then $\mu_y = \mu'_y$ - ν a.e. (refer to [1, Section 2]).

Theorem C.1.17 (Model)

A direct integral representation for the operator M_ϕ acting on $L^2(X, \mu)$ is the operator M of multiplication by the coordinate function y on the direct integral space $\int_Y \oplus L^2(\mu_y) d\nu$.

Clearly, combining the Theorems C.1.16 and Theorem C.1.17, we have the following corollary, see [55, Theorem 1.24].

Corollary C.1.12

Two multiplication operators M_ϕ and M_ψ on $L^2(Y, \mu)$ and $L^2(Y', \mu')$ represented as $\int_Y \oplus L^2(\mu_y) d\nu$ and $\int_{Y'} \oplus L^2(\mu'_y) d\nu'$ are unitarily equivalent if and only if

(1) $Y = Y'$, *that is, the spectra are equal;*

(2) $\dim L^2(\mu_y) = \dim L^2(\mu'_y)$, *that is, the multiplicities match; and*

(3) $\nu \sim \nu'$, *that is, the scalar spectral measures ν and ν' are mutually absolutely continuous.*

References

[1] M. B. Abrahamse and T. L. Kriete, Spectral multiplicity of a multiplication operator. *Indiana University Mathematics Journal*, **22** (1973), 845–857.

[2] J. Agler, A disconjugacy theorem for Toeplitz operators. *American Journal of Mathematics*, **112** (1990), 1–14.

[3] J. Agler and M. Stankus, m-isometric transformations of Hilbert spaces I. *Integral Equations and Operator Theory*, **21** (1995), 383–429.

[4] J. Agler and M. Stankus, m-isometric transformations of Hilbert spaces II. *Integral Equations and Operator Theory*, **23** (1995), 1–48.

[5] J. Agler and M. Stankus, m-isometric transformations of Hilbert spaces III. *Integral Equations and Operator Theory*, **24** (1996), 379–421.

[6] J. Anderson, A C^*-algebra $\mathcal{A}$ for which Ext($\mathcal{A}$) is not a group. *Annals of Mathematics*, **107** (1978), 455–458.

[7] P. Ara, F. Lledó and D. Yakubovich, Følner sequences in operator theory and operator algebras. In *Operator Theory, Operator Algebras and Applications* (Basel: Birkhauser/Springer, 2014), pp. 1–24.

[8] J. Arazy, A survey of invariant Hilbert spaces of analytic functions on bounded symmetric domains. *Contemporary Mathematics*, **185** (1995), 7–7.

[9] W. Arveson, A note on essentially normal operators. Proceedings of the Royal Irish Academy Section A **74** (1974), 143–146.

[10] W. Arveson, Notes on extensions of C^*-algebras. *Duke Mathematical Journal*, **44** (1977), 329–355.

[11] W. Arveson, Subalgebras of C^*-algebras, III, Multivariable operator theory. *Acta Mathematica*, **181** (1998), 159–228.

[12] W. Arveson, The Dirac operator of a commuting d-tuple. *Journal of Functional Analysis*, **189** (2002), 53–79.

[13] W. Arveson, p-summable commutators in dimension d. *Journal of Operator Theory*, **54** (2005) No. 1, 101–117.

[14] W. Arveson, Quotients of standard Hilbert modules. *Transactions of the American Mathematical Society*, **359** (2007), 6027–6055.

[15] W. Arveson, D. Hadwin, T. Hoover and E. Kymala, Circular operators. *Indiana University Mathematics Journal,* **33** (1984), 583–595.

[16] M. F. Atiyah, Algebraic topology and operators in Hilbert space. In *Lectures in Modern Analysis and Applications I* (Berlin, Heidelberg: Springer, 1969), pp. 101–121.

[17] S. Axler. *Linear Algebra Done Right,* Third Edition, Undergraduate Texts in Mathematics (Cham, Heidelberg: Springer, 2015).

[18] B. Bagchi and G. Misra, Constant characteristic functions and homogeneous operators, *Journal of Operator Theory,* **37** (1997), 51–65.

[19] B. Bagchi and G. Misra, Homogeneous operators and projective representations of the Mobius group: a survey. *Proceedings of the Indian Academy of Sciences (Mathematical Sciences),* **111** (2001), 415–437.

[20] I. D. Berg, An extension of the Weyl-von Neumann theorem to normal operators. *Transactions of the American Mathematical Society,* **160** (1971), 365–371.

[21] I. D. Berg and K. R. Davidson, Almost commuting matrices and the Brown–Douglas–Fillmore theorem. *Bulletin (New Series) of the American Mathematical Society,* **16** (1987), 97–100.

[22] I. D. Berg and K. R. Davidson, Almost commuting matrices and a quantitative version of the Brown–Douglas–Fillmore theorem. *Acta Mathematica,* **166** (1991), 121–161.

[23] S. Biswas, O. Shalit, Stable division and essential normality: the non-homogeneous and quasi homogeneous cases. *Indiana University Mathematics Journal,* **67** (2018), 169–185.

[24] B. Blackadar. *K-theory for Operator Algebras,* Second edition. Mathematical Sciences Research Institute Publications, (Cambridge: Cambridge University Press, 1998).

[25] L. G. Brown, R. G. Douglas and P. A. Fillmore, Extensions of C^*-algebras, operators with compact self-commutators, and K-homology. *Bulletin of the American Mathematical Society,* **79** (1973), 973–978.

[26] L. G. Brown, R. G. Douglas and P. A. Fillmore, Unitary equivlence moduló the compact operators and extensions of C^*-algebras. *Proceedings of a Conference on Operator Theory,* Springer Lecture Notes, **345** (1973), 58–128.

[27] L. G. Brown, R. G. Douglas and P. A. Fillmore, Extensions of C^*-algebras, and K-homology. *Annals of Mathematics,* **105** (1977) No. 2, 265–324.

[28] J. Bunce, The joint spectrum of commuting nonnormal operators. *Proceedings of the American Mathematical Society,* **29** (1971) 499–505.

[29] S. Chavan, On operators Cauchy dual to 2-hyperexpansive operators. *Proceedings of the Edinburgh Mathematical Society,* **50** (2007), 637–652.

[30] S. Chavan, Essential normality of operators close to isometries. *Integral Equations Operator Theory,* **73** (2012) No. 1, 49–55.

[31] S. Chavan and D. Yakubovich, Spherical tuples of Hilbert space operators. *Indiana University Mathematics Journal,* **64** (2015), 577–612.

[32] M. D. Choi and E. G. Effros, The completely positive lifting problem for C^*-algebras. *Annals of Mathematics,* **104** (1976), 585–609.

[33] K. Clancey. *Seminormal Operators*, 742 Lecture Notes in Mathematics, (Berlin: Springer, 1979).

[34] J. Conway. *The Theory of Subnormal Operators*, Mathematical Surveys and Monographs, Vol. 36, (Providence, RI: American Mathematical Society, 1991).

[35] J. Conway. *A Course in Functional Analysis*, (New York: Springer-Verlag, 1997).

[36] J. Conway. *A Course in Operator Theory*, Vol. 21 of Graduate Studies in Mathematics, (Providence, RI: American Mathematical Society, 2000).

[37] J. Cuntz, R. Meyer and J. M. Rosenberg. *Topological and Bivariant K-theory*, (Birkhäuser: Verlag, 2007).

[38] R. E. Curto, Fredholm and invertible n-tuples of operators: The deformation problem. *Transactions of the American Mathematical Society,* **266** (1981), 129–159.

[39] R. E. Curto, Applications of several complex variables to multiparameter spectral theory. In *Surveys of Some Recent Results in Operator Theory,* Vol II, (Harlow: Pitman Research Notes in Mathematics Series 192, Longman Sci. Tech., 1988), pp. 25–90.

[40] K. R. Davidson. *C*-algebras by Example*, Fields Institute Monographs, Vol. 6, (Providence, RI: American Mathematical Society, 1996).

[41] K. R. Davidson, Essentially normal operators. *A Glimpse at Hilbert Space Operators,* (Basel: Springer, 2010), pp. 209–222.

[42] A. M. Davie, Classification of essentially normal operators. In *Spaces of Analytic Functions (Sem. Functional Anal. and Function Theory, Kristiansand, 1975),* (Berlin, Heidelberg: Springer. Lecture Notes in Math., Vol. 512, 1976), pp. 31–55.

[43] J. A. Deddens and J. G. Stampfli, On a question of Douglas and Fillmore. *Bulletin of the American Mathematical Society,* **79** (1973), 327–330.

[44] R. G. Douglas. *Banach Algebra Techniques in Operator Theory*, (New York: Academic Press, 1972).

[45] R. G. Douglas. *Banach Algebra Techniques in the Theory of Toeplitz Operators*, CBMS, No. 15 (Providence, RI: American Mathematical Society, 1973).

[46] R. G. Douglas, The relation of Ext to K-theory. *Symposia Mathematica,* **20** (1976), 513–529.

[47] R. G. Douglas. *C*-algebra extensions and K-homology*, (New Jersey: Princeton University Press, 1980).

[48] R. G. Douglas, Evolution of Modern Analysis: In Shing-Tung Yau, ed., *The Founders of Index Theory: Reminiscences of Atiyah, Bott, Hirzebruch and Singer,* (Massachusetts: International Press of Boston, 2003).

[49] R. G. Douglas, Essentially reductive Hilbert modules. *Journal of Operator Theory,* **55** (2006), 117–133.

[50] R. G. Douglas, Essentially reductive Hilbert modules II, *Hot Topics in Operator Theory,* Theta Series in *Advanced Mathematics,* **9** (2008), 79–87.

[51] R. G. Douglas, K. Guo, Y. Wang, On the p-essential normality of principal submodules of the Bergman module on strongly pseudoconvex domains, arXiv:1708.04949.

[52] R. G. Douglas and G. Misra, Some calculations for Hilbert modules. *Journal of the Orissa Mathematical Society,* **12** (1993) No. 15, 75–85.

[53] R. G. Douglas and G. Misra, Equivalence of quotient modules-II. *Transactions of the American Mathematical Society,* **360** (2008) No. 4, 2229–2264.

[54] R. G. Douglas, G. Misra and C. Varughese, On quotient modules: the case of arbitrary multiplicity. *Journal of Functional Analysis,* **174** (2000) No. 2, 364–398.

[55] R. G. Douglas and V. I. Paulsen, Hilbert Modules over Function Algebras. *Longman Research Notes,* **217** (1989).

[56] R. G. Douglas, X. Tang and G. Yu, An analytic Grothendieck Riemann Roch theorem. *Advances in Mathematics,* **294,** (2016), 307–331

[57] R. G. Douglas and D. Voiculescu, On the smoothness of sphere extensions. *Journal of Operator Theory* **6** (1981), 103–111.

[58] M. Engliš and J. Eschmeier, Geometric Arveson–Douglas conjecture. *Advances in Mathematics,* **274** (2015), 606–630.

[59] Q. Fang and J. Xia, Analytic aspects of Drury-Arveson space. In *Handbook of Analytic Operator Theory,* (Boca Raton: CRC Press, 2018).

[60] J. Faraut and A. Korányi, Function spaces and reproducing kernels on bounded symmetric domains. *Journal of Functional Analysis,* **88** (1990), 64–89.

[61] S. H. Ferguson and R. Rochberg, Description of certain quotient Hilbert modules. *Operator Theory* **20** (2006), 93–109.

[62] P. A. Fillmore, J. G. Stampfli, and J. P. Williams, On the essential numerical range, the essential spectrum, and a problem of Halmos. *Acta Scientiarum Mathematicarum (Szeged),* **33** (1972), 179–192.

[63] S. Ghara and G. Misra, Decomposition of the tensor product of two Hilbert modules, In *Operator Theory, Operator Algebras and Their Interactions with Geometry and Topology,* (Cham: Birkhäuser, 2020), pp. 221–265.

[64] S. Ghara, S. Kumar, and P. Pramanick, K-homogeneous tuple of operators on bounded symmetric domains. To appear, *Israel Journal of Mathematics,* arXiv:2002.01298.

[65] J. Gleason and S. Richter, m-isometric commuting tuples of operators on a Hilbert space. *Integral Equations and Operator Theory,* **56** (2006), 181–196.

[66] J. Gleason, S. Richter and C. Sundberg, On the index of invariant subspaces in spaces of analytic functions of several complex variables. *Journal für die Reine und Angewandte Mathematik,* **587** (2005), 49–76.

[67] K. Guo, Essentially commutative C^*-algebras with essential spectrum homeomorphic to $\mathbb{S}^{2n-1}$. *Journal of the Australian Mathematical Society,* **70** (2001), 199–210.

[68] K. Guo, J. Hu, and X. Xu, Toeplitz algebras, subnormal tuples and rigidity on reproducing $\mathbb{C}[z_1,\ldots,z_d]$-modules. *Journal of Functional Analysis,* **210** (2004), 214–247.

[69] K. Guo and Y. Duan, Spectral properties of quotients of Beurling-type submodules of the Hardy module over the unit ball. *Studia Mathematica,* **177** (2006), 141–152.

[70] K. Guo and K. Wang, Beurling type quotient modules over the bidisk and boundary representations. *Journal of Functional Analysis,* **257** (2009), 3218–3238.

[71] K. Guo and P. Wang, Essentially normal Hilbert modules and K-homology III, Homogenous quotient modules of Hardy modules on the bidisk. *Science in China Series A: Mathematics,* **50** (2007), 387–411.

[72] P. Halmos, Ten problems in Hilbert space. *Bulletin of the American Mathematical Society,* **76** (1970), 887–933.

[73] P. Halmos, Continuous functions of Hermitian operators. *Proceedings of the American Mathematical Society,* **31** (1972), 130–132.

[74] P. Halmos, Limits of shifts. *Acta Scientiarum Mathematicarum (Szeged),* **34** (1973), 131–139.

[75] P. Halmos. *A Hilbert Space Problem Book,* Graduate Texts in Mathematics, Vol. 19, (New York, Berlin: Springer-Verlag, 1982).

[76] P. Halmos, BDF or the infinite principal axis theorem. *Notices of the American Mathematical Society* **30** (1983), 387–391.

[77] N. Higson and J. Roe. *Analytic K-homology,* Oxford Mathematical Monographs, (Oxford: Oxford University Press, 2000).

[78] W. Ingram and W. Mahavier. *Inverse Limits: From Continua to Chaos,* Developments in Mathematics, Vol. 25 (New York: Springer, 2012).

[79] Z. Jabłoński, Il Bong Jung, and J. Stochel, Weighted shifts on directed trees. *Memoirs of the American Mathematical Society,* **216** (2012).

[80] K. Jänich, Vektorraumbündel und raum der Fredholm-operatoren. *Mathematische Annalen,* **161** (1965), 129–142.

[81] N. Jewell and A. Lubin, Commuting weighted shifts and analytic function theory in several variables. *Journal of Operator Theory,* **1** (1979), 207–223.

[82] R. V. Kadison and J. R. Ringrose. *Fundamentals of the Theory of Operator Algebras,* Vol. I. Elementary Theory [Reprint of the 1983 original, Graduate Studies in Mathematics, Vol. 15], (Providence, RI: American Mathematical Society, 1997).

[83] A. Kirillov. *Elements of the Theory of Representations,* (New York: Springer-Verlag, 1976).

[84] A. Kirillov and A. Gvishiani. *Theorems and Problems in Functional Analysis,* [Translated from Russian by Harold H. McFaden], Problem Books in Mathematics (Berlin, New York: Springer-Verlag, 1982).

[85] A. Korányi and G. Misra, Homogeneous operators on Hilbert spaces of holomorphic functions. *Journal of Functional Analysis,* **254** (2008), 2419–2436.

[86] A. Korányi and G. Misra, A classification of homogeneous operators in the Cowen–Douglas class. *Advances in Mathematics,* **226** (2011), 5338–5360.

[87] T. L. Kriete, An elementary approach to the multiplicity theory of multiplication operators, *The Rocky Mountain Journal of Mathematics,* **16** (1986), 23–32.

[88] S. T. Kuroda, On a theorem of Weyl-von Neumann, *Proceedings of the Japan Academy,* **34** (1958), 11–15.

[89] J. S. Lancaster, Lifting from the Calkin algebra. Dissertation, Indiana University (1972).

[90] R. N. Levy, Algebraic and topological *K*-functors of commuting *n*-tuple of operators. *Journal of Operator Theory,* **21** (1989), 219–253.

[91] R. N. Levy, Spectral picture and index invariants of commuting *n*-tuples of operators, Contemporary Mathematics. *American Mathematical Society,* **185** (1993), 219–236.

[92] G. W. Mackey. *The Theory of Unitary Group Representations,* (The University of Chicago Press, 1976).

[93] M. Mandelkern, Metrization of the one-point compactification. *Proceedings of the American Mathematical Society,* **107** (1989), 1111–1115.

[94] V. Manuilov and K. Thomsen, Relative *K*-homology and normal operators. *Journal of Operator Theory,* **62** (2009), 249–279.

[95] A. Mukherjee, *Atiyah-Singer Index Theorem,* (Hindustan Book Agency, 2013).

[96] J. Munkres, *Topology: A First Course,* (Englewood Clis: Prentice-Hall, 1975).

[97] G. J. Murphy, Diagonalising operators on Hilbert spaces. *Proceedings of the Royal Irish Academy, Section A: Mathematical and Physical Sciences,* **87** (1987), 67–71.

[98] G. J. Murphy, Extensions and K-theory of C^*-algebras. *Irish Mathematical Society Bulletin,* **18** (1987), 18–29.

[99] G. J. Murphy. *C^*-algebras and operator theory,* (Boston: Academic Press, 1990).

[100] E. Nelson. *Topics in Dynamics. I: Flows,* Mathematical Notes. (Princeton: Princeton University Press, 1969).

[101] J. von Neumann. *Charakterisierung des Spektrums eines Integraloperators,* (Paris: Hermann, 1935).

[102] J. Newburgh, The variation of spectra. *Duke Mathematical Journal,* **18** (1951), 165–176.

[103] D. P. O'Donovan, Quasidiagonality in Brown–Douglas–Fillmore theory. *Duke Mathematical Journal,* **44** (1977), 767–776.

[104] D. P. O'Donovan, Unitary equivalence of essentially normal operators modulo the compacts. *Proceedings of the Royal Irish Academy, Section A: Mathematical and Physical Sciences,* **80** (1980), 147–153.

[105] C. Olsen, A structure theorem for polynomially compact operators. *American Journal of Mathematics,* **93** (1971), 686–698.

[106] V. I. Paulsen, Weak compalence invariants for essentially *n*-normal operators. *American Journal of Mathematics,* **101**, 979–1006.

[107] V. I. Paulsen, A covariant version of Ext. *Michigan Mathematical Journal,* **29** (1982) No. 2, 131–142

[108] V. I. Paulsen. *Completely Bounded Maps and Operator Algebras*, Cambridge Studies in Advanced Mathematics, No. 78, (Cambridge: Cambridge University Press, 2002).

[109] C. Pearcy. *Some Recent Developments in Operator Theory*, CBMS Regional Conference Series in Mathematics, No. 36, (American Mathematical Society, 1978).

[110] C. Pearcy and N. Salinas, Operators with compact self-commutator. *Canadian Journal of Mathematics*, **26** (1974), 115–120.

[111] G. K. Pedersen. *C*-algebras and Their Automorphism Groups*, (Academic Press, 1979).

[112] L. Peng and G. Zhang, Tensor products of holomorphic representations and bilinear differential operators. *Journal of Functional Analysis*, **210** (2004), 171–192.

[113] M. Reed and B. Simon. *Methods of Modern Mathematical Physics: I. Functional Analysis*, 2nd edition, (New York: Academic Press, 1980).

[114] C. E. Rickart. *General Theory of Banach Algebras*, The University Series in Higher Mathematics D, (Princeton: van Nostrand Co., 1960).

[115] S. Richter, Invariant subspaces of the Dirichlet shift. *Journal für die Reine und Angewandte Mathematik*, **386** (1988), 205–220.

[116] S. Roman. *An Introduction to the Language of Category Theory*, Compact Textbooks in Mathematics, (Cham: Birkhäuser-Springer, 2017).

[117] I. Rosenholtz, Another proof that any compact metric space is the continuous image of the Cantor set. *The American Mathematical Monthly*, **83** (1976), 646–747.

[118] R. A. Martínez-Avendaño and P. Rosenthal. *An Introduction to Operators on the Hardy-Hilbert Space*, (New York: Springer-Verlag, 2007).

[119] W. Rudin. *Principles of Mathematical Analysis*, (New York: McGraw-Hill, 1976).

[120] W. Rudin. *Real and Complex Analysis*, (New York: McGraw-Hill, 1987).

[121] N. Salinas, Extension of C^*-algebras and n-normal operators. *Bulletin of the American Mathematical Society*, **82** (1976) 143–146.

[122] N. Salinas, Homotopy invariance of Ext($\mathcal{A}$). *Duke Mathematical Journal*, **44** (1977) 777–794.

[123] N. Salinas, Hyperconvexity and n-normal operators. *Transactions of the American Mathematical Society*, **236** (1979), 325–351.

[124] D. Sarason, The multiplication theorem for Fredholm operators. *The American Mathematical Monthly*, **94** (1987), 68–70.

[125] O. Shalit, Operator theory and function theory in Drury–Arveson space and its quotients. In D. Alpay, ed., *Operator Theory*, (Basel: Springer, 2015).

[126] O. Shalit, Stable polynomial division and essential normality of graded Hilbert modules. *Journal of the London Mathematical Society*, **83** (2011), 273–289.

[127] A. Shields, Weighted Shift Operators and Analytic Function Theory. *Mathematical Surveys and Monographs, American Mathematical Society*, **13** (1974): 49–128.

[128] S. Shimorin, Wold-type decompositions and wandering subspaces for operators close to isometries. *Journal für die reine und angewandte Mathematik*, **531** (2001), 147–189.

[129] W. Sikonia, The von Neumann converse of Weyl's theorem. *Indiana University Mathematics Journal*, **21** (1971/1972), 121–124.

[130] B. Simon. *Real Analysis, A Comprehensive Course in Analysis*, Part 1, (Providence, RI: American Mathematical Society, 2015).

[131] B. Simon. *Operator Theory, A Comprehensive Course in Analysis*, Part 4, (Providence, RI: American Mathematical Society, 2015).

[132] E. Stein and R. Shakarchi. *Complex Analysis,* Princeton Lectures in Analysis, No. 2., (Princeton, NJ: Princeton University Press, 2003).

[133] E. Stein and R. Shakarchi. *Real Analysis: Measure Theory, Integration, and Hilbert Spaces,* (Princeton and Oxford: Princeton University Press, 2005).

[134] V. S. Sunder, The Riesz representation theorem. *Indian Journal of Pure and Applied Mathematics*, **39** (2008), 467–481.

[135] V. S. Sunder, Fuglede's theorem. *Indian Journal of Pure and Applied Mathematics*, **46** (2015) No. 4, 415–417.

[136] J. L. Taylor, The analytic-functional calculus for several commuting operators. *Acta Mathematica*, **125** (1970), 1–38.

[137] V. Varadarajan. *Geometry of Quantum Theory*, Second edition, (New York: Springer-Verlag, 1985).

[138] D. Voiculescu, A non-commutative Weyl-von Neumann theorem. *Revue Roumaine des Mathematiques Pures et Appliquees*, **21** (1976), 97–113.

[139] P. Wang and C. Zhao, Essential normality of homogenous quotient modules over the polydisc: distinguished variety case. *Integral Equations and Operator Theory*, **90** (2018): 1–24.

[140] P. Wang and C. Zhao, Essentially normal homogeneous quotient modules on the polydisc. *Advances in Mathematics*, **339** (2018), 404–425

[141] H. Weyl, Über beschränkte quadratischen Formen deren Differenz vollstetig ist. *Rendiconti del Circolo Matematico di Palermo*, **27** (1909), 373–392.

[142] K. Zhu. *Spaces of Holomorphic Functions in the Unit Ball*, (Springer, 2005).

Subject Index

Index of Symbols